The Zodiac of Paris

The Zodiac of Paris

How an Improbable Controversy over an Ancient Egyptian Artifact

Provoked a Modern Debate between Religion and Science

Jed Z. Buchwald & Diane Greco Josefowicz

Princeton University Press • Princeton and Oxford

Published by Princeton University Press, 41 William Street,
Princeton, New Jersey 08540
In the United Kingdom: Princeton University Press,
6 Oxford Street, Woodstock, Oxfordshire OX20 1TW

Library of Congress Cataloging-in-Publication Data

Buchwald, Jed Z.
The zodiac of Paris : how an improbable controversy over an
ancient Egyptian artifact provoked a modern debate between
religion and science / Jed Z. Buchwald and Diane Greco Josefowicz.
p. cm. Includes bibliographical references and index.
ISBN 978-0-691-14576-1 (hardcover : alk. paper)
1. Science—Europe—History. 2. Science, Ancient.
3. Science—Egypt—History—To 1500. 4. Science—Philosophy.
5. Science and astrology. 6. Religion and science.
7. Napoleon I, Emperor of the French, 1769–1821.
8. France—History—Restoration, 1814–1830.
9. Egypt—Antiquities. I. Josefowicz, Diane Greco. II. Title.
Q127.E8B83 2010
201'.65—dc22   2010002409

British Library Cataloging-in-Publication Data is available

This book has been composed in ITC Caslon 224
Design by Marcella Engel Roberts

Printed on acid-free paper. ∞
press.princeton.edu
Printed in the United States of America
1 3 5 7 9 10 8 6 4 2

CONTENTS

vi   Contents

The Zodiac of Paris

INTRODUCTION

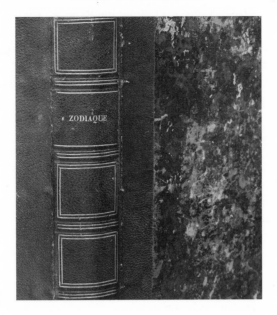

Seven years ago one of us wandered into a small book shop near the Luxembourg Palace in Paris. A volume half-clad in red Morocco sat intriguingly apart on a lower shelf. Its spine, stamped in gold with one word, *Zodiaque*, held a bound selection of pamphlets, heavily annotated in a contemporary hand. Each pamphlet concerned something called the "zodiac of Dendera." Why had these obscure articles been collected? Indeed, why were they written in the first place, and why had someone devoted such a considerable amount of time to commenting on them? One thing was certainly clear: whatever the subject of the pamphlets was, it had created a sensation in Paris at the beginning of the nineteenth century.

The "zodiac of Dendera" was easy enough to find. Consisting of two large sandstone blocks, it is embedded in the ceiling of a small room of its own in the Egyptian antiquities section of the Louvre. Some of the figures that cover the stones look like the familiar symbols for the zodiacal constellations; along with many hieroglyphs, they are arrayed in an apparently regular fashion around the center. First seen, and drawn, by Europeans during Napoleon's Egyptian expedition in the late 1790s, the stones had landed in Paris in 1821 after a Frenchman by the name of Lelorrain journeyed to Egypt with explosives and saws to steal the artifact for personal profit and national glory—glory being at a rather low ebb even six years after Napoleon's fall and the country's occupation by foreign troops.

Book illustrations, drawings, and eventually prints of this and three other Egyptian "zodiacs" produced during and shortly after Napoleon's campaign circulated widely two decades before the Dendera ceiling made its physical appearance in France. Many of the *savants*, as the scholars, engineers, scientists and mathematicians who accompanied the expedition were known, claimed that the placement of figures on these objects was intended to represent the state of the sky at the time they were designed, if not carved. If this were true, then astronomy could be used to prove that Egypt's history reached back much further than anyone had hitherto suspected, back at least fifteen thousand years, nine millennia before the biblical date of Creation itself.

In the early years of Napoleon's consulateship and empire, revived religiosity coupled to continuing royalist sentiment contended with republican ideology. Both fought the empire's efforts to establish its legitimacy, producing a precarious polity that the regime took considerable trouble to control. The great age claimed for the zodiacs thrust them into the midst of these conflicts. For they threatened to resurrect the dangers of impiety and the disintegration of traditional social hierarchies that repatriated reactionaries associated with the *philosophes*, the free-wheeling eighteenth-century intellectuals who were blamed for the excesses of the revolution and the Terror. The zodiacs also gave mathematically trained savants the opportunity to challenge their historical and philological colleagues for control of the past, since many of the savants were certain that the zodiacs testified to an Egypt of profound sophistication, the primordial source that had flooded the ancient world with scientific and mathematical knowledge. When, years later, the Dendera artifact was ripped from its home and brought to a Restoration Paris where angry royalist Ultras busily policed print and speech, the zodiacs again exerted their power, summoning a demonology of greed, anger, reaction, calculation, Machiavellian power-plays, intrigue—and a vaudeville comedy with political overtones.

Not surprisingly, the zodiacs were from the beginning stamped with contemporary French attitudes to Egyptian civilizations, past and present, attitudes that rested on specific beliefs about history. These views had taken a particular cast during the invasion. Napoleon's rank-and-file soldiers neither tolerated the savants well, nor did they demonstrate much understanding of Egyptian customs. The savants themselves had considerable sympathy for the *fellahin*, or peasants, and the Egyptian literate classes, but even they had expected to find a people oppressed by centuries of alien domination. Moreover, the French Arabists who accompanied the expedition apparently had little sense of the language's modern form, to the annoyance of educated Egyptians, who reciprocated French disdain. Al-Jabarti, an astute and sophisticated Cairene chronicler of the invasion, deplored the colonizers' crude intrusions into local government, not to mention their irreligious views and their inelegant Arabic. There were few grounds here for a respectful meeting of minds, improbable though that would have been in any event, given the violence of the invasion.

Admiration, even awe, for Egypt's past grew among the French invaders in direct proportion to their disdain for Egypt as they found it. The discovery of what seemed to be four ancient zodiacs carved into the ceilings of temples buried nearly to their roofs in sand fit neatly into this vision, for they were taken to be evidence of a glorious civilization, now lost, that dated from millennia before the Greeks were believed to have developed astronomy. The zodiacs, two at Esneh and two at Dendera, were first found by General Desaix as he led his army up the Nile near Luxor, the site of ancient Thebes. The artist Vivant Denon, who was traveling with the company, rapidly drew several of them. Denon's drawings, along with a romantic version of his adventures with Napoleon's army, circulated soon after the campaign in a newspaper account and then in a massive folio printing. Smaller editions followed, making Denon's *Voyage* a bestseller in France and in England, propelling the debates on zodiacal antiquity as well as fueling a rage for all things Egyptian during the empire. Egyptian motifs soon appeared on wallpapers, dinner services, architecture, even clocks, charging the period's esthetic culture with exoticism and politics.

Two decades later, the advent of the Dendera stones seven years into the Restoration reignited the incendiary power of Egypt in France. The zodiac awakened submerged, and politically dangerous, memories of Napoleonic glory, not to mention republicanism's notorious hostility to religious dogma. Scholarly battles raged and overlapped with politics, religion, and popular culture as the Dendera zodiac conjured Egypt's ancient and France's recent past. Aristocratic Ultras and a reempowered Catholic clergy fought every sign of renascent *philosophie*, which was to them the evil fruit of eighteenth-century intellectual license. The zodiac seemed a particularly dangerous object, good only for would-be philosophes to mount new attacks on the organic unity of throne and altar, ever-threatened by materialistic republicanism.

The foundation of the zodiac debates had been laid years before Napoleon's expedition, for events during the revolution had turned questions about the age and character of antique civilizations into sites of political contention. A claim made by one Charles Dupuis during the 1780s concerning the origins of all forms of religion, including monotheism, and which interpreted myths in astronomical and agricultural terms, was particularly potent. It had spread so widely and proved so durable that in later years it was nearly impossible for scholars to engage with questions about ancient Egyptian astronomy and religion without implicitly or explicitly referring to Dupuis. His theory enraged reactionaries, and it is not hard to see why. For Dupuis connected his claims to speculations about the origins of the zodiac, whose birthplace he located in an Egypt older by far than any chronology based on textual arguments, and especially on the Books of Moses, could possibly allow. During the revolution, Dupuis' scheme played out in extraordinary ways. It was well known to, and admired by, many of the savants who accompanied Napoleon to Egypt, men who regarded all forms of religion as remnants of superstitious awe hijacked by scheming priests to hoodwink an ignorant populace.

Although the details of Dupuis' argument were new, and its feverish antireligiosity breathed the atmosphere of prerevolutionary France, Egypt had been considered the original source of knowledge as early as the time of Plato. Scholars in the seventeenth century had provided countervailing arguments. In 1614 Isaac Causabon had demonstrated that one group of texts, the influential Hermetic Corpus, actually dated from about 200 CE and not, as had been claimed, from Egypt before the time of Moses. Nevertheless, Egypt and the mysteries of its hieroglyphs continued to capture the European imagination throughout the eighteenth century. One widely read French work of the time by Constantin Volney argued that history amounted to a succession of continually reemerging ancient civilizations, notably Egypt's. Influenced by that vision, Napoleon conceived his invasion as the latest act in this grand historical drama. For Napoleon expected to be greeted as a liberator by native Egyptians, descendants of a wise and graceful past, who had been subjected for centuries to the oppression of the Ottoman Turks and their Mameluke satraps. He was neither the first, nor certainly the last, to cloak naked conquest in the guise of benevolent assistance to a tyrannized populace.

The Napoleonic expedition was, in the end, a debacle. Native Egyptians had little love for the Mamelukes, but neither did they greet the French invaders as liberators. Revolts and resistance to the occupation were frequent, and the French responded with great brutality. The English fleet under Admiral Nelson destroyed the French flotilla not long after its arrival at Alexandria, effectively isolating the army in Egypt. Napoleon returned clandestinely to France a year later, leaving in charge General Jean-Baptiste Kléber, who was assassinated nine months afterward by a Syrian who detested the presence of non-Muslims in the region. The French occupiers were forced to capitulate by the end of August 1801, though not to the local rulers but to the English, who had managed to mount an Egyptian invasion of their own.

A great deal has been written about Napoleon's military expedition, and about the savants who accompanied it. We will not trod these well-worn paths.[1] Instead, we will ask how the savants' views and prejudices shaped, and were shaped by, their understanding of the artifacts, sketches, drawings, engravings, models, and finally the Dendera carving itself, that they encountered in their efforts to make sense of the zodiacs. Images and words sent back to France, and the eventual return of the savants to a country under the effective control of the man who had led them into Egypt and had abandoned them there, molded early reactions to the zodiacs. These reactions continued to evolve over the decades following Napoleon's expedition, often in response to prevailing political winds.

The arrival of the stones in France provoked impassioned arguments in many quarters: in the pages of popular and scholarly journals, in salons, around dinner tables, at the meetings of learned societies, and even on the Parisian stage. These debates, nominally about the age of antique civilization as revealed by the zodiacs, open a window onto deeper issues of the day, such as the comparative values of different ways of knowing about the past, including astronomy and calculation versus historical philology, and the nature and power of scientific against religious authority. Profoundly motivated by either religious conviction or its opposite, the zodiac debates hinged in part upon the ways in which numbers could be trumped by words and architectural styles, or vice versa. They mark a moment at which these two worlds began strongly and overtly to separate, in environments saturated with conflict over the proper place for, and the very nature of, belief in the unseen.

The hub around which the arguments swirled concerned the age of the world—more precisely, how many centuries had elapsed since the catastrophe that had erased most, if not all, of humanity in a universal Flood. By the late eighteenth century many students of Earth's geologic history had concluded that the world had existed long before, perhaps immensely long before, the Flood, which meant that Creation itself had not occurred a mere six millennia ago. Nevertheless, in almost every such discussion two elements were not up for reinterpretation: the nearly complete extermination of humanity in a universal catastrophe about four millennia or so ago, and the creation of humans by a deity not many centuries before that.

The antireligious philosophes and their underground followers and competitors in the decades before the revolution had challenged nearly every aspect of religious belief, generating a powerful local reaction from the Gallican church and its hangers-on. The revolutionary years had had further dramatic effects, culminating in the establishment of several cults devoted to the extirpation of churchly "fanatics" and their replacement with devotees to icons of "reason." These events markedly affected the attitudes of the young savants who invaded Egypt with Napoleon, and especially the controversies that erupted upon their return, since the savants were certain that Egyptian civilization predated the biblical date of the world's origin, to say nothing of the Flood. These themes remained a constant, if occasionally subterranean, factor in the zodiac debates throughout the empire and the first decade of the Restoration. Religious conservatives received the savants' zodiacal calculations with contempt, propelled in part by their attitude to science, which many among them both feared and hated.

The diaries, reminiscences, books, newspapers, pamphlets, and remnant ephemera of these years provide ample evidence of the often vicious and threatening scribblings of the religious press that followed the return of émigrés to France. Not that their efforts were entirely new, since the reaction had begun years before the revolution itself. The horrific events of the Terror, when prelates had either to leave, to renounce their vows, or to die, cemented the Right's long-standing conviction that freedom of expression and the abolition of social hierarchy lead inevitably to the abyss. The years after the Restoration saw the emergence of repeated, though only partially successful, efforts to establish a thoroughly repressive regime based on the Right's vision of an idyllic past that had never existed, a past in which a faithful and dutifully submissive populace knew its proper place in a social hierarchy that united throne with altar. The "infernal" stone from a pagan temple, as some reactionaries called the Egyptian zodiac, generated great fear and anger for dread of its power to subvert the historical ground of religious belief. Yet the zodiac's inflammatory potential also provided opportunities for influence and profit that could be, and were, taken advantage of.

Among those who became entangled in the later debates was Jean-François Champollion, a young man recently arrived in Paris and well connected with one of the major savants on Napoleon's expedition. A superb linguist, Champollion was convinced that hieroglyphs should be treated neither as cryptographic codes nor as mystical talismans, as had mostly been done until quite recently. Rather, these mute symbols could only be made to speak if one first understood how ancient Egyptians themselves spoke as they went about their daily lives. On September 22, 1822, Champollion wrote to Bon-Joseph Dacier, the permanent secretary of the Académie des Inscriptions et Belles-Lettres, where philology, linguistics, and antiquity increasingly mixed with one another. That famous missive provided a first explanation of the essential principles for reading the hieroglyphs. A signal element in Champollion's decipherment was his penetrating analysis of signs appearing on parts of the Dendera stones that had been left behind but that were available to him, as to many others, through drawings made by the expedition's savants.

When Champollion visited Egypt many years later, he made excitedly for Dendera. Standing among the remnants of the ruined temple, he was confronted, perhaps for the first time, with the limits of his own philological genius. What he found near the void that had once contained the zodiac was sufficiently explosive that it remained private for a dozen years following his death in 1832, two years after the expiration of the Bourbon monarchy. Although by the end of the 1830s the zodiacs had lost much of their power to excite political and cultural passions, arguments concerning their astronomical character continued to reappear over the years, as they do even today.

1

All This for Two Stones?

On October 1, 1820, an engineer named Jean Lelorrain left Marseilles for Alexandria on a ship under heavy sail, laden with saws, chisels, jacks, and a sledge with wooden rollers made especially for transporting a large object over rough terrain. Lelorrain had been commissioned to remove an immense circular zodiac from the ceiling of an ancient temple near the village of Dendera on the west bank of the Nile. The zodiac, one of only four still extant, had excited tremendous interest and controversy when it had been discovered during Napoleon's Egypt expedition two decades before. Lelorrain's employer, an antiquities speculator named Sebastian Saulnier, hoped to ensure a good return on his investment by resurrecting the ceiling, now all but forgotten, as a particular cause celébre: a fresh symbol of French national glory, then in dire need of a boost.

Saulnier was born in Nancy on February 28, 1790, the son of Pierre Dieudonné Louis Saulnier, a deputy of the Chamber and secretary general of Napoleon's Ministry of Police in 1810 under its inquisitorial minister, Anne Marie Savary, the duc de Rovigo. The younger Saulnier, due no doubt to his father's influence, had been police commissioner in Lyons, and then, during Napoleon's hundred-day return from exile in Elba, prefect in Tarn-et-Garonne and the Aude, both in southern France. Stripped of his official responsibilities during the Restoration, he turned his hand to literary and scientific matters, publishing fourteen volumes of the *Bibliothèque historique* before shuttering the enterprise in April 1820, as well as the somewhat longer-lived and more widely read *Minerve française*, a journal of current events and opinion.[1] Saulnier was well connected in Egypt, having befriended Youssef Boghos Bey, the Egyptian pasha's closest adviser, a canny polyglot who effectively controlled the issuance of firmans, or permissions for excavation, throughout Egypt (and who reportedly received one-thirtieth the value of each item exported).[2]

As a publisher Saulnier knew how to influence public opinion. To coincide with the zodiac's Parisian debut, he produced a book promoting

Figure 1.1. Dendera today and as photographed by Francis Frith circa 1863

the ceiling as an archeological artifact comparable in importance to the Rosetta Stone. With its inscriptions in the undeciphered hieroglyphs, in demotic,[3] and in Greek, the Rosetta Stone was among the greatest trophies seized by Napoleon's Egyptian campaign—which made its loss, after Napoleon's defeat by the British navy at Aboukir, a profound humiliation. The British poured salt into the wound by inscribing the captured antiquities with jibes against the defeated French.[4] While acquainting his readers with the Dendera ceiling as a singular archeological find, Saulnier fanned the flames of French resentment, accusing the British of the grossest philistinism in their careless treatment of the magnificent remnants of Egyptian civilization.[5]

Saulnier's assertion of the zodiac's centrality to French national pride rested on a single breathtaking assumption: that the zodiac, like the lost Rosetta Stone, was French property in the first place. Saulnier described the zodiac as a treasured reminder of one of the most glorious episodes in French military history, General Desaix's rout of the Mamelukes in the Thebaid, the ancient desert region of Egypt that stretches from Abydos to Aswan. It was during this campaign, Saulnier pointed out, that Napoleon's troops first entered the temple at Dendera, where they found a chamber whose ceiling was adorned with what appeared to them to be an intricately carved depiction of the heavens.[6] The ceiling was sketched for the first time by Vivant Denon, one of the artists on Napoleon's expedition.[7] Denon's sketch (Figure 4.3), Saulnier averred, had brought the zodiac to the world's attention, a fact that he adduced as further evidence to support his claim for French ownership. For Saulnier, to discover, describe, publish, and promote something was to own it.

Saulnier's advertisement of the zodiac as a symbol of the lost glories of France's recent past connected the mysterious object to popular resentment against the British. As Saulnier well knew, the value of the zodiac—a matter about which he, the excavation's financier, was decidedly not neutral—would naturally rise along with French pride at its having been snatched from the English captors of the Rosetta Stone. But there was more to Saulnier's position than mere nationalism. Rather, Saulnier's proprietary attitude was typical of a view of Egypt and its antiquities common to cultured Europeans of the period. The astonishing ease with which Saulnier could claim the zodiac as a French possession attested to the ubiquity and familiarity of this point of view, which was the product of a constellation of European attitudes toward the East known today as "orientalism."[8] In nineteenth-century

Paris, booksellers' stalls were crowded with volumes devoted to the lore of Egypt, as well as with travelers' accounts that fueled dreams of treasure hunting in the Valley of the Kings. The mysteries of ancient Egypt, of which the enigmatic hieroglyphs were the perennial emblem, had for centuries fascinated Western seekers and mystics, linguists and scholars. For over two centuries they had produced a general narrative about this most ancient of civilizations, a narrative that demanded material objects to complete and enliven it. To them, Egypt was a warehouse for antiques that belonged in European collections and museums, to be studied for whatever they might have to say about the origins of European civilization. (Although the civilization of ancient Egypt was considered thoroughly alien, it was often thought to be the unfathomable source of Europe's own.) The very frontispiece to the atlas of the monumental account of the Napoleonic expedition's activity—the *Description de l'Égypte*—communicated this attitude precisely, by depicting the Dendera ceiling set carelessly amidst statues and other objects like items for sale in a flea market (Figure 4.4). The haphazard arrangement made it clear that the value of Egypt's antiquities inhered less in the objects themselves than in their abundance and apparent availability.

Pasha and Firman

Although Saulnier had ensured that Lelorrain was materially well prepared for his expedition, removing the zodiac from Egypt required more than ready cash and equipment: it required diplomacy. The politics of antiquities excavation were tricky and competitive, and Egypt's stability was precarious. Decades of struggle against foreign invaders had left the country in social and political disarray, its treasures vulnerable to the incursions of vandals like Lelorrain. Egyptians after all had greater problems than the fate of ancient stones, no matter how important they seemed to be to foreigners. During his campaign of 1798, Napoleon had defeated the Mamelukes—a foreign slave-soldier caste that had for centuries alternately supported and struggled against the Ottomans in Egypt. But after the British forces routed the French, they allowed the surviving Mamelukes to regroup, laying the ground for a civil war between the Mamelukes and the Ottomans. The French invasion had reoriented Egyptian trade toward Europe and away from its traditional centers on the Red Sea, impoverishing Egypt's artisan class, who could neither compete with the influx of cheap textiles from industrial Europe nor profit from the substantially increased demand for Egyptian raw materials by those same European factories.[9] At the same time, Ottoman soldiers continued to pillage Alexandria and Cairo, so that daily commerce all but halted, the streets became thoroughly unsafe, and Egypt's prized silk and cotton trades were abandoned, as was all learning.[10] The civil conflict began to wind down in 1806 when the Mamelukes were evicted from Cairo at the hands of Mehmet Ali, a former coffee dealer born in Kavala, in what is now Macedonia, who by canny manipulation of Cairo's complicated politics had lifted himself to prominence within the Ottoman power structure in Egypt.[11] Ali's final victory over the Mamelukes in the Cairo massacre of 1811 consolidated his position as ruler of Egypt, which meant, among other things, that it was to Ali that Saulnier had to apply for permission to excavate, and to Ali that Lelorrain would have to address any problems that might crop up during his sojourn at Dendera.[12]

Ali had been made viceroy of Egypt in 1805, but his sphere of real influence was limited primarily to Cairo until the Mameluke massacre of 1811. He remained in control until 1848 and died the following year.[13] Ali's European visitors found him somewhat preoccupied but also congenial, dignified, and down-to-earth. He certainly disappointed Europeans who arrived at his quarters expecting to meet a so-called oriental despot like those caricatured in then-popular travelogues about the East.[14] Traveling through Egypt on a trip to India in 1806, George Annesley, the Viscount of Valentia, struggled, as many Europeans in Egypt did, to integrate his experiences with his expectations. Annesley described Ali as "a little man of intelligent countenance, with a reddish brown beard of moderate dimensions" that, Annesley conjectured, must have been a source of "pride" for Ali, "as he was continually stroking it." Annesley later noted that when coffee was served, "the cup out of which [Ali] drank was set with diamonds," an opulent detail that contrasted sharply with Annesley's appraisal of Ali's residence as "remarkable neither for its size nor its richness."[15] Dr. R. R. Madden, who visited in 1826, observed that Ali was a restless person with a "ruddy fair complexion and light hazel eyes, deeply set in their sockets, and overshadowed with prominent eyebrows," and given to insomnia.[16] The French traveler Edouard de Montulé, in Egypt from 1816 to 1819, admired Ali for having "a physiognomy characteristic of his conduct; it is cold, like that of the Turks [musulmans], but noble, majestic, and dignified according to the rank he has attained."[17] Still others praised Ali as a talented horse- and swordsman, a crack shot who excelled at throwing the *djerid* (javelin), a person of "prompt, decisive" character who was also a shrewd and entertaining conversationalist.[18] Even Saulnier, who had little to gain from flattering the pasha by the time his book on the Dendera zodiac appeared in Paris, praised Ali to the skies, aligning him with Ptolemy I, the first Greek successor to the pharaohs, in virtue of the fact that both men happened to be born in Macedonia. He applauded Ali's efforts to modernize Egypt; and he invited the reader to compare Ali favorably with great leaders of the past by sharing the list of books that Ali had reportedly asked Saulnier to obtain for translation, which included biographies of Peter the Great and Frederick II of Prussia, together with the writings of Napoleon and Plutarch's *Lives*, a selection that seemed calculated to suggest that the pasha was—like Saulnier's implied reader—an educated person with a taste for the classics and an interest in Europe's storied military and political figures.[19]

Less admirable aspects of Ali's character also attracted notice—although, again, special care must be exercised since the exoticizing tenor of European travel writing made exaggerations de rigueur. Robert, Richardson, a British traveler who accompanied the Earl of Belmore up the Nile in 1817–1818 noted that, at the age of forty, Ali remained illiterate (notwithstanding Saulnier's list of books); when Ali finally learned to read and write, Richardson claimed that he was not especially good at either; and, astonishingly, he never learned to speak Arabic.[20] His sexual appetite was a matter of some interest as well. In his travelogue, De Montulé reported that Ali "adores the pleasures of women" and claimed, moreover, that "his harem is composed of more than five hundred females."[21] Finally, in a time and place often noted for its violence, Ali could be exceptionally duplicitous, ruthless, and bloody-minded. In 1811, after inviting between four hundred and five hundred Mamelukes to a banquet at the Cairo citadel, Ali secretly ordered an attack on the assembly. As the killing commenced, he looked on, crying "Vras! Vras!" (Kill! Kill!). Nearly a thousand people died in the carnage, which continued for six days, spilling into the harems. Ali was forced to kill several of his own men in order to halt the violence.[22]

These peccadilloes notwithstanding, Europeans generally admired Ali. He was courted by European consuls, who preferred him to the recalcitrant Mamelukes, and who had their eyes on Egypt as a strategic spot for trade with the Middle East and southern Asia.[23] In addition, he was generous with subsidies and positions for foreigners, who ran the businesses and factories that he founded and who trained his army.[24]

Ordinary Egyptians liked him much less. That Ali was more generous to foreigners than to the people he ruled was just one example of
the policies that made him unpopular. Although he is remembered for
ridding Egypt of the Mamelukes, Ali's efforts to modernize the country came at a tremendous cost to his subjects, beginning with a bitterly resented conscription policy that blighted the young adulthood
of two generations of Egyptian men, the imposition of corvée labor,
and extraordinarily heavy taxation. Ali's monopolization of production included surveillance of peasants, who were watched closely to
ensure that not a single olive or boll of cotton was diverted from Ali's
coffers. Although Ali did undertake useful improvements to the Nile
irrigation systems, as well as the construction of the Mahmoudiyah
Canal between Alexandria and the Nile (linking Cairo more closely to
the Mediterranean), his success depended on the subjugation of legions
of workers who labored for little pay under the threatening eyes of his
troops. Although the Mameluke alternative was generally thought to be
worse, ordinary Egyptians—upon whom Lelorrain would soon rely for
the efforts required to remove the zodiac from the ceiling of the temple
at Dendera—were hardly fond of Ali Pasha.[25]

Lelorrain was not the only European treasure hunter on the Upper
Nile. He had two chief rivals on the ground in Egypt: Henry Salt, the
British consul, and Bernardino Drovetti, who had been consul for
the French in Egypt until 1815, but who had remained thereafter in
the country. Both were colorful figures, and each had some control
over which areas could be excavated, as well as by whom. An intense
rivalry had developed between the two men, who vied fiercely for the
Pasha's favor—and for the excavation firmans that flowed from it. If
either man were to learn of Lelorrain's designs on the Dendera zodiac,
he would certainly take countervailing measures—whether or not the
pasha's permission had been obtained, which indeed Saulnier had
arranged before Lelorrain's departure.

The Piedmontese Drovetti, "the doyen of the diplomatic corps in
Egypt," had served as colonel during Napoleon's Egyptian expedition,
and then as consul general during Ali's rise to power, positions that
involved him in many diplomatic adventures and that made him both
a favorite of the pasha and a formidable challenge to the British occupiers, even after he was relieved of his duties following Napoleon's
downfall.[26] By this time, the career change represented no hardship,
for Drovetti had also established himself as a reputable antiquarian.
He spent his early postconsul years traveling around Egypt, amassing

antiquities and sending what he didn't keep to colleagues in Europe; Jean Du Boisaymé, for instance, received from Drovetti a shipment of mummified animals including ibises, a monkey, some snakes, and a loaf of ancient bread.[27] Drovetti's debut collection of papyri, statues, and everyday objects earned high marks from the Swiss orientalist Johann Burkhardt, who declared it "the finest in all Europe," and he was reportedly approached by Henry Salt to see about purchasing the collection for the British Museum.[28] (Eventually Drovetti sold it to the Turin Museum.) De Montulé wrote of Drovetti in 1818: "I aspire to the lot of M. Drovetti, who can count on new discoveries brought to him daily by his Arab workers, who adore him."[29] Admiration of Drovetti was apparently general. Visiting Drovetti's camp at Karnak, De Montulé portrayed him as an archaeological prince: "In the midst of the ruins of Karnak, the loveliest of Thebes, of Egypt and the world, rises a portal sixty feet high; from the small earthen house he has constructed here, M. Drovetti appears to command the precious relics of antiquity that surround him. . . . He is an educated man, revered and loved throughout the country."[30]

In contrast Drovetti's British counterpart, Henry Salt, was an unlikely power broker, with neither Drovetti's charm, nor his good looks, nor his long experience in local Egyptian politics. Nevertheless, he cut such an enduring figure that even E. M. Forster, in his guide to Alexandria (1922), was moved to describe him as "vigorous but rather shady . . . with an artistic temperament" (an ambivalent characterization that Salt's most recent biographer is eager to discredit).[31] Bookish and often in poor health, Salt had trained in London as a fine artist, at the Royal Academy Antique School, where he made little progress despite the efforts of his teacher, the landscape painter Joseph Farington.[32] His artistic skills were sufficient, however, to support a work-for-hire career, traveling with an entourage of the sort that touring nobility typically employed on their trips abroad. As secretary to George Annesley, Viscount Valentia, Salt traveled around the Cape of Good Hope to Sri Lanka, where they surveyed the African coast by dhow en route to India. Salt visited Egypt in 1806, where he met the pasha for the first time, returning in 1815 as the British consul. Thanks to the close relationship he cultivated with Ali, Salt was granted a cornucopia of firmans, and he generally kept himself and his fellow travelers in comfortable style even while on excavations. Forbin, a painter and traveler, wrote in his *Voyage au Levant* (1823) that "Salt, English consul, was established with a numerous suite, under tents in the Valley of the Kings. . . . Plenty of money, plenty of presents have won for him the affection of the Arabs."[33]

The pasha extended excavation privileges to both Salt and Drovetti, an even-handedness that, perhaps predictably, brought these two ambitious men into constant conflict. Richard Burton wrote in 1880 in the *Cornhill Magazine* that, under the pasha's stewardship, "the archaeological field became a battle plain for armies of Dragomans and Fellanavvies. One was headed by the redoubtable Salt; the other owned the command of Drovetti, whose sharp Italian brain had done much to promote the Pasha's interests."[34] Eventually the rivalry became so intense that it threatened to end in a shootout over a disputed obelisk. Drovetti invited Salt to his chambers at Karnak where he plied his guest with sherbet and lemonade. Recognizing that a certain amount of cooperation might benefit everyone, the two consuls arrived at a gentlemen's agreement to divide the entire Nile Valley into zones over which each had exclusive control, and they made sure that locals in their employ warned off any interlopers. Those who persisted would find themselves without laborers for their excavations—a problem that Lelorrain, sailing for Alexandria, would soon encounter for himself.[35]

In nineteenth-century travelogues, European observers of Egypt's antiquities trade routinely claimed that valuable artifacts were available in striking abundance. Such descriptions encouraged travelers in Egypt to believe that the country was full of souvenirs ripe for the picking. Arriving in Alexandria just ahead of Lelorrain, in October 1820, Sir Frederick Henniker, the first recorded European to scale the Second Pyramid at Giza, wrote: "I have been on shore; the very stepping stones at the water's edge are a mass of antiquities about to quit their native country." Henniker, who was no archaeologist, nevertheless derided a shipment of relics that had come to his attention as "defaced hieroglyphics and noseless statues sent [to Europe] for no visible reason, unless for ballast." He pointed out that, despite their apparent lack of value, they were leaving "with strong letters of recommendation" from the consuls, destined for sale into the museums and private collections of England and France.[36] One common variation on this theme excused the wholesale plunder by claiming that Egyptians were themselves unaware of the value of these goods in European eyes. According to the Baroness von Minutoli, who accompanied her husband, the Prussian officer and archaeologist Heinrich Menu von Minutoli, to Egypt in 1820, artifacts were so plentiful, and the Egyptians were apparently so unaware of their value, that wooden antiquities were even burned when other sources ran short. "Being in want of wood," she wrote, "the Arabs supplied us with a considerable quantity, consisting of the remains of mummy cases, among which were some very valuable pieces which my husband saved from the auto da fé."[37] Illustrating the truism that there are two sides to every culture clash, Egyptian laborers employed by Europeans did not hide their surprise when confronted with Europeans' unfamiliar habits and prejudices. Workers on excavation sites viewed their employers with particular skepticism. Jean-Jacques Rifaud, one-time architect to Ali Pasha and an excavator employed by Drovetti, noted that local laborers laughed at him because "the trouble and expense [of digging] seem absurd." Each one, Rifaud claimed, had a different explanation for the activities of European antiquities hunters. "Some [Egyptians] take them for pagans who carry infamy so far as to caress statues. They have seen statues being moistened with the tongue to see what kind of stone they were made of, and they have concluded that the statues were being kissed. According to others, the marbles that we take from them contain gold, that we alone have the secret of extracting." Rifaud's description, while not free of duplicity, conveys an idea of the surprising lengths to which travelers would go in order to extract profit from Egyptian antiquities. If Europeans found it easy to deny their greed, they could not hide it from the Egyptians they encountered (nor did they trouble to). Rifaud noted uneasily that "not a single European move or gesture escaped" the Egyptian laborers in his employ.[38]

Europeans in search of relics might have been figures of fun, but their enormous demand for Egypt's antiquities was a matter of concern at the highest levels of government. The antiquities themselves were valuable; firmans, or permissions to dig for them, were also a source of income. But here, again, the question of stewardship was subordinated to other interests. Ali Pasha, for instance, was interested in selling Egypt's antiquities, as well as the rights to dig for them, in order to subsidize his armaments and public works projects. While one might see in the pasha's attitude a perfectly rational response to the great demand for antiquities, it offended Europeans who felt that Egypt's treasures should be handled differently, perhaps by keeping them together in their original home, under the watchful eye of a trustworthy (i.e., European) guardian. In his *Travels*, de Montulé loftily declared that "the Pasha's government knows even less than the Arabs about how to profit by the antiquities of the country; he permits everything to be taken piecemeal, by different individuals, while the whole enterprise could be delegated to a single company, which might propose to found a museum at Cairo or Alexandria."[39] De Montulé's attitude was typical; Europeans generally believed that the Egyptians were inadequate stewards of their artifacts and monuments. For Saulnier, this conveniently self-serving belief amounted to an apology for plunder. Making his case for bringing the Dendera zodiac to France, Saulnier explained that the Egyptians themselves—Ali Pasha's efforts to modernize Egypt notwithstanding—were not truly ready to handle the responsibility. According to Saulnier, the civilization of modern Egypt was "still imperfect" and its people "half-savage"—so that Egypt's antiquities, whose fate was less pressing than the immediate needs of ordinary Egyptians, should be left to Europeans, whom he presumed to have the appropriate resources and the requisite sophistication.[40] Curiously, those who sought Egyptiana in the museums and shops of Paris and London rarely, if ever, questioned their right to see, handle, and own these goods; nor did travelers in Egypt hesitate to deface ancient structures and objects by inscribing them with their names and other graffiti. This greed and destructiveness was not lost on Ali, who used it to deflect his own responsibility. When asked why he was not doing more to save the antiquities, he replied: "How can I do so, and why should you ask me, since it is the Europeans themselves who are their chief enemies?"[41]

Lelorrain and Saulnier's interest in the stewardship of antiquities extended only to the preservation of their own stakes in the expedition. Their mercenary attitude was, again, typical. Two generations later, the discourse surrounding European plunder of antiquities would shift, as scientific archaeologists blamed tourists and the consuls from whom they received excavation firmans for the destruction of artifacts and the integrity of the sites in which those artifacts were originally found. As much as later scholars benefited from the papyri and other objects that these expeditions brought to Europe, they faulted Drovetti, Salt, and the explorers whose work they directed for putting the accumulation of precious objects ahead of the interests of scientific knowledge.[42] Forty years after Lelorrain, Henry Rhind, a Scottish lawyer turned archaeologist working at Thebes, complained that the consuls of France and England in Egypt had failed to subordinate profit to scientific interests. "The [destructive] turn" of the putatively archaeological digs "has in great degree depended on the fact that even or indeed chiefly when under the auspices of governments, the economics of a mining speculation rather than the scope of a scientific survey have been imported into fields of research—the condition being imposed or implied that for so much expenditure so many tangible returns were expected."[43] In an 1865 issue of the *Revue des deux Mondes*, the orientalist Ernest Renan, who had spilled much ink identifying a putative Semitic mentality averse to science and philosophy, echoed Rhind's view, venting his characteristic spleen against the plunder of Egypt with consular permission. "For more than half a century Egyptian antiquities have been pillaged. Purveyors to museums have gone through the country like vandals; to secure a fragment of a head, a piece of inscription, precious antiquities were reduced to fragments. Nearly always provided with a consular instrument, these avid destroyers treated Egypt as their own property." In winding up his indictment, Renan came to the same conclusion Ali had reached years before: "The worst enemy, however, of Egyptian antiquities is still the English or American traveler. The names of these idiots will go down to posterity since they were careful to inscribe themselves on famous monuments across the most delicate drawings."[44]

Purloining the Zodiac

Lelorrain debarked in Alexandria in early November 1820. As the second city of Egypt and the country's most important port, Alexandria should have been a bustling, cosmopolitan place. Yet in 1820 the city was struggling to emerge from a long period of decline, and it was difficult for foreign travelers to sense the grand Alexandria of historical legend, with its famous library and its centuries of intellectual flowering. Arriving in the city the same year as Lelorrain, the Baroness von Minutoli echoed the disappointment of many European visitors to Alexandria when she described it as run-down, noisy, filthy, and dull. Despite the city's cosmopolitan reputation, Alexandria's intellectual life was marred by provincialism and, as the sore Baroness put it, "the gossiping of a little country town."[45] The orientalist William Lane, who was in Alexandria in 1825, complained that "it is a poor, wretched town; its climate is unhealthy; and nothing but sea and desert meets the eye around it."[46]

Perhaps for these reasons, Lelorrain did not linger in Alexandria. Immediately after recuperating from his passage, he struck out for Cairo. As he headed up the Nile, he no doubt saw many of the sights described by Anne Katherine Elwood, wife of Colonel Elwood of Hastings, in her account of travels on the Nile in 1825: buffaloes floating in the water; storks rising, in their ungainly way, when approached by boats; crocodiles sunning themselves on the Nile's banks or splashing suddenly into the river.[47] He arrived at Cairo in January.

The "city of a thousand minarets" must have been an imposing sight, with its ancient wall, its famous citadel, and the pyramids of Giza rising above the horizon. Cairo sprawled on the right bank of the Nile, below the Mukattam Hills, perpetually fogged by the dense smoke from the house fires of its 250,000 inhabitants.[48] As the terminus of long-distance caravan routes through North Africa and the Near East, Cairo's bazaars overflowed with the products of the caravan trade: cotton, flax, grain, ivory, salt, spices, exotic nostrums like rhinoceros horn (touted as an aphrodisiac), silks, precious metals, leather goods, ceramics, slaves. Sailors and soldiers rubbed shoulders with religious pilgrims, merchants, and travelers from Europe and elsewhere; and, five times a day, all were summoned by the *adhan*, or call to prayer, as it rang out from minarets across the city, sung by one muezzin after another. According to one traveler, Eliza Fay, who visited the year before Napoleon's invasion, Cairo "smelled of hot bricks." Another described it as full of "every species of insect, crawling, creeping, jumping, flying, buzzing and humming about one to a tormenting degree."[49] Solid wooden doors closed the European quarter off from the rest of the city. The pasha and other wealthy notables lived in the shady garden district of Bulaq. Despite the bustle, noise, and heat, Saulnier informed his readers, the city was so safe that "one could, without fear, walk at night in the streets of Cairo with both hands full of gold."[50] Lelorrain must have found it at least somewhat congenial, for he stayed until the middle of the following month.

On February 12 Lelorrain left the minarets of Cairo for Dendera, traveling on a chartered boat with an interpreter and one of the pasha's Janissaries, for protection from brigands who roamed the Nile preying on unwary travelers.[51] Halfway there, as the boat maneuvered through a tight spot on the river, Lelorrain noticed a man loitering on deck and asked what he was doing. The man replied, with surprising candor, that Salt had sent him to observe the expedition. He was ushered hastily ashore.

Dendera is located on a bend in the Nile, on the west bank opposite Qena, four hundred miles south of Cairo, sixty miles north of Thebes. The town's name descends from the Greek Tentyris; it had been the principal seat of the cult of Hathor, goddess of love and joy, corresponding to the Greek Aphrodite. Rising at dawn on his first day there, Lelorrain went directly to the temple. An impressive rectangular structure consisting of several chambers, the monument was fronted by a portico of two dozen enormous columns, each twenty-one feet in circumference, covered with relief sculptures and hieroglyphics, and topped with sculpted heads of Hathor that still bore bright traces of their original paint. The temple was surrounded by a wall of mud brick and piles of debris that nearly touched the roof.

Lelorrain found the circular zodiac on the ceiling of a small chapel on the temple's roof. Twelve feet long, eight feet wide, and three feet thick, the chapel's ceiling weighed some sixty tons and was composed of three separate stones. One of these three and nearly half of its neighbor were decorated with a large medallion in which were arranged bas-reliefs of human and animal figures as well as sections of hieroglyphics. The remainder of the neighboring slab was inscribed with two parallel columns of hieroglpyhs between which stretched a nude female figure. The regions above and below the medallion proper were filled with a zig-zag pattern (Figure 10.1 depicts the medallion, nude figure, and zig-zags as drawn by Jollois and Devilliers during Napoleon's expedition).

Saulnier and Lelorrain had assumed that the zodiac was incised into a single block of stone; they had accordingly thought that it would not be difficult to detach the ceiling medallion from the roof and slide it down the pile of rubble that had collected to the sides of the temple.[52] Lelorrain soon found that two stones had to be removed and transported, not just one. To make matters worse, a group of English travelers had appeared outside the site. Although they said they were there to make drawings, Lelorrain worried that, if he tried to remove the ceiling, the English artists would spread the news as they traveled to other places along the Nile. And then it would not be long before Drovetti and Salt moved to foil the venture.

Determined to keep his intentions to himself, Lelorrain started back up the Nile to Thebes, where he learned just how long a shadow Drovetti and Salt cast over Egypt's antiquities trade. When he tried to hire laborers, Lelorrain discovered that the two consuls had intimidated the locals, who steadfastly ignored his offers.[53] He did, however, manage to pick up a few antiquities—several painted wooden amulets, a few sarcophagi, and two sycamore coffin lids, one of which Saulnier speculated might have held a priest, as it depicted its inhabitant holding an ankh and a nilometer. In addition to their intrinsic interest, these items would support the cover story Lelorrain had concocted: if asked, he would present himself as just another amateur treasure-hunter. To ensure further secrecy, Lelorrain created a false itinerary, according to which his next stop would be the Red Sea, where, he let it be known, he intended to collect seashells. This news traveled back to Cairo along with (false) reports that Lelorrain had fallen sick and was now laid up in a village near Thebes. Lelorrain had by this time ceased direct communication with Saulnier because he feared that any letter might fall into the wrong hands. Back in France, the news of Lelorrain's illness nevertheless reached Saulnier via others "with the speed that envy and other vile passions of the human heart never fail to give evil tidings."[54] Meanwhile, the now-incommunicado but quite healthy Lelorrain moved quietly downriver, only a little behind schedule due to the inconvenient English artists.

Lelorrain returned to Dendera on April 18, the worrisome artists having decamped. At last free to begin, he assembled a team of twenty workers who had not been contacted by the consuls, unaware as they were of his aims at Dendera. He arranged for a foreman to supervise the workers and led the group, together with an interpreter, to the temple. Given the huge weight of the slabs, Lelorrain decided to remove only the central portion of the medallion on the first stone, without the bordering zig-zags. He decided as well to cut through the second slab at the point where the medallion proper ended, leaving behind the bordering columns of hieroglyphs and the female figure sandwiched between them.

The first task was to build a scaffold that would support the stones once they had been freed. Having accomplished this, the team was directed to cut a hole in the three-foot-thick ceiling, through which they would introduce a saw to remove the blocks that bore the zodiac. Using the saw risked significant damage. But Lelorrain recognized immediately that his chisels, which he would later need to thin the stones and lighten them for transport, would be ruined if they were used instead.

Boring the hole proved difficult. Frustrated with the slow pace of the work, Lelorrain resorted to blasting the ceiling with gunpowder, a delicate job that required several days of testing to see what formula would do the least damage. After two days of working with high explosives in hundred-degree heat, Lelorrain achieved his objective. But when the workers began to operate the saw, he found that it could cut through only one foot per day. And there were three sides to be cut—twenty-four feet in all. Lelorrain reckoned a single saw would require more than three full weeks of work—plenty of time for Drovetti and Salt to catch the scent of his undertaking and move to thwart it. Fearing discovery more than damage to the ceiling, Lelorrain blasted two more holes, so

Figure 1.2. Lelorrain's handiwork, showing the plaster cast put in the place of the zodiac. The line visible in the cast marks the boundary between the two unequal stone slabs on which the original was carved. Note the fractured stone into which an iron bar has been set on the right to hold the replacement.

that three saws could be used simultaneously. As long as the equipment held out, the job would require only slightly more than a week.

The crew worked hard and fast, and Lelorrain kept continual watch on the saws until, overcome by the heat and sun, he succumbed to fever and paralysis. His illness lasted eight days, during which, again fearing discovery, he refused to send for a doctor. His interpreter saved the project, keeping the saws moving and making sure the work was not interrupted. Meanwhile, a local dosed Lelorrain "with the juice of a plant whose name" he claimed not to know, and he recovered.[55]

After twenty-two days, the ceiling was freed, and the workers hoisted the two stones onto the roof in preparation for their next task, hauling them to the river four miles away. They loaded the blocks onto the large sledge that had been built in France, which was then set in motion on wooden rollers. The first day of dragging went passably well; on the second, they covered more than a mile. But the weight of the stones destroyed the rollers. Although Lelorrain managed at considerable cost to obtain logs that might be used instead of the ruined rollers, these too were destroyed in a single day. After that he was forced to rely on brute force and mechanics, using levers, tackles, and fifty workers' worth of sheer elbow grease to drag the sledge. Even these mighty efforts bore progressively less fruit. At the expedition's lowest point, it took twelve hours to cover sixty steps, and Lelorrain himself, though still convalescing, helped to pull on the ropes. It took sixteen days to reach the Nile, where a fresh setback awaited.

Just as Lelorrain and his crew arrived, the Nile's annual retreat reached its lowest ebb, leaving a twelve-foot vertical cliff above the water, where a barge awaited the zodiac. Undaunted, Lelorrain directed his men to construct a sixty-foot ramp made of earth, onto which they positioned soaped planks. The first stone was hoisted on top. To prevent slips, the workers secured the stone to a palm tree with a cable and roped it to thirty men who were responsible for controlling the stone's descent. The barge was positioned at the bottom of the ramp. As they hoisted the first, massive block onto the ramp, the cable snapped, sending the stone down the ramp at a speed that flipped the men holding the ropes into the air. The stone just missed the barge—which would have sunk on impact—and fell sideways into the soft mud six feet below. The workers dug it out and managed to hoist it onto the barge, which promptly began to sink. By bailing on one side and caulking on the other, the crew managed to keep the barge afloat. The smaller stone was loaded without incident.

At this point, Lelorrain's problems seemed to be over, but respite proved fleeting, for the boat's owner refused to budge, claiming the waters were now too low to set out. Lelorrain argued but to no avail. Puzzled by the owner's recalcitrance, which seemed to him to be unjustified by the Nile's condition, Lelorrain recalled seeing an American tourist at Dendera just as he had finished removing the ceiling from the temple. Moreover, Lelorrain had recently seen a prowler hanging around the boat, someone who looked like a man he had noticed while shopping for antiquities in Thebes, where the consuls had alerted their spies. In a quiet moment, Lelorrain's interpreter informed him that he had overheard the boat's owner boasting that one of Salt's agents—the same man Lelorrain had seen prowling around—had paid him a considerable sum to delay the transport for three weeks. Salt at least had learned Lelorrain's true goals.

Figure 1.3. The Dendera circular zodiac, now in the Louvre (compare with Figure 1.2)

Lelorrain's next move was clear: he made a counteroffer, in the same amount as the original bribe, if the man would cast off right away. The boat's owner, now two thousand piastres richer, was so overcome with gratitude that he fell to his knees, swearing eternal fidelity. Lelorrain set off for Cairo in the hope that no additional business opportunities would come the captain's way. Just outside the city the boat was hailed by an envoy, yet another of Salt's men, who claimed he had an order signed by Kaya Bey, the grand vizier, forbidding the removal of the zodiac. Lelorrain asserted that his authority came from the pasha himself and declared that if anyone wanted the zodiac they would have to board the ship and take it by force. His gamble paid off, and the envoy retreated.

Arriving at last in Cairo, Lelorrain learned that Salt had discovered his activities from Luther Bradish; this was the American Lelorrain had encountered earlier, who was on tour in Egypt as Mehmet Ali's guest after having completed a diplomatic mission in Constantinople for John Quincy Adams, then the American secretary of state under James Monroe. On hearing Bradish's news, Salt flew into a rage. Evidently he had long harbored dreams of capturing the zodiac for himself. He had even developed a plan for which, according to Saulnier, the wealthy antiquarian William Bankes had already sent equipment from London. Salt complained about Lelorrain's activities to the pasha, but Ali was occupied with a more pressing matter—the disruption of a plot to massacre all the Christians in Alexandria. After failing to get the pasha's ear, Salt applied to the grand vizier and sent his envoy down the Nile with the document that Lelorrain had scoffed at when stopped outside Cairo. In addition to this ploy, and perhaps because he knew it had little chance of success, Salt chased the pasha to Alexandria, where he prosecuted his case, arguing that the zodiac was rightfully his because he had excavated at Dendera before Lelorrain had arrived. After hearing his arguments, the pasha asked rhetorically whether it was not the case that he had himself authorized Lelorrain's project. Of course, he had, and that, he informed Salt, was the end of the matter. After the decision was handed down, several Turks in the pasha's entourage expressed wonderment at the extraordinary fuss that had been made over "two stones" for they lived in a country that seemed to contain, as Saulnier put it, "all the stones in the world."[56]

2

Antiquity Imagined

By the time the purloined zodiac arrived in France, the country had been scarred by a quarter century of political upheaval, as a series of bloody events transformed the nation from monarchy to republic to empire and back again. Egypt was to play a central part in the iconography of Napoleon's empire, but it had long before evoked deep and conflicting reactions among the reading public of the late ancien régime, as arguments flourished about the earliest civilizations. The years immediately preceding the revolution were especially fertile in generating novel, occasionally outlandish, and often heretical theories concerning remote antiquity and the origins of religion; many of them looked to science for support. When the revolution turned from legislation to terror, academic discussions about remote antiquity degenerated into ideological justifications for violence. For Egypt—place, mystery, idea—had armed the Jacobins with weapons to slay religion, or what they termed *fanaticism*, which included almost any form of belief in the supernatural. When the Egyptian zodiacs captured the French imagination in the early years of the empire, controversy and even fear swirled about them as conservatives remembered the uses to which Egypt had been put scarcely a decade before.

The Origins of Civilization

In 1760 the sixty-four-year-old epitome of philosophes, Voltaire, received a mysterious manuscript from a military adventurer by the name of Laurent de Féderbe, Chevalier de Maudave. Written in French, it purported to be the translation by a Brahmin of an ancient Sanskrit document. Believing it to date from before the time of Alexander's conquests, Voltaire had the document, known as the *Ezourvedam*, copied and presented to the Royal Library in Paris.[1] In this peculiar manuscript, Voltaire spied the origins not only of specifically Christian precepts but of an altogether superior form of civilization and morality. Although the antique Brahmins were as subject as other ancient peoples to the absurdities of theology and metaphysics, Voltaire thought, they had nevertheless developed a society free from the sorts of corruptions epitomized for him by the French clergy. India, not Greece, and certainly not Egypt or ancient Israel, had produced the closest humanity had come to the ideal civilization. Although the *Ezourvedam* turned out many years later to be a missionary forgery designed to show that Vedic precepts were derived from Christian sources rather than the reverse, Voltaire never suspected anything and chose in any case to ignore remarks about Indian civilization by his contemporaries that did not fit his idealized image.[2]

Voltaire's interest in remote antiquity would eventually come back to haunt him. In his old age, he entered into an extensive correspondence with the astronomer Jean-Sylvain Bailly, who had developed an unusual theory about remote antiquity that he pushed the aged philosophe to accept.[3] According to Bailly, civilization began not in India or China (the two places Voltaire had favored), much less among the ancient Hebrews, but about five hundred miles from the North Pole. He placed the first people in Plato's apocryphal Atlantis, which he eventually, if rather inconveniently, located on the island of Spitzbergen. Thoroughly frozen over, at least until recent times, Spitzbergen had been discovered in 1596 by the Dutch explorer Willem Barents, who perished the next year when his ship was trapped in ice. The clear, cool air of the far North, warmed in remote times by Earth's inner fire, had according to Bailly stimulated its ancient inhabitants to develop agriculture and, thereafter, astronomy and the arts of civilization, which they then spread through colonization. These were the happy Hyperboreans of Greek myth, living in perpetual sunlight beyond the north wind, among whom Apollo wintered.[4]

Bailly's route to this peculiar result followed his encounters with two of France's great astronomers, Nicolas de Lacaille and Alexis Clairault, both of whom he met while keeper of the king's paintings, a position inherited from three generations of paternal ancestors, and which he held until 1783. The astronomers' influence encouraged Bailly to begin his own research in 1759, focusing on Halley's comet. A year later he had an observatory at the Louvre, and in 1763 Bailly replaced Lacaille at the Académie des Sciences. After 1771 Bailly began a history of ancient astronomy that appeared in 1775. He followed this four years later with a supplement that took the story to 1730, which he then continued through 1782. Along the way Bailly engaged Voltaire's views on remote antiquity, publishing in 1777 a tome on the origins of science and another of letters detailing his own remarkable speculations. His reputation expanded during the 1780s, leading to membership in both the literary Académie Française and the philological and historical Académie des Inscriptions et Belles-Lettres in 1783. This was not merely unusual, it was almost unprecedented either before or after; only Bernard de Fontenelle had previously acquired the triple crown.[5] The now famous Bailly became mayor of Paris on the day following the fall of the Bastille, but the dangerous and fast-moving events of the next two years overtook him. On the Champ-de-Mars in July 1791 he ordered troops to fire on a crowd clamoring for the king's elimination, killing fifty people. This order brought Bailly to the guillotine two years later—doomed, wrote Carlyle, "for leaving his Astronomy to meddle with Revolution."[6]

The aged Voltaire must have been exhausted by the relentless Bailly's twenty-eight letters on the northern Atlantis, the last sent on May 12, 1778, just eighteen days before the sage of Ferney died. "I don't dispute anything against you," Voltaire wrote Bailly in 1777, "I only seek to learn. I'm an old blind man who asks you to point out the way. No one is more capable than you to rectify my ideas about the Brahmins."[7] Though Bailly's books were "treasures of the most profound erudition"—this of a man who traced the path of Jason and the Argonauts all the way to a polar Atlantis—"the sick old man of Ferney," as Voltaire signed himself, remained persuaded that philosophy and science originated in India among the Brahmins. Despite Voltaire's skepticism, Bailly continued to expand his pursuit of the northern Atlanteans. In a lengthy series of letters in 1782 to the poetess and muse of Ferney, Madame du Bocage, Bailly even went so far as to outline an entire history of myths. The letters were collected and published posthumously in 1799 as *Essai sur les Fables*.[8] The volume's anonymous editor ("P****"—probably Pierre Baudin) lamented the unfortunate Bailly's death and praised his assiduous research as "one of the grandest enterprises of the human spirit," which "lit the night of time" itself.[9]

In the two decades or so before the revolution, Bailly was hardly the only person to club religion with nature by inventing schemes that, for all their purportedly scientific qualifications, gave away little in fancifulness to the biblical creation myth. Odd though they may seem in retrospect, his views exhibit the particularly febrile interest during those years in the origins of religious beliefs and practices of every kind. Joseph de Guignes provides a case in point. A specialist in Chinese, de Guignes had a sufficiently solid reputation that he had been admitted to the Royal Society of London in 1752, and to Inscriptions et Belles-Lettres the following year. He had obtained the chair in Syriac at the Collège Royal in 1757, whereupon he took the opportunity to argue, in Latin, that the French kings were "much better made for letters" than were the "princes of Asia," according to a nineteenth-century biographical notice that is replete with the odor of exoticism and condescension that so often characterized European attitudes toward the East at the time (and not only then).[10] Nonetheless, he believed that Egyptian culture had had an immense influence. Indeed, so great had it been that, de Guignes argued in a lecture given at the Académie, China had been colonized by Egyptians in remote antiquity. Chinese culture therefore originated in Egypt, and Chinese writing was itself a transformation of the unreadable Egyptian hieroglyphs. Not surprisingly, de Guignes thought that the Chinese zodiac was essentially Egyptian in origin, and that the zodiacal images proper represented agricultural labor.[11]

Another theory located the origins of civilization in Plato's Atlantis—or, rather, in an Atlantis transplanted to different climes from Bailly's. In 1770 Jean Baptiste Claude Izouard, writing under the pseudonym Delisle de Sales, produced his own *Philosophie de la Nature*, a bizarre and fractured work that countered revelation with nature, arguing from geological evidence that the earliest civilization (Atlantis) had been located in Sardinia. The book was condemned in France, which ensured its popularity (it went through no fewer than seven editions before 1804).[12] Meanwhile, Izouard's property was confiscated, and he left for exile in London. Like many other provincials who had come to Paris seeking to make their names as philosophes, Izouard had plied his trade in the fermenting literary underground in the last years of the ancien régime. Angered by their inability to penetrate the academy and the powerful cultural milieu known as *le monde*, with its lucrative government pensions and honors, and which by the 1770s had absorbed the now-aging philosophes of the first generation, fringe writers like Izouard exploded the boundaries of discussion. Their works, which were often printed outside France, in Holland or in Switzerland, circulated clandestinely through networks of dealers connected to the foreign printers.[13]

Although the 1770s and 1780s were especially fertile periods for theories concerning the origins of credence in invisible powers, the nature of pagan religion had been an important theme in the seventeenth century as well. Pierre Bayle had, for example, rejected the allegorical interpretations of religious belief common among Renaissance humanists and insisted that pagans took their myths to be literally truthful. For Bayle there was no essential difference between Greek ways of thinking and those of the African and Micronesian fetish-worshippers brought to European attention through exploration and colonial exploitation.[14]

Fables and mythologies continued to occupy French scholars and philosophes during the first half of the eighteenth century. Fontenelle saw in fable a recourse to the supernatural in the face of the world's complexity, while the Abbé Banier vigorously upheld the notion, which dates to Euhemerus in the fourth century BCE, that mythologies are transfigurations of historical events. Banier's colleague at Inscriptions et Belles-Lettres, Nicolas Fréret, at first himself inclined to Euhemerism, evolved a new position, according to which fables must neither be dismissed as uninformative fantasies nor overrationalized but treated as "a confused mélange of imagination's reflections, of philosophy's dreams with the debris of ancient history."[15] After 1745 the pursuit of mythologies and fables abated, only to be revived with vigor in the early 1760s with the iconoclastic work of Charles de Brosses and Nicolas Boulanger.

De Brosses, a French magistrate and scholar, journeyed with a friend to Italy in 1739 to observe the newly discovered ruins of Herculaneum, about which he wrote the first work to reach print, *Lettres sur l'État Actuel de la Ville Souterraine d'Herculée* (Letters on the Present State of the Underground City of Herculaneum).[16] Having coined the word "fetishism" to describe the magical use of wooden idols in West Africa, de Brosses became convinced that Egyptian religion had operated in the same fashion—the sphinx was, for example, not an enigma but a fetish, an object thought to be invested with magical powers. His further insistence on the extreme antiquity of animal worship verged on heresy because it suggested that monotheism may itself have emerged directly out of idolatry. To avoid this problem, de Brosses creatively suggested that the Noachian deluge marked an absolute break with the past. Whereas the postdiluvian world was infected with fetishism, a pure form of monotheism had prevailed universally before the Flood.[17]

An engineer who had served in military campaigns under the Baron de Thiers, Boulanger was a close friend of the philosophe Denis Diderot. Having built bridges and observed the geological actions of water, Boulanger went beyond de Brosses to construct a hydrological history of all religions, including monotheism. For him, the deluge marked the signal moment in natural history, and all myths and fables were distant echoes of the vast displacements caused by this universal flood. "One must," he wrote in the posthumously published *L'Antiquité dévoilée* (Antiquity Unmasked),

> begin the history of societies and of present nations with the Deluge. If there have been false and harmful religions, it's to the Deluge that I will go back to find their source; if there have been doctrines inimical to society, I'll see their principles in the consequences of the Deluge; if there has been vicious legislation and an infinity of bad governments, I will accuse only the Deluge; if a crowd of usages, ceremonies, customs and bizarre prejudices introduced themselves among men, I will attribute them to the Deluge; in a word the Deluge is the principle of everything that over the centuries produced the disgrace and misfortune of nations. . . . The fear which thereafter seized man's heart prevented him from discovering and following the true means to reestablish the destroyed society.[18]

Though the Deluge existed for both de Brosses and Boulanger, for the one it marked the divide between antediluvian monotheism and postdiluvian polytheism, while for the other it marked the moment after which humanity, terrified by nature, succumbed to supernatural explanations, whether of single or multiple deities. According to Boulanger, liberty can arrive only once religion vanishes, and the memory of the Deluge evaporates. His work was published in Amsterdam to avoid censorship and was considered to be sufficiently outside acceptable bounds that it was confiscated at least nine times at French customs and three times by the Paris police between 1771 and 1786.[19]

Boulanger had also written an unpublished work entitled *Les anecdotes de la nature* (Nature's Anecdotes) in which he used geological facts to illuminate the postdiluvian epoch.[20] The historian of geology Martin Rudwick remarks of this and similar works that they "created a substantive link between human history and the history of nature."[21] These, he explains, are indicative of a new genre that was emerging during the eighteenth century. Instead of skipping like Gulliver over a featureless timescape, in this new genre earth historians examined local geological details in much the same way that eighteenth-century antiquarians immersed themselves in the details of a region's history, details that included material evidence such as architecture. "The sciences of the earth," notes Rudwick, "became historical by borrowing ideas, concepts, and methods from human historiography."[22] Moreover among geological investigators the grip of Mosaic history, at least for the period of Creation proper, had greatly loosened during the century, and not only among them. The Flood became a link that was often used to join human to natural history, which eventually had the effect among many of reducing its substantive reality to a question of history as well as of belief. This "lively concern to understand Genesis in scientific terms, and more particularly an interest in identifying the physical traces of the Flood, facilitated just the kind of thinking that was needed in order to develop a distinctly *geohistorical* practice within the sciences of the earth."[23]

Protestant and Catholic belief that the deity could have done otherwise accordingly enabled a focus on the contingencies of geological history. "Given this sense of the sheer contingency of human history" among believers, Rudwick continues, "a newly *historical* science of nature, stressing a similar contingency in the deep prehuman past, could well be built on foundations borrowed not only from erudite antiquarianism but also from the radically historical Judeo-Christian religion that was so scorned by most of the philosophes of the Enlightenment."[24] Deists, and especially the French philosophes, were in contrast gripped by visions of an utterly noncontingent history of Earth, and for that matter of humanity itself. A vision in which natural law alone, having perhaps been set into action in some remote period by a thereafter absent power, progressed rigidly of its own accord. That way of thinking tended to make the details of history, whether natural or human, subservient to an extremely broad theory, a "geotheory," as Rudwick terms it.

Although the power of a literal reading of the Mosaic account of Creation had already waned by the middle of the eighteenth century, the Flood remained for some time a significant linchpin connecting human with geological history. Even here the biblical account was not read altogether literally. Instead, the general idea was appropriated from it that human history dates back only five or six thousand years. The cataclysm of a universal flood wiped out any of humankind's creations that existed up to that moment (if any), but the preceding periods of geological time extend back vastly further than that: The restriction of human history to several thousands of years was widely ridiculed by philosophes, who immediately linked any such claim to religious beliefs.

Ruminations about antiquity and the origins of religion strongly exercised conservatives during the two decades or so before the revolution, and we will see that the attitudes that they developed during these years returned in particularly virulent forms in the early years of the Restoration. Claims that either implicitly or explicitly treated the Bible as myth seemed in their view to strike at the foundations of society and of morality itself. An uncompromising vision of perfect good versus absolute evil came to dominate their discourse, as conservatives deplored intellectual and social license, praised the family, and denigrated abstract rights. Stimulated especially by the arch-enemy Voltaire's triumphant return in 1778 to Paris, antiphilosophe works spread like wildfire under the auspices and subsidies of the Committee on Religion and Jurisdiction of the Assembly General of Clergy. Jesuits with moderate views were thrown out of editorial positions, and subtlety vanished as positions hardened into simplistic assertions of orthodoxy. Sacred texts were plundered for visions of a philosophe-engendered apocalypse, Louis XVI was accused of compromise with the forces of evil, and the vision of a premodern, profoundly religious society that had never actually existed took shape. During the early years of the Napoleonic empire many of these themes reemerged in the anti-Enlightenment attacks of such authors as Chateaubriand, Bonald, and de Maistre, foreshadowing the vengeful, utopian hopes of the implacable Catholic Ultras who came to power and influence during the Restoration of the Bourbon monarchy after Napoleon's fall.[25]

Partly as a result of this situation, increasing numbers of pamphlets, books, and periodicals were formulating an extraordinarily vituperative attack against the social and cultural predominance of those who espoused one or another form of Enlightenment philosophie, with particular anger directed at attempts to undercut sacred chronology. Elie-Cathérine Fréron was, for example, a vocal enemy of Voltaire, whose journal, the *Année littéraire*, had for some time opposed philosophes. He also supported young aspirants to literary fame who could be used to further his battles, among them, for example, Nicolas Gilbert. Born to a peasant family, Gilbert sought like many others to gain entry to the world of Parisian philosophes, having written a small book of verse that he hoped would do the trick. And, again like so many, he failed. D'Alembert, and no doubt others, refused help. Gilbert did not, however, turn to the bubbling underground literary press of the Parisian equivalent to London's Grub Street, there to produce scandalous pamphlets and satires. Supported by Fréron, Gilbert instead directed his anger entirely at the world of the philosophes.[26] Nor was he alone, for others also contributed to a growing chorus that bracketed together the many strands of philosophe discourse, various though they certainly were. Family, throne, and altar had to be defended from philosophe attacks, attacks furthered by the infectious spread of licentiousness of all sorts, from the literary to the social. In 1785 Marc Antoine Noé, the bishop of Lescar, warned that unless the noxious plague of philosophes was extirpated, France would be steeped in a "sea of blood" and subject to a "flood of fire."[27]

A Voyage to Egypt and Syria

Two years after Noé's warning, a young man returned to France after an extended visit to Egypt and Syria, soon to produce two extraordinarily influential books that exemplified both contemporary attitudes to the "Orient" and the conviction that the Bible was merely myth. In 1775 the young Constantin François Chasseboeuf Boisgirais took advantage of the 1,100 livres per year left to him by the estate of his mother, who had died when he was only two, and moved to Paris where he intended to study medicine. Introduced by his fellow student, Pierre Cabanis, to the salon of Mme Helvetius, and eventually frequenting that of the vocal atheist Baron d'Holbach as well, he became acquainted with the day's luminaries, including Voltaire, Diderot, and the American Benjamin Franklin. Having lost interest in medicine, Chasseboeuf pursued the study of Hebrew, an interest he had brought with him to Paris from the provinces, and took up Arabic as well. Enamored of ancient history, Chasseboeuf studied it with the aim of uncovering links between beliefs and mores, especially religious beliefs, and the rules that govern the natural world. He soon authored a memoir on the chronology in Herodotus and conceived the idea of journeying to Egypt and to Syria, thereafter to write an account patterned on the Greek historian's (though there are indications that he was sent for some reason by the French Ministry of Foreign Affairs).[28] He left in December 1782 and did not return for over two years. Shortly before leaving, Chasseboeuf added "Volney" (a contraction of Voltaire and Voltaire's village, Ferney) to his name. He was always known by this adulatory addition, one that would inevitably, and no doubt deliberately, antagonize conservatives, for whom Voltaire epitomized the evils of philosophie.

Volney limited his excursions in Egypt for the most part to Alexandria and Cairo, remaining there for seven months. In September 1783 he left for Syria, where he lived for eight months among the Druzes before ensconcing himself in a Coptic monastery. His *Voyage en Égypte et en Syrie*, a two-volume account of the journey, was published in

1787. In it he surveyed the mores and habits of the peoples he had visited and offered as well a theory about their present state—namely, that what he thought to be the parlous and oppressed condition of the region's inhabitants reflected the invidious impacts of despotism and religion. After disparaging the Turks, whose Ottoman Empire had since 1517 ruled Egypt through the medium of the Mamelukes, Volney helpfully remarked how exposed to invasion the coast was.

> The Turkish spirit ruins the works of the past and hope for the future, because there is no tomorrow within the barbarity of ignorant despotism. Alexandria can hardly be considered a town fit for war. One sees there no fortification, not even the lighthouse with its high towers. There aren't even four canons in a fit state, and not a cannoneer who knows how to point them. The five hundred Janissaries who are supposed to constitute the garrison, reduced by half, are workers who know only how to smoke their pipes. The Turks are happy that the French are interested in running the town. One frigate from Malta or Russia would be enough to reduce it to cinders.[29]

French merchants had for years been established in Alexandria, and Volney was apparently convinced that they ran the town with the indifferent connivance of its indolent rulers.

Volney's remarks evoked an image of the "Orient" that held powerful sway over the European imagination for many decades to come, and that continues to color Western attitudes to the present day. Ignorance, despotism, and torpor conspired with a retrograde religiosity to make the region ripe for exploitation—in the interests, naturally, of leading its inhabitants out of medieval darkness into the light of modernity. So, for instance, Volney warned that travel in Egypt by Europeans was difficult and dangerous because "the superstitious people look upon them as sorcerers come to steal by magic the treasures guarded under the ruins by genies."[30] He attributed this primitive mindset to the influence of Islam, claiming that the "ignorance of the peoples in this part of the world is the more or less immediate effect of the Koran and its morals."[31] Despite its evident Enlightenment expectations and prejudices, Volney's treatment of the region's habits, mores, and governments was detailed and vivid, and it proved quite popular. It was translated into English and printed in London the next year, then reprinted in France in 1792 and again in 1799.

Though hardly a political radical, Volney welcomed the convening in France of the Estates-General and then the Constituent Assembly, at which he represented Anjou. In 1791, having left Paris following the dissolution of the Assembly, Volney thought to put his economic views into practice in Corsica. Like his friend, the American Thomas Jefferson, Volney thought that a country's prosperity is proportional to the number of its independent farmers. The government was, it seems, sufficiently impressed by the venture that it appointed him director of agriculture and commerce for the island, though Volney's scheme eventually proved fruitless. Shortly before leaving for Corsica, he had published the most influential of his several books, the *Ruines, ou Méditations sur les révolutions des empires* (Ruins, or Meditations on the Revolutions of Empires). Beginning with the contemplation of the detritus of fallen kingdoms, Volney wondered whether all such are destined to tumble into ruin.

Reflecting the increasingly visible antireligious fervor of early revolutionary France, Volney imagined a convocation of all faiths, during which their inherent contradictions would be exposed through rational debate. Following an astronomical theory based on Egyptian sources that was current at the time (and about which we will have more to say), Volney explicitly targeted monotheism, intending to "demonstrate that the tenth chapter [of *Genesis*], among others, which treats of the pretended generations of the man called Noah, is a real geographical picture of the world, as it was known to the Hebrews at the epoch of the captivity. . . . All the pretended personages from Adam to Abraham, or his father Terah, are mythological beings, star constellations, countries. Adam is Bootes: Noah is Osiris."[32] At Volney's imagined conclave of religions, the Brahmins protested the conceit of the monotheists. "What are these new and almost unheard of nations," they cried, "who arrogantly set themselves up as the sources of the human race, and the depositories of its archives. To hear their calculations of five or six thousand years, it would seem that the world was of yesterday; whereas our monuments prove a duration of many thousands of centuries. . . . And is not the testimony of our fathers and our gods as valid as that of the fathers and the gods of the West?"[33]

Volney combined a heady reduction of all forms of religious belief to misplaced worship of the heavens with an attempt to construct "the social virtues" entirely out of a form of "natural law." Deities of any kind, whether plural or singular, had no place in this system. He went into particular detail in an attempt to connect Christian belief to the

constellations Taurus, Aries, and Pisces. As the well-known phenomenon of celestial precession shifted the stars around the ecliptic (of
which more below), religion moved in lockstep. "The worship of the
bull of Egypt—the celestial Taurus—has given place to that of the lamb
of Palestine—the celestial Aries; and under the astronomical emblem
Pisces—the twelfth sign of the zodiac—the dominant faith of today
[Christianity] was appropriately taught by the twelve apostolic fisher-
men." Indeed, in the worship of Pisces, represented by two fish tied
together, "may be found the true secret of the origin of the rite of bap-
tism." And since astronomical precession ever continues, and inas-
much "as the masses still continue credulous and devout, they may in
succeeding ages be again called upon to worship the god Apis, when the
sign of Taurus shall again coincide in the zodiac and the ecliptic; and
Aries, 'the lamb of God,' may again be offered in the fullness of time as
a sacrifice for mankind, again be crucified, and again shed his redeem-
ing blood to wash away the sins of a believing world."[34]

A work that incarnated a profound disdain for all forms of belief, but
especially for monotheism, the *Ruines* proved so popular that a second
printing appeared the year after the first, and it was soon translated
into English (1792), Dutch (1796), and German (1792), with separate
American editions in 1796 and 1799. In America, the English exile,
chemist, and pamphleteer Joseph Priestley, who held rather unorthodox religious views of his own, nevertheless attacked Volney in print
for unwarranted skepticism and atheism. Priestley had long cast a cold
eye on the French, whom he suspected of rampant disbelief. On a visit
to Paris in 1774 he had found "all the philosophical persons to whom I
was introduced at Paris, unbelievers in Christianity, and even professed
Atheists. As I chose on all occasions to appear as a Christian, I was told
by some of them, that I was the only person they had ever met with, of
whose understanding they had any opinion, who professed to believe
in Christianity."[35] Priestley hardly veiled his criticisms, and not only
of Volney, whom he accused of pretending to linguistic knowledge that
he surely lacked. Volney responded in kind, bringing in Locke, whom
Priestley admired. "If you admit, with Locke," Volney replied, "and with
us infidels, that every one has the right of rejecting whatever is contrary
to his natural reason, and that all our ideas and all our knowledge are
acquired only by the inlets of our external senses; What becomes of the
system of revelation, and of that order of things in times past, which is
so contradictory to that of the time present? Unless we consider it as a
dream of the human brain during the state of superstitious ignorance."[36]

The fervent antireligiosity of the revolution made it extremely difficult for anyone to make claims, however well grounded in evidence, that seemed to square with religious belief. A particularly significant case for our purposes is provided by the geologist Déodat de Dolomieu who, though an aristocratic Catholic, at first backed the revolution. Dolomieu had argued for the geological significance of a series of great tsunamis, which he applied to local details. The most recent such event demarcated known human history and had occurred around six thousand years ago. And here he ran into problems with his contemporaries. "This truth would not perhaps have been attacked so fiercely and been so strongly combated," he complained in 1792, "had it not been related to religious opinions that one wanted to destroy. . . . It was believed to be an act of courage, showing oneself exempt from prejudice, to increase—by a kind of bidding up—the number of centuries that had elapsed since our continents were given over to our [human] industry."[37.]

Volney himself did not fare especially well during the tumultuous years after the revolution. Upon his return to Paris in 1793, he published a pamphlet that set out his views of natural law based on a form of Locke's sensationalism. Like many others deemed insufficiently radical or otherwise undesirable, he fell victim to the rapidly escalating Terror and was arrested in November. Accused of an aversion to the cleansing power of anarchy, Volney spent the next ten months in prison, no doubt expecting at any moment to hear the cartwheels turning that would convey him to the guillotine. He managed to survive until Robespierre's fall on the 9th of Thermidor, Year 2 in the new Revolutionary Calendar (July 27, 1794). His fortunes recovered thereafter, and Volney became one of the professors at the newly established, but temporary, École Normale in 1795. Journeying to the United States later that year, he ran afoul of the government under John Adams, which thought him to be spying for France. We will encounter him again when we consider Napoleon's early career.[38]

As the revolutionary imagination became increasingly concerned with shaping a new French culture purged of all forms of "fanaticism" (especially Catholicism), the Jacobins drew on prevailing ideas about remote antiquity to forge an ideology that was, ironically if not surprisingly, itself steeped in a form of idolatry. Now, however, the idol was *reason*, incarnated in human form and presented in dramatic, carnivalesque stagings.

The Cult of Reason

Jacobin fervor reached a climax on October 23, 1793. On that day, the Council of the Commune of Paris ordered that "the Gothic simulacra of the kings of France that are placed in the portal of the church Notre Dame will be overturned and destroyed."[39] The cathedral's doomed sovereigns were in good company, for the revolutionary iconoclasts also aimed to obliterate all images of saints, Virgin, and Christ child, emblems that reeked of priestly intrigue and superstition. Early the previous month, the Terror had begun to gather its victims, stimulated in part by Charlotte Corday's assassination of Jean-Paul Marat in July. Among the most ardent of the Terror's proponents were the militant spokesmen for the sansculottes, Pierre Chaumette (the president of the Paris Commune who had styled himself "Anaxagoras" Chaumette) and his fellow member of the radical Cordeliers Club, Jacques Hébert.[40] Hébert was also editor of the newspaper *Le Père Duchesne*, in which he sarcastically attacked aristocrats, lawyers, and priests—a group that Hébert singled out for especially harsh treatment. It was Hébert's supporters who inaugurated the Terror, and it was Hébert who insisted on transmuting churches into Temples of Reason. These new temples were not, however, devoid of a reigning deity.

At ten in the morning of November 10, the goddess of Liberty visited the Temple of Reason. Draped within to hide its religious symbols, the cathedral of Notre Dame that day contained a platform whose top bore a small Grecian temple evocatively labeled "À la Philosophie." At the platform's foot burned the Flame of Truth. Martial music played for the assembling crowd. Two young girls emerged from behind the temple. Descending solemnly to the Altar of Reason, they stopped before it and, crossing themselves, bowed before the flame. From the temple emerged a woman draped in a white robe and blue cloak, capped by a red bonnet: Liberty personified. To her the crowd sung Marie-Joseph Chénier's lyrics, set to music by François-Joseph Gossec: "Descend, oh Liberty, daughter of the Nation; The people have recovered their immortal power; On the pompous debris of ancient imposture, their hands raise up your altar; Come, conqueror of kings, Europe beholds you; Come, swell your success to envelop these false gods; You, saint Liberty, come, dwell in this temple; Become the goddess of the French."[41]

Arriving after the ceremony at the National Convention, Chaumette announced that "the people have just sacrificed to Reason in the former metropolitan church; they come now to offer one as well in the sanctuary of the Law." Whereupon a group of young musicians followed by equally young republicans filed in, singing a patriotic hymn; swift on their heels came a group chanting "Vive la République! Vive la Montagne." At last, the goddess of Liberty herself entered, carried aloft in a chair borne by four devout citizens. "We have not offered our sacrifices to empty images, to inanimate idols," Chaumette declaimed. "No: We chose a masterpiece of nature to stand for her, and this sacred image has enflamed every heart. One sole vow, one sole cry was everywhere heard. The people have declared: no more priests, nor other gods than those offered by nature herself. We, her judges, we have gathered this vow, we bring it to you from the temple of Reason."[42] Carried forward by the power and emotion of the moment, and no doubt by the presence of so many armed Parisians, the convention rebaptized Notre Dame as the Temple of Reason, upon which the goddess of Liberty herself ascended next to the president. A goddess indeed: in a previous incarnation, Chaumette's "masterpiece of nature" had sung at the Opéra. "A woman fair to look upon, when well-rouged" wrote Carlyle decades later, "she, borne on palanquin shoulder-high; with red woolen nightcap; in azure mantle; garlanded with oak; holding in her hand the Pike of the Jupiter-Peuple, sails in: heralded by white young women girt in tricolor . . . *Goddess of Reason*, worthy. And alone worthy of revering. Her henceforth we adore."[43] Half the assembled members of the convention reconvened at Reason's temple, where the morning's ceremony was repeated. The other half chose discreetly to absent themselves.

Two weeks later, on November 23, Chaumette proposed to shutter all the churches in Paris. Anyone demanding their reopening was to be arrested as a traitor.[44] Additional "Festivals of Reason" cropped up throughout France on December 12, a day on which the Vendée rebellion killed seven thousand republicans, and even Voltaire's niece, Madame de Villette, climbed the scaffold. Yet it was none other than this same Chaumette who pleaded, in the name of "arts and philosophy," that the images at Notre Dame should be preserved.[45] One of the three carvings of the Virgin on the portal of Notre Dame (the one on the central axis, or trumeau, between the doors, depicting her with the infant Christ) had already been wrecked, but two others remained. Why should they be saved? Because, Chaumette intoned, the images were not the abominable figments of priestly imagination—not at all, for they represented instead the Egyptian goddess Isis and her child, Horus, god of the day! Of even greater significance, the façade was

Figure 2.1. The apse of Notre Dame—the "Temple of Reason" in 1793—as it appears today

replete with explicit symbols of the zodiac and of agriculture (Figure 2.2). Together they represented features of the natural order, not the invidious superstitions of despotic prelates. Indeed, Chaumette argued, these very figures had recently led the prominent scholar Charles François Dupuis to a new understanding of Christian imagery that fit Chaumette and Hébert's "Cult of Reason." Although Dupuis had not by then fully worked out his system, his basic ideas had already spread quite far. Three years earlier, one François Delaulnaye had even published a popular prospectus that framed Dupuis' system squarely as an antidote to religious belief.[46] Chaumette's assimilation of the Virgin to Isis must

Figure 2.2. Signs of the zodiac (*top*) and of agriculture (*bottom*) on the façade of Notre Dame

have been particularly convincing, since the Paris Council thereupon appointed Dupuis to the administration of public works, in charge of conserving worthy monuments.

Though Chaumette himself may have been more mystic than atheist,[47] at the time anticlericalism combined readily with political radicalism. In this potent brew of ideas, any form of religious belief amounted to a concealed monarchism. The self-described "orator of the human race," Anacharsis Cloots (née Jean-Baptiste du Val-de-Grace, baron de Cloots) explained to the convention that "a people of theists necessarily become revelationists, that is to say, slaves of priests, who are but religious go-betweens, and physicians of damned souls. . . . We shall instantly see the monarchy of heaven condemned in its turn by the revolutionary tribunal of victorious Reason; for Truth, exalted on the throne of Nature, is sovereignly intolerant."[48]

Maximilien Robespierre feared both the destabilizing power of these antireligious fulminations as well as their political strength. He was, moreover, hardly an atheist, though by this time he may have had trouble distinguishing himself from an incarnation of divine virtue.[49] Six days after Chaumette's proposal to shut the churches of Paris, Robespierre declaimed against attacks on religion. The following May he proposed instead a new "Cult of the Supreme Being" (Figure I.1). Though a deity still, Robespierre's new Being was untainted by priestly hands.[50] Chaumette and Hébert met the guillotine that April and March. In June Robespierre held a festival in honor of his newfound Being, while falling himself in mid-July to the Thermidorean reaction. Dupuis, for his part, survived both his embrace by Chaumette and the fall of Robespierre.

By 1795, Year 3 in the Revolutionary Calendar, and a year after the demise of Robespierre, attitudes had changed. Baudin, the man who would write so fulsomely of Bailly's fabulous and dangerously irreligious history in 1799, spoke before the convention, not merely in favor of religious toleration, but of religion's decidedly positive virtues. A cleric who had retained his bishopric, Baudin argued that only those with leisure enough to read the philosophes had become irreligious. The true salt of the earth, France's farmers, artisans, and workers, had "neither the means to buy these voluminous collections, nor the time to bother with them"[51]—a decided exaggeration in view of the extensive, and cheap, anticlerical literature of the underground press. Nowadays, he went on, "only bewigged boys mock those who go to mass." True, there had been abuses on the part of the clergy, but that was not due to the intrinsic character of the class. It reflected instead the real source of intolerance and fanaticism: the court and its powerful allied families. They had passed into history. Privilege was the source of all problems, but privilege had been abolished. It was now time to "complete our work" by giving "to those what they demand, religious liberty, which they ought never to have lost." With, however, one small exception: the public espousal of atheism must not be tolerated. The very first article of Baudin's proposed decree emancipated les cultes from proscription and specified that "the French nation denies the exercise of political rights to those who publicly profess atheism."[52] Perhaps a prudent measure in light of the horrors to which the extremes of Chaumette and Hébert had led, the decree never formed part of the constitution voted in that year. Nevertheless, Baudin's proposed exclusion of atheism from the public sphere foreshadowed the reappearance of religious discourse, with concomitant strictures on words that might have doctrinal implications, such as remarks about remote antiquity.

3

The Origin of All Religions

A Reluctant Scholar in Public Life

In Year 3 of the French Republic, Dupuis published his controversial, and hugely influential, magnum opus, the *Origine de tous les cultes* (Origin of All Religions), which laid out at stupefying length the evidence for his thesis: that every religion originated in the connections between the heavens, agriculture and the seasons. Fifty-three in 1795, Dupuis had established himself quite well long before the revolution.[1] His interests in both science and religion were long-standing. At the time of Dupuis' birth, his father was an instructor in mathematics and surveying in the town of Trie-le-Chateau near Gisors in the Oise, about forty-five or so miles northwest of Paris. A somewhat improved position moved the family two dozen miles to the village of Roche-Guyon, near Mantes. The local seigneur, the duc d'Anville and (by marriage in 1732) de la Rochefoucauld, one day found Charles-François, then about twelve, attempting to measure the height of an old tower with a surveying instrument. Impressed, Rochefoucauld paid for him to attend the ancient Collège d'Harcourt in Paris as soon as he was fit to do so. This apparently took little time as the young Dupuis threw himself into the study of Latin, leaving behind for a while the interests in measurement and instruments that he had absorbed from his father.

Established originally in 1280 by the canon of Paris for poor students, Harcourt included among its graduates such luminaries as the dramatist Jean Racine, the poet Nicolas Boileau, and the fabulist Charles Perrault. Dupuis' father died during his years at Harcourt, whereupon Rochefoucauld undertook the support not only of the boy but also of his mother and siblings. Dupuis distinguished himself to such an extent in Paris that he was granted a master of arts without formal examination. His triumphant march through the system continued as, now a twenty-four-year-old student in theology, he was similarly allowed to dispense with the statutory examination normally required of an incipient professor when he took up the influential chair of rhetoric in the Collège de Lisieux of the University of Paris in 1766 without (or so it was said) having asked to be nominated.

Dupuis had arrived in Lisieux just two years after a period of religious upheaval, during which the Jesuit order was expelled from France.[2] The ejection of the Jesuits between 1762 and 1764 had a number of complex causes, some of which dated back more than a century. As one of the ten major colleges of the University of Paris, Lisieux had been moved to the buildings of the now-vacated Jesuit college of Louis-le-Grand in 1762, which had not formed part of the University of Paris. After two years of increasing difficulty occasioned by conflicts between the teachers and officers with the new organization, Lisieux separated and was replaced by the Collège de Dormans-Beauvais.[3] Dupuis arrived not long after the collège had returned to its previous structure. Once there, Dupuis rose rapidly in reputation, eventually becoming professor of Latin eloquence at the Collège Royale with the help of the Abbé LeBlond.[4] In 1783 he was chosen to give the funeral oration in the name of the University for Maria Theresa, Archduchess of Austria, the mother of Louis XVI's queen, Marie Antoinette.

In 1778 Dupuis had begun to evolve a theory concerning the origins of mythologies for which he needed to know something about astronomy. He accordingly began to follow the course given by the astronomer Joseph de Lalande at the Collège Royale. Lalande was an extraordinarily difficult personality at a time and in a place that hardly lacked the type. An enthusiastic astronomer avid for fame, praised by Voltaire, extremely ugly, obsessed with pornography, and peculiarly addicted to the consumption of insects, Lalande was not only an excellent teacher (if not a first-rate astronomer); he was also an ardent and quite vocal atheist.[5] Dupuis was by this time reasonably well-known himself; he and Lalande became close friends. According to his first published account, Dupuis conceived his theory connecting the heavens, agriculture, and religion on May 18, 1778, at which point, his wife later wrote, his "head became a volcano" as he immersed himself in the scheme.[6] The theory had developed out of Dupuis' interest in settling the relations between the present names of the months and their names in Attic Greece. "I imagined," he wrote, "that they might have been born out of the [zodiacal] signs through which the sun passes . . . or out of the stars whose rise fixes the sun's passage in each of the signs." Lalande himself made the resulting system public eight months later in the prestigious *Journal des Savants*, which Dupuis followed (in the same journal) with his own account. The whole was

Figure 3.1. Charles Dupuis in the late 1790s and early 1800s

included by Lalande in the continuation in 1781 of his massive *Astronomie*.[7] There, Dupuis announced that he aimed to produce a further "considerable work" on the subject.

Dupuis aimed high, for he intended to find the very time, the very place, and indeed the very people who first designed the zodiac and who had connected their religion to it—as, indeed, he thought true for all religions. How was this to be done? By means, we shall see, of astronomical computation, by the millennial motion, or precession, of the solstices and equinoxes among the constellations. The use of precession for historical purposes was hardly new, for even Isaac Newton had used it a century before to redate the Trojan War, indeed to develop an entire theory of the origins of civilization among the ancient Hebrews.[8] Unlike the religious (if unorthodox) Newton, Dupuis used astronomy as a weapon to attack supernatural belief by cracking the very foundations of time itself. There were, however, barriers to a full public exposition of his views, for due regard to his position and the realities of the day dictated caution.

In the years before the fall of the Bastille, Dupuis had done quite well for himself. Already holder of the chair in Latin eloquence at the

Collège Royal, he wished especially to become a member of the prestigious Académie des Inscriptions. There was opposition, probably due in substantial part to what was quickly seen as his work's implications for sacred chronology, apparently to such an extent that Dupuis at one point decided to burn all his writings on antiquity, which his wife forestalled by having them copied. He finally did succeed in being elected in 1788 as a replacement for the deceased Rochefort (translator into modern French verse of the *Iliad* and *Odyssey*), but only as a result of pressure from several "grandes personages." These included his friends Louis Alexandre, duc de la Rochefoucauld (the son of Dupuis' original patron and himself French translator in 1783 of the "Constitution of the Thirteen United States of America") and Jean-Jacques Barthélemy (antiquarian and archaeologist). A proviso attached to his election required that he not press too hard against the prudent limits of wisdom. In part, this surely reflects unease among the Académie members with the claims concerning the origins of myths that Dupuis had already made (and which could easily target Christianity itself). Still, their objections at the time had as much to do with the odor that he emitted of being overly enamored of the "spirit of system," something that was not then considered to be de rigueur. Dupuis prudently decided not to expose his theories before his new colleagues.

His views were nevertheless being discussed, not least in the pages of the *Histoire de l'Académie des sciences* by the astronomer Le Gentil de la Galasière. A series of mishaps and adventures following a failed attempt to observe the transit of Venus in 1761 had landed de la Galasière far from France for over eleven years, a good part of the time in India, where he had learned much about Indian astronomical lore from the Brahmins. Though he was impressed by the structure of Dupuis' arguments, de la Galasière countered that the original zodiac could as well have been developed in India as in Egypt, where, we shall see, Dupuis placed its origin.[9] His opinion does not seem to have gained any adherents at the time. Indeed, Dupuis' fame spread sufficiently far that, on the recommendation of no less than the Marquis de Condorcet, Frederick the Great, king of Prussia and sometime friend of Voltaire, offered him a chair of literature and a position in the Berlin Academy. The negotiations apparently fell through when Dupuis asked to remain in Paris long enough to qualify for his pension. In any case, shortly thereafter, in 1788, Dupuis became involved in public affairs when he was named one of the four commissioners of public instruction, which brought his name before a wider audience than historians, astronomers, and philologists. The very next year the revolution broke out with the storming of the Bastille on July 14.

At first Dupuis remained unengaged with the revolution's rapidly moving developments, though he was hopeful that the new freedom of the press from censorship would enable him at last to move ahead with the full-scale publication of his theories. Events, however, caught up with him, and in the spring of 1792 he sent the National Assembly a letter that would later bring him directly into the increasingly dangerous whirlpool of public affairs. "The sacred duties that tie every citizen to his country," he wrote, "require that he come to her assistance when danger threatens and contribute to the support of those who defend her. Imbued with this feeling, I offer one hundred gold tokens to the country, the fruits of my works at the Académie des Belles-Lettres."[10] Sincere though Dupuis may have been, the offer was also prudent in a period when loyalty to the new dispensation had become increasingly imperative.

Early the following fall, Dupuis' sense of security suffered a devastating blow. His close friend, Louis Alexandre (Rochefoucauld), who had been a Paris deputy to the revolutionary National Constituent Assembly, was assassinated in front of his mother and his wife in the September massacres of 1792. These had been prompted by the news of the Prussian army's successful invasion of France under the Duke of Brunswick, with his demand for an end to the revolutionary developments. On August 10 the king and his family had fled the Tuileries in fear for their lives. For two days in early September, mobs of enraged Parisians attacked and often killed whomever seemed to them to be enemies of the revolution, in particular aristocrats and nonjuring priests (those who had not signed the 1790 Civil Constitution of the Clergy). The National Convention, elected by male citizens over twenty-five, was formally constituted on September 20, and the next day it abolished the monarchy.

After learning of Rochefoucauld's murder, Dupuis fled Paris for a friend's house in a different department. Along the way, he "was informed" that his own department, Seine-et-Oise, had actually nominated him as a member of the convention. Filled with trepidation and indecision, Dupuis eventually decided he had no choice but to accept and returned home. Few details are known about his early activities at the convention, except that he seems to have kept his head reasonably low, with one exception. In January 1793 the convention voted on the fate of Louis XVI. Found "guilty of a conspiracy against the liberty of the nation" by 693 of the 745 members (with 25 absent or sick and 27 offering a variety of views), Louis faced permanent detention, banishment, or some related way to remove him altogether from the body politic, or death. This time 721 members voted; of these, 366 voted unconditionally for death, 319 for "confinement"—a majority of only 5 for execution.[11] Though he was among those who opposed death, Dupuis nevertheless voiced his hope as he voted that, whatever the result, it "will bring luck to all my fellow citizens, and that it will do so if it can withstand the severe examination of Europe and of posterity, which will judge the king and his judges."[12] Louis met the guillotine on January 21.

On December 21 Dupuis had been named by the convention's Committee of Public Instruction (of which he was himself a member, although he apparently spoke little) to a commission assigned the task of developing a calendar that would appropriately represent the new revolutionary dawn. The chair of the commission, Gilbert Romme, had voted for Louis' death and would eventually commit suicide before he could be guillotined following his support of a Jacobin insurrection the spring after Robespierre's own fall. Although Romme was chair, according to his later recollection Dupuis himself developed the idea for and the basis of the new calendar. "When in 1793," he recalled thirteen years later, "I proposed to the Committee . . . the calendar project which it adopted, and for which my colleague Romme did the report, I took for my guide the ancient sages of the East, who made the year consist of twelve months, each of thirty days, plus five epagomenal ones. Like the Athenians and the Chinese I divided each month in three parts, called *Decades*. This was the calendar of ancient science."[13] The new calendar would begin with Year 1 on the day that the convention had abolished the monarchy (September 21, 1792). The day was particularly appropriate because it coincided with the autumnal equinox. Romme presented the commission's report the following September 20, on the "day of opinion," Year 1, this being one of the five end-of-year days that were added to the twelve thirty-day months.

The system mimicked the month scheme of the ancient Egyptian calendar, albeit with its own naming conventions, tied to the seasons, and with the months each divided into three ten-day "*décades*" instead of the Egyptians' seven-day weeks. Romme insisted that the new system ripped away the deleterious remnants of religion and despotism. "The Vulgar Era arose among an ignorant and credulous people amid the troubles preceding the fall of the Roman Empire," Romme told the convention. "For eighteen centuries its designations of the duration of time have been entwined with the progress of fanaticism," he continued, "the debasement of nations, the scandalous triumph of pride, vice, and stupidity, the persecution and humiliation visited on virtue, talent, and philosophy by cruel despots or by those permitted to act in their name."[14]

Little more than a month later, the Council of the Paris Commune issued the order for royal iconoclasm, followed by Chaumette's insistence that all of Paris's churches be closed. Chaumette may have known for some time about Dupuis' theory, though Dupuis had not as yet seen his grand scheme through for presentation to a wide audience. He had, however, continued to work on the system even after his election to the convention and his appointment to the calendrical commission. During the nine-month gestation of the new calendar, it seems, Dupuis' theory became known to the convention's members. Chaumette relied directly on it to argue that the remaining statues on Notre Dame had nothing to do with Christianity but reflected instead the Egyptian deities, which were themselves linked to astronomical and agricultural events. The Egypt of remote antiquity consequently underpinned both the Revolutionary Calendar and much of what seemed otherwise to be "fanaticism" carved in stone (Figure 3.2)—or at least what was acceptable from ancient Egypt, since not all of its calendrical practices fit revolutionary views. Romme and Dupuis had accordingly abandoned the seven-day week, which the Egyptians had used, on the grounds that it had been produced by "astrologers" to represent their belief in the synchrony between the number of planets and the number of days of the world's creation. That made the scheme utterly unacceptable as a relic of 'fanaticism.'[15]

Romme's presentation of the new calendar occurred just weeks after the convention, pressed by the sansculottes, proclaimed the Terror, fearful as they were for the republic's survival in the face of Prussian and British military successes at Mainz and Toulon. Dupuis himself not only managed to navigate the turbulent waters of the period, he was elected secretary of the convention and member of the new Council of the Five Hundred after its proclamation in August 1795 following Robespierre's fall and the Terror's end. His contemporaries thought Dupuis a better scholar than politician, though his ability to survive and even prosper in these dangerous years attests either to a chameleon-like ability to change his colors or (more probably) to a prudential capacity to keep out of the limelight as he continued work on his magnum opus, the full-scale account of the origins of all religions.

Figure 3.2. Dupuis' depiction of the carvings on the façade of Notre Dame, which he interpreted as the "calendar of Isis, or of the Virgin . . . here the Virgin represents the year, as Isis was its symbol among the Egyptians." Compare with the photographs in Figure 2.2.

Religion Exposed

Dupuis' revised chronology for antiquity formed the main part of the *Origine*'s argument. It depended on the astronomical phenomenon of precession, in which Earth's axis rotates slowly in a conical motion that causes it to point, over time, to different places among the heavens. At the end of the eighteenth century, the period for this precession was taken to be about 25,960 years (as compared to the 36,000 given by the Alexandrian astronomer Ptolemy in the second century CE).

Precession can affect chronology because it causes the position of the sun among the stars at particular times of the year to change over the centuries. To understand the effect, imagine that the plane of Earth's orbit cuts a great circle, called the ecliptic, in the sphere on which the stars appear to lie. On this circle lie the star groups, or asterisms, that form the twelve zodiacal constellations. During the course of the year, the sun appears to move along the ecliptic, traveling bit by bit through the zodiac. Imagine now that the plane that contains Earth's equator itself cuts a great circle on the stellar sphere; this great circle will intersect the ecliptic at two points. The sun in its yearly motion will pass at six-month intervals through each of these points. When it does, the hours of day and night are equal, and so they are called the equinoxes. The two points that lie on the ecliptic at right angles to the equinoxes are the solstices, and at these points (in the Northern Hemisphere) the hours of daylight are longest (at the summer solstice) or shortest (at the winter solstice). Because of precession, the position of the sun at the equinoxes and the solstices with respect to the zodiacal constellations changes over time (Figure 3.3). If this apparent motion could be connected to human activity in the distant past, astronomy could be used to construct chronology.

Dupuis' flash of insight did precisely that: by assimilating the figurative characters of the constellations to the rhythms of planting and sowing, he could use precession to date the origins of both agriculture and the zodiac. The basic association between the sky and planting was in itself hardly an original suggestion. Many before him had, he admitted, suspected that the constellations of the zodiac were related to agriculture and the seasons—including de Guignes and, especially, the Protestant preacher and advocate of religious tolerance, Gébelin, to whom Dupuis referred. Gébelin had made the connection by linking the constellations through the labors of Hercules to the seasons. For example, he interpreted the constellation Leo as the Nemean Lion killed by Hercules, providing its meaning for ancient Greeks. Since the constellation rose there in antiquity during the heat of July, it could be related to growing crops.[16]

Dupuis was much more systematic than Gébelin had been, whose several associations lacked completeness, for he aimed to produce a comprehensive system that could ultimately be extended to explain the nature and origin of every religion. To find, first of all, the "primitive people who traced this symbolic calendar, written in character of fire, in the sky" (the zodiacal constellations), he "proceeded in the

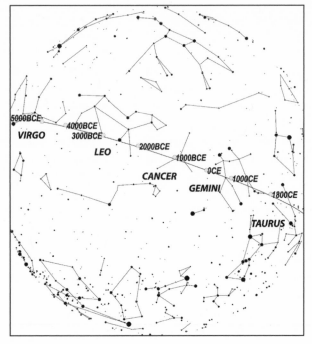

Figure 3.3. Precession of the summer solstice from 5000 BCE to 1800 CE

simplest manner": the inventors had to be those people for whom the positions of the stars at significant times of the year uniquely fit an appropriate natural interpretation. His scheme worked to assign astronomical or agricultural meanings to the constellations by associating characteristics of the object represented by the figurative image with a seasonal event.

To effect the correspondences, Dupuis turned first to what he deemed the most significant astronomical moments for an agricultural people, namely, the times of the year when the periods of daylight are longest and shortest (the solstices), and the times that divide these two extremes (the equinoxes). The longest and shortest days were particularly important because after the longest day, on the summer solstice, the sun is visibly lower each succeeding day at noon and retreats ever more to the regions beneath the horizon until, on the shortest day (the winter solstice), the sun reverses its path. Dupuis further assumed that his cultivating early astronomers would have paid close attention to the asterisms that, after a period of invisibility, rise just before the sun, what astronomers called their heliacal risings.

Here Dupuis' knowledge of Latin and Greek texts stood him in good stead. The fourth-century CE neo-Platonist Macrobius's *Saturnalia* contained apposite remarks, for he had supposed that the signs of the zodiac symbolize the course and effects of the sun. Though he demurred from several of Macrobius's associations, Dupuis fastened on the two that linked a pair of constellation figures to the solstices. According to Macrobius, Capricorn, the constellation symbolized by a goat, represented the winter solstice because goats always seek lofty ground, as does the sun in climbing ever higher in the sky *after* the winter solstice. Moreover, because Cancer, the crab, crawls backwards and sideways, it should represent the summer solstice, when the sun reverses its direction and begins its southerly descent to the Capricorn-governed depths of winter.[17] Dupuis generalized from Macrobius, linking Cancer directly to a reversal of the sun's motion; Capricorn he linked to height proper and not merely (as Macrobius had) to northerly ascent. The constellations at the equinoxes then had equally natural interpretations. Libra, symbolized by a balance, is "the most expressive and simple symbol of equality," and so it should correspond to a time when the hours of day and night become equal, as indeed it does if Cancer and Capricorn mark the solstices, with Cancer governing summer and Capricorn winter.

These particular associations are reasonably consistent with Greek antiquity, provided that Libra corresponds to the autumnal equinox. Dupuis had considerable leeway here because it takes about 2,160 years for a constellation to traverse its entire extent past a given point on the ecliptic (see Figure 3.4). On this basis alone it was possible that the original zodiac had been delineated at some time during this period. However, these identifications could not specify just where the zodiacal inventors lived. Dupuis' particular genius aimed to solve this problem by associating other zodiacal constellations with the agricultural seasons. If the period of the zodiac's invention did fall somewhere in the two millennia BCE, then, Dupuis reasoned, it should be possible to identify a place at which others among the constellations could be appropriately interpreted.

Much more followed, as Dupuis canvassed, interpreted, and twisted texts in his quest for support, eventually claiming not only that the ancient Egyptians had delineated the zodiac, but that their efforts had passed to India, China, and even the Americas through some unknown

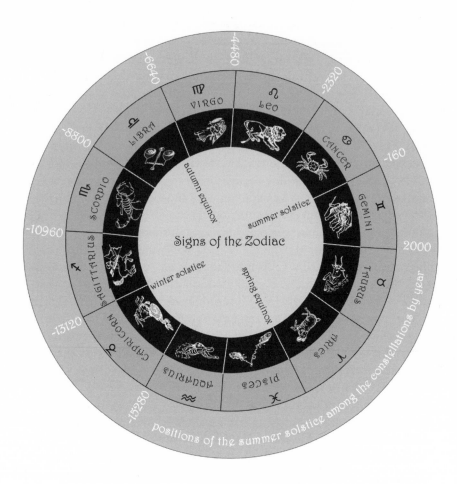

Figure 3.4. The second ring in contains the names of the fixed zodiacal signs
and the symbols for them. Each of the signs extends exactly thirty degrees. The
sequence begins with Aries, which is located at the intersection of the ecliptic
with the equator, where the spring equinox always occurs. The locations of all
four equinoxes and solstices among these fixed signs are marked on the central
disk. The black ring contains representations of the correspondingly named
constellations. As a result of precession, the black ring rotates clockwise over
the millennia, thereby turning the constellations through the fixed signs. The
outermost ring marks very approximately the positions of the summer solstice
among the precessing constellations every two millennia from about 15,000 BCE
to the present day. The black ring's position represents the configuration of the
constellations among the signs in about 300 BCE. The spring equinox occurred in
the constellation Aries from about 2000 BCE for two millennia or so.

route. In this, his first publication on the subject in 1781, he argued that all religions were connected to the associations primitively established in the original zodiac. Sabeanism (the worship of the sun, moon, stars, and planets), Dupuis asserted, underpinned every form of polytheism. There is, he wrote, "only one physico-astronomical system that can explain the theology, the monuments, and the fables of these religions"—namely, the one that Dupuis himself had found in the zodiac.[18] At the time he excepted only the religion of the ancient Hebrews (or at least he excepted it in print). Their prohibitions against idolatry, he argued, followed "naturally" from their rejection of the "connection which these symbolic animals in Egypt and Syria had with the stars themselves . . . because the Jews lived in the midst of the nations whose religion was entirely astronomical and represented divinity in the form of the animals painted in the constellations."[19]

There was, however, a problem: Dupuis did not actually specify the period for the zodiac's invention. Nevertheless, on the basis of his identifications, which placed the zodiac's origin at a time when the summer solstice occurred in Capricorn, the zodiac must have been devised no later than thirteen thousand years ago, and perhaps as much as fifteen thousand (depending on whether the zodiac was first delineated when the solstice occurred, respectively, at the end or beginning of Capricorn), predating by perhaps as much as nine thousand years any biblically based date for the world's creation. "The period of this invention," Dupuis admitted without providing a specific number, "reaches far beyond the time fixed by our chronologists for the creation of the world, and this would be a great objection to our hypothesis"—it would be one, that is, provided that we were certain the precession of the equinoxes had always occurred at its present rate. Perhaps, he suggested, the tremendous disturbances produced by the Noachian deluge had radically slowed precession. This we cannot know, Dupuis admitted, though we must conclude it to be so on the basis of accepted chronology. Or so Dupuis averred in print in the years before the revolution.[20]

In the 1770s Dupuis and his family were living in the town of Belleville, northeast of Paris. Although by 1860 Belleville had disappeared into the nineteenth and twentieth districts of Paris proper as a result of Haussmann's rebuilding of the city, in Dupuis' day it was a large and lively country town. Here Dupuis not only delved into the astronomical origins of religion, he also pursued more material projects, such as the development of a semaphore telegraph based on the optical system that had been created a century before by Amontons. This telegraph (which, ever prudent, Dupuis dismantled after the onset of the revolution, fearing with good reason that it would be thought a tool of subversion) combined his early interest in devices with his training in the art of rhetoric. That endeavor brought him into close contact with someone who would also aid him by contriving a mechanism to represent his astronomical theories. Not far away, in the town of Bagneux just to the south of Paris, was the country home of Nicholas Fortin, a successful instrument maker, manufacturer of celestial and terrestrial globes, and sometime scholar who in 1776 had published an abridgement of the English royal astronomer John Flamsteed's 1729 celestial atlas. He also fabricated apparatus for the unfortunate chemist Lavoisier (an eventual victim of the guillotine), among others. He and Dupuis used the semaphore system to communicate telegraphically with one another.

According to Dupuis, Fortin followed a "method already used by Lalande" in order to produce for his friend an extraordinarily rare sort of celestial globe, one that could be adjusted to represent precession. This was hardly a simple matter because the device required a pair of mutually inclined axes of rotation for the entire stellar sphere: one to turn the equator around, and another to rotate the stars about the ecliptic. Ptolemy had discussed how to make one in his *Almagest*, though not without ambiguity,[21] and even in the late Renaissance and through the eighteenth century few devices of the sort seem to have been made. The Museum of the History of Science at Oxford University does, however, have one that apparently dates to the mid-sixteenth century and that illustrates the degree of complexity required (Figure 3.5).

Fortin's version of a precession sphere was visually impressive because it carried a globe painted with stars and perhaps constellations as well. Since it could certainly be adjusted for latitude (place) as well as for precession (time), Dupuis could use the mechanism to test various locations and periods for their fits to the constellations. This was more than a convenience because the calculation of stellar risings and settings at a given date in the far past is not a simple matter for even a single star, much less for the members of a group that compose a constellation. But with a precession globe to hand, Dupuis had only to choose some latitude and then rotate the device about its ecliptic axis until some part of Capricorn lay at the winter solstice (the low point of the ecliptic below the equator), with the other constellations in their proper places according to his theory. If it could not be done then he could try another latitude.[22]

Figure 3.5. Armillary sphere, Flemish circa 1550. The second ring in is for precession.

Dupuis' elaborate system, backed by a device produced specifically for him by the foremost Parisian instrument maker of the day, boldly bound together the worlds of erudition and science in a way that had been essayed only by Isaac Newton himself a half century earlier. Newton had used an astronomical poem by Aratus (third century BCE) that had been immensely popular in Greco-Roman antiquity and about which the astronomer Hipparchus (mid- to late second century BCE) had produced a barbed critique, the only Hipparchan work that has survived.[23] Unlike Dupuis, Newton was not particularly interested in developing a theory of pagan religious belief (though he did have views about the matter that had been common since late antiquity). He was instead concerned to demonstrate that all civilization postdated the death of Solomon in order to prove the importance of the ancient Hebrews in their special relationship with the monotheistic deity. This had led Newton to move the date of the Trojan War several hundred years past the accepted period of the day, which in turn produced considerable dispute, not only in Britain but especially in France.[24] The disputes had several aspects, but among them was the sense among historians and philologists in the first half of the eighteenth century that calculating science had intruded dismissively into their world of texts and inscriptions.

Dupuis came from nearly the opposite pole to Newton. He had risen to prominence in essentially the very universe that Newton had barged into with his numbers and his astronomical claims—the universe of letters, albeit one that had seen considerable change in the half century since posthumous the publication in 1728 of Sir Isaac's *Chronology of Ancient Kingdoms Amended*. A master of Latin rhetoric, familiar with the panorama of ancient Greek and Latin texts, as well as with the works of his eighteenth-century French predecessors, including Fréret, who had spent much effort critiquing Newton's *Chronology*, Dupuis moved from history to astronomy and not the other way around. He could not be accused of antagonism to texts, or indeed to erudition proper, unlike his philosophe contemporaries.

Early Reactions to Dupuis' *Origine*

Dupuis' particular merger of philology with astronomy captured considerable attention and admiration during this last decade before the revolution. The philologist Dacier, who became perpetual secretary of Inscriptions et Belles-Lettres, and who, no doubt wisely, had refused the offer of the Ministry of Finance by Louis XVI in 1790, recalled the contemporary effect in his obituary for Dupuis. "We have often seen the Heavens peopled at the expense of the Earth," Dacier remarked, "but no one before M. Dupuis undertook to show that it was on the contrary the Heavens alone that peopled the Earth with this multitude of imaginary beings whose symbolic origin having been forgotten metamorphosed into princes, warriors, and heroes; and that the simple theory of the risings and settings of the stars . . . was the origin of this immense number of marvelous feats, of mythology's astonishing chimerical adventures."[25]

When the *Origine* was printed in 1795, the massive, nearly unreadable work could be bought in September at Henri Agasse, its publisher, in the rue des Poitevins. The work required three large quarto volumes as well as a separate one for plates, at the cost—unbound—of 600 livres. For 3,000 livres (about a month's rent for a good apartment in Paris in 1795) it could be had on vellum, and a cheaper edition in octavo was also available. Rumblings of trouble ahead for the *Origine* could be heard almost immediately after its appearance in print.

Claude le Coz, priest and principal of the Collège de Quimper, had taken the constitutional oath for the clergy in 1791, became a member of the legislative assembly that September, was elected Bishop of Ille-et-Villaine—and had his election nullified by the powerless pope. A comparative moderate in a period of extremes, le Coz perceived the direction of the winds and hurriedly left Paris in September 1792.

Imprisoned for fifteen months at Mont St. Michel, he managed to make it back to his diocese in Rennes in 1794. Disliked by conservatives for having exhibited revolutionary fervor before 1792 and for having joined the juring clergy, and by the Left for insufficient fervor and for having protected nonjuring priests, le Coz was hardly moderate in his defense of Catholic belief, his dislike of religious toleration, and, above all, his profound loathing for Dupuis' several works.

Le Coz vocally abhorred the new calendar, which had thrown out Sunday as a sanctified day. This alone brought him close to deportation. And then he heard about the publication of the *Origine*. "If this new production of impiety seems to you to require that we occupy ourselves with it," he wrote on December 7, 1795, to the juring Abbé Grégoire (whom we will encounter again, and who did not share le Coz's antagonisms), "and if it's sold for 600 livres in paper, you will oblige me by sending a copy. Since these frenzied ones never cease to multiply the forms of their poisons, we must also have the courage to chase them through all their disguises and show the people the bad faith or ignorance of these poisoners."[26] Two weeks later he wrote again, his tone more urgent: "The work of Dupuis is triumphantly announced in the journals: do you think of opposing some good David to this new Goliath? Have you sent me a copy of this book, as I asked? Since some impious people in this town already extol it, I intend to combat it, even from the pulpit, if the sensation it makes seems to demand it."[27] The following April he wrote a third time. Grégoire still hadn't sent the book, and by now it had become prohibitively expensive. By September le Coz was still at it, though Grégoire had clearly sent him neither the book nor anything much by way of remarks. If you don't have the time, le Coz pleaded, then have one of your friends send me a summary, for "I would like an idea of this new Goliath, who has just insulted our Lord."[28] Five years later, with Napoleon now first consul, le Coz was even more troubled because Dupuis, a "monstrous man who knows no God other than the passions and interests of the people," was about to be made a senator. "This scandal," he cried to Grégoire, "has already echoed in our countrysides, and it may cause the death of more than ten thousand men." Le Coz's loathing for Dupuis burns through these remarks, and he was a moderate in comparison with the extreme religious reactionaries of the period. Over time, conservatives would spew out increasing hate for views like Dupuis', as we will see. For them his *Origine* ever remained a red-letter symbol of everything that had gone wrong with France during the Enlightenment.

Conservative revulsion intensified when Lalande, that convinced and vocal atheist, hyperbolically praised the *Origine* in the *National Gazette* on September 16, 1795, remarking that the work "contains the finest discovery ever made in the study of antiquity."[29] He also noted that one could find a second edition of Condorcet's posthumous *Esquisse de l'histoire des progres de l'esprit humaine* (Sketch of the Progress of the Human Spirit) at the same bookseller. The appearance of both works at the same publisher was hardly coincidental because the two books were consistent with a doctrine that came to be known as *idéologie*—a philosophy that became pervasive, and even politically dominant, in the years after Robespierre's fall, during which the country was run by an executive that consisted of a five-member group called the Directory (1794–1799).

The partisans of *idéologie*, the *idéologues* (so-called by Napoleon in later years), who counted Volney among their members, espoused an extreme form of sensationalist philosophy due to Étienne Bonnot, the Abbé de Condillac. According to him, all true knowledge derived ultimately from the senses and could be captured only imperfectly through language. The idéologues were particularly interested in tracing errors of understanding to their sources in the misrepresentation of original sensations by language. Religious belief, or at least the sort of belief imposed by an organized clerisy, exemplified just that sort of error. Their journal, the *Décade philosophique*, began publication on April 29, 1794, at the very height of the Terror and ran until 1807. Deeply anticlerical and fervently republican, the *Décade* aimed to forward the aims of *idéologie*, namely, to renovate the human sciences by building them, and especially language, on the basis of a set of simple ideas. The group's identity was forged by, among others, Volney himself, Ginguené, Chamfort, Daunou, Chénier, Lakanal, and de Tracy. The extraordinary Chamfort—playwright, essayist, onetime secretary to Louis XVI's sister, early republican, defender of Charlotte Corday, sardonic critic of Robespierre, and failed suicide—founded the *Décade* (but did not live to see it succeed) together with Ginguené, who had himself been imprisoned during the Terror. Daunou, originally an Oratorian, member of the convention, and prisoner from 1793 until the Terror's end, was primarily responsible for writing the Constitution that established the new, governing Directory; he had also originated the plan for the Institut National des Sciences et des Arts, which replaced the academies of the ancien régime. De Tracy, who had also been imprisoned during the Terror, became the principal spokesman

Figure 3.6. Frontispiece to Dupuis' *Origine*

of the group, eventually authoring the lengthy *Élémens d'Idéologie* (1803).[30] He was particularly enchanted with Dupuis' *Origine*, of which he wrote an *Analyse* that, published in 1799, soon became the primary point of entry into the elaborate system for those who did not have the fortitude to plow through its turgid volumes, though Dupuis himself had written a less compelling, one-volume *Abrégé* that appeared in 1797 and was reprinted twice the next year.[31]

Although the idéologues hardly formed a cohesive group, and neither did they ever produce anything like a Newtonianism of the moral sciences, nevertheless they did exercise considerable influence during the Directory. Volney lectured on history at the short-lived École Normale, which had been established in 1794 for the training of instructors in the "art of teaching," who would attend lectures in a variety of areas. In his, Volney aimed to interpret historical developments as the actions of large forces, such as major changes in technology, working their effects in the social world. To do so he relied not only on texts or artifacts, but also on customs, religions, artisanal and skilled activities, morality, and languages. Eschewing the particularities of individual events and persons, Volney characterized the "habits, customs, and character of nations."[32] Combining Dupuis' *Origine*, made especially well-known by de Tracy's *Analyse* and Dupuis' own *Abrégé*, along with Volney's *Ruines*, his *Voyage*, and his public lectures, Volney forged a vision and an aim that would soon resonate with the political ambitions of a young general by the name of Napoleon Bonaparte.[33]

Napoleon

Napoleone Buonaparte was born in 1769, the son of a Corsican lawyer descended from minor Italian nobility. While his father, Carlo, represented Corsica in Paris for several years from 1778, the young Napoleone was sent to the Collège Militaire Royale de Brienne, near Troyes, where he developed a particular taste for mathematics and geography and his name changed to the more French-sounding Napoléon Bonaparte. In 1784 he entered the École Royale Militaire, from which he graduated after a year, having learned a great deal about artillery and ballistics. After he and his family left Corsica altogether for France in June 1793, Napoleon, who had supported the Jacobins in Corsica, achieved his first military success while in charge of the artillery at Toulon, then under siege by the British. Taking advantage of his training, he succeeded in repelling the British fleet and recapturing the city. This first military success earned him Robespierre's favor, though the connection landed Napoleon in prison for a fortnight after Thermidor. Nonetheless, a year later he defended the National Convention in Paris against an uprising by royalist sympathizers and others, which gained him the support of Paul Barras, one of the five directors of the republic. Barras had a hand in arranging the marriage of his former mistress, Josephine de Beauharnais, to Napoleon, and it was Barras who nominated Napoleon to command the army in Italy in the early spring of 1796. His successes there were followed by further triumphs in Austria in 1797 and after that in Venice.

In September 1797 a coup d'état against the Directory was planned by a group of moderate and conservative members of the legislative Council of Five Hundred, who intended to cede most of the Republican conquests in exchange for peace. The coup of 18 Fructidor (September 4) was foiled by Barras and two other directors, Reubell and

Revelliere-Lepeaux. The failed uprising led to a closer connection between the Directory and the conquering generals, including Hoche and Moreau, as well as Napoleon. It was during this period that Napoleon first articulated the notion of an expansionist "Grande Nation," which would swell the republic beyond the boundaries of Europe.

The bulk of the French populace, hostile by this time to republican government, saw in Napoleon a man who could guarantee the decade's territorial and economic gains while ending the social disruptions of revolution. His main rival for their affection, Hoche, died of tuberculosis or pneumonia on September 19. In the summer of 1797, Napoleon had come into epistolary contact with the supple diplomat Talleyrand, newly returned from America, who had left France for England just before the September Massacres, and who had succeeded in becoming the French foreign minister on July 16 of that year. Talleyrand had met Volney in the United States, who had told him much about the young general. Volney no doubt provided particular insights as a result of his agricultural sojourn among the Corsicans.

Talleyrand felt that the foundation of England's power lay in its access to India, which had become an increasingly important source of imports, especially cotton, as well as a locus for the sale of British manufactures. To sap British power required the destruction of their entrée to India. The route to the subcontinent lay through Egypt, around which now coalesced the dreams of both Talleyrand and Napoleon. On August 16 Napoleon wrote the Directory that "the times are not far off when we will sense that Egypt must be seized in order truly to destroy England. The vast, perpetually dying Ottoman Empire obliges us soon to find ways to preserve our commerce with the Levant." Talleyrand replied on August 23 that "nothing is more important than to put us on a good footing in Albania, Greece, Macedonia, and the other provinces of the Turkish empire in Europe, and even all those places that are bathed by the Mediterranean, especially Egypt, which can become very useful to us."[34]

Napoleon arrived in Paris in December, preparations having been undertaken the previous two months to organize the apparatus for an invasion of England itself using especially Napoleon's forces in Italy. On December 10 he sealed his military success with the presentation of the formal Treaty of Campo-Formio to the Directory. They, however, feared the growing power of the generals, and with good reason, since Moreau, Kléber, and Cafarelli had already suggested a coup, one that Napoleon thought premature. In late February he reported to the

Directory his conclusion that an invasion of England could not succeed. Instead, British maritime power could best be countered by an
expedition to the Levant that would block their access to India. Talleyrand also pressed for an expedition to Egypt: "Egypt was a province of
the Roman Republic, it must become one of the French Republic. The
conquest by the Romans marked the beginning of this beautiful country's decline; conquest by France will mark that of its prosperity."[35]
The Directory agreed to the expedition on March 5, 1798.

Like many others in mid- to late eighteenth-century France, Napoleon had visions of the "Orient" as a place of decayed opulence and
sensuous indulgence. Although the most elaborate of these fantasies
had not yet evolved (they would do so, not least, as a result of the Napoleonic expedition itself), nevertheless many elements of the vision had
long been in place. In 1741, for example, Voltaire had published a play
in five acts entitled *Mahomet*. In it Voltaire had Mohammed himself
lament the lost glories of Arabia, which he would now renew:

Every people has its turn on Earth to shine,

Through laws, through arts, through war above all;
Arabia's moment has arrived at last.
This generous people, too long unknown,
Glory abandoned, in her deserts entombed;
Come now new days, by victory marked.
Behold, from north to south a desolate world,
Persia still bloodied, her throne unsettled,
India enslaved and timid, Egypt abased,
The splendor of Constantine's Walls eclipsed;
Behold, the Roman Empire in pieces fallen,
Its great body dismembered, whose parts
Dispersed languish, bereft of honor and life;
On this worldly debris will Arabia rise.[36]

If the adolescent Napoleon had not yet read Voltaire's evocation of
an Arabia risen, now again debased, he had certainly perused Marigny's
1750 *Histoire des Arabes*, du Tott's *Mémoires sur les Turcs et les Tar-
tares*, and, especially, Volney's *Voyage* and *Ruines*.[37] Moreover, Napoleon had apparently encountered Volney himself in Corsica during the
latter's ventures there. He would likely have heard of Dupuis' *Origine*,
if only through Volney; he certainly knew it in later years, as we shall

see. From it, he might have learned of the birth of religion in Egypt, and "the pretended history of a God, born of a Virgin at the winter solstice, who resuscitated at Easter or at the equinox of spring, after having descended into hell; of a God, who has twelve apostles in his train . . . of a God-conqueror of the Prince of Darkness, who restores to mankind the dominion of Light, and who redeems the evils of Nature— is merely a solar fable."[38] By the time of the Egyptian expedition, or shortly thereafter, Napoleon had read Voltaire's *Mahomet*. By then he objected to it, primarily because he saw in Mohammed an early vision of himself, "a great man who changed the face of the earth" and not, as Voltaire had depicted him (regretfully, as it turned out), "an abject villain, worthy of being hanged."[39]

Egypt: a land the French imagined as despoiled for centuries by the oppressing Mamelukes, gateway to India, key to British power, and the original home, not only of religion, but also, perhaps, of science itself. A land whose inhabitants were just waiting to be liberated intellectually, socially, and economically by the enlightened French bearers of the fruits of civilization. Where religion had been born, Republican virtue and science would now reign.

4

On Napoleon's Expedition

In the spring of 1799, a baby crocodile died while in the care of the twenty-first semibrigade of Napoleon's army, camped at Qena outside Dendera.[1] Led by generals Desaix and Belliard, the army was in the sixth of an eight-month inland excursion to survey (and subdue) the countryside. The crocodile's fate would have been forgotten in the press of events, had it not caught the attention of a civilian attached to the division, Dominique Vivant Denon, the fifty-one-year-old artist, writer, and courtier who would shortly make the sketches that introduced the zodiacs to Europe.

"Arriving at Qena," Denon wrote, "I had occasion to regret the death of a crocodile, which some peasants had surprised while it was sleeping." The image of the defenseless crocodile seized while asleep dramatized the unfairness of the attack, emphasizing the animal's small size and the use of what Denon considered excessive force to overpower it. Although the crocodile was not fully grown and so, Denon acidly pointed out, "could not have been especially formidable," it had been thoroughly restrained, "chained with a circle of iron around its shoulders and abdomen."[2]

Although the crocodile's death was unremarkable both to the soldiers and to the villagers who had made them a present of the animal, it disturbed Denon. Ten violent months in Egypt with Napoleon had made it difficult to sustain whatever faith he might originally have had in the official aims of the expedition: that French interests in Egypt were to do purely with the liberation of the Egyptians from Mameluke oppression, and that the French were in Egypt to export the values of the Enlightenment to the Near East. Commissioned by Napoleon's scientific institute at Cairo to write an account of a military expedition in Upper Egypt, Denon's role as an observer required him to be merely an unobtrusive part of the scenery. But as part of General Desaix's army, he was also an instrument of the occupation. As Desaix and his men chased the Mameluke leader Murad Bey throughout Upper Egypt, there was no shortage of bloodshed, fear, and death, such as the gruesome discovery, some weeks earlier, of the desiccated remains of a French convoy in the Redifi Pass, near Chenubis.[3]

Denon's disquiet in the face of the reptile's death reflected the uneasy dependence of scientific knowledge on the violent realities of invasion, occupation, and war—a war without which the scientific work would never have taken place. Unlike earlier French encounters with Egypt, which occurred primarily in the pages of books and in the imaginations of scholars, this one was starkly real and full of baffling surprises. Arriving in Egypt, the French were intellectually and imaginatively primed by received fantasies about the lost grandeur of the country's past—a grandeur that they were keen to resurrect by, among other activities, seizing antiquities and conveying them to museums and private collections in Europe. This, it was hoped, would reflect the light of ancient glory on the French "liberators." But the innocent nobility of motive that the French tried to project, and that many among them certainly believed, at least initially, was repeatedly undermined by wartime realities of cruelty, violence, and greed.

These contradictions were not lost on Denon as he watched the infant crocodile expire at Qena. The injustice and futility of the affair of the crocodile—the show of force against a relatively harmless animal that was going to die anyway—provoked him to meditate on what had been lost in this encounter between the French occupiers and the least of the inhabitants whose country they had subjugated. An opportunity to learn more about crocodiles had been lost, to be sure. Perhaps, Denon suggested, from a French scientist's perspective, that was all. He linked the lost chance to French greed for knowledge itself, a greed that made one oblivious to the sufferings it provoked, however unintentional they might be. Had the crocodile lived, Denon mused, it might have been presented to Napoleon's savants, who would have been "interested to learn" certain things about it. Chief among them, he thought, would have been the animal's horrific eating habits: "how the amphibian ate, and what it ate; whether chewing was necessary, and, if so, how this is accomplished when all its teeth are incisors." Perhaps even the crocodile's storied appetite could itself be put to use, harnessed in the service of taming the amphibian. It is rather hard to envision what immediate practical value this might have produced. Neither does Denon's image of a crocodile socialized into a reptilian facsimile of Jean-Jacques Rousseau's *Émile* spring easily to mind. Nonetheless, these retrospectively odd conceptions suggest that Denon had evolved a complex view of the occupation. Had the Egyptian crocodile made it alive to Paris, Denon wrote, it not only would have been "submitted to the observations of naturalists, and the curiosity of Parisians," but would also have provided "an homage to the nation as a trophy of the conquest of the Nile."[4]

That a crocodile's appetite should have captured Denon's imagination is no surprise, for energies of greed and accumulation characterized much of Napoleon's adventure in Egypt, from the humdrum imposition of new taxes and fees on local owners of businesses and property to the spectacular seizure of the Rosetta Stone and other trophies of antiquity. Moreover, Denon was himself a longtime student of various appetites. Arriving in Egypt, he brought along the philosophe's characteristic preoccupation with the contradiction between the good and the natural: how can virtue, a good thing, be reconciled with appetite—which, if it was not good, was certainly natural and therefore must, in some way, be moral. Twelve years before, in 1777, Denon had anonymously penned the novella *Point de Lendemain* (No Tomorrow), in which a naive young man succumbed to the charms of a certain marquise only to find, at the conclusion of their night together, that she had used him to divert her husband's attention from yet a third man, with whom she had been carrying on an affair. The point of the book was relatively modest: it did not explicitly attack the political system, as other pornographic works of the time did, but merely tested the limits of what could be said without provoking the censors of the ancien regime.[5] This exploration of the connections between appetite and virtue grounded the latter in materialist philosophy and defense of liberty without slipping into a defense of licentiousness. More recently, Denon had deepened his repertoire as a producer of salable erotica by writing and illustrating the *Oeuvre Priapique* (1793), a fanciful guide to the sexual mores of the ancient city of Pompeii. This volume, with its quite explicit twenty-odd engravings, was far racier than anything he had written in *Point de Lendemain*, and the first edition promptly sold out in the bookstalls of Paris.[6]

Denon came to Egypt via a circuitous route that had taken him from a consular post in Italy to the offices of the revolutionary government to the most fashionable Parisian drawing rooms. By the time he arrived in Egypt, Denon had accumulated a number of interests, talents, and influential friends: he was an artist, a diplomat, a man of letters, an intimate of philosophes like Diderot and Voltaire, and a popular presence in Parisian salons. Before the revolution, he was a favorite of Louis XV, who sent him to St. Petersburg as a diplomatic attaché after Denon resigned his appointment, also made by order of the king in 1769, as the royal curator of a collection of gems, coins, and medals that had been a gift from the king to his mistress, the fashionable Madame de Pompadour.[7] Unsurprisingly, during the revolution Denon's property and financial holdings were confiscated. But thanks to his connections (in addition to Diderot and Voltaire, Denon claimed the friendship of the artist David, who was a great favorite of the revolutionary government), he escaped the worst of the period's horrors. After a miserable period in the Montmartre slums, where he sold sketches to make ends meet, he found regular work, through his patron David, as the National Engraver of Republican Uniforms, where he engraved David's pictorial efforts to transform French dress according to a style befitting the revolutionary regime.[8] He also attended meetings of the Tribunal, where he sketched such leading figures as the fiery orator Danton and the notorious public prosecutor Fouquier-Tinville. There is a story of uncertain truth that, while waiting for official approval of his job, Denon was summoned by the Committee for Public Safety to report on his progress in a midnight meeting with none other than Robespierre. He arrived on time but was made to wait for hours. Through the walls he could hear another meeting in vociferous progress. The door opened; a tense Robespierre emerged and, upon seeing Denon, flew into a rage, demanding to know what he was doing in the hallway eavesdropping on official revolutionary business in the middle of the night. Undaunted, Denon quietly introduced himself. They retired to Robespierre's office, where he assuaged Robespierre's anger by conversing about fine art until dawn. By the end of their meeting Denon's estate had been restored.[9] He was, in short, a complicated man living in a complicated time: a talented, diplomatic, and tactful person with a flair for elegant prose, who also possessed enormous physical courage and stamina.

In Upper Egypt Denon lived as ordinary soldiers did, sleeping rough with his sword at the ready. He complained of exhaustion, boils, sunstroke, irritations of lung, eye, and intestine, and the relentless *khamsin* wind. Through it all his enthusiasm for sketching never flagged. He often drew in the heat of battle, with a box of art supplies slung over his shoulder, marching at the head of the column as shots whizzed past. Once, "amid bugled alarms," Denon found himself in the middle of a skirmish. As the soldiers returned fire, "Denon encouraged them to the fray by waving his drawing paper." The immediate danger past, he returned to his sketch, trying to capture the scene as he experienced it, perched on the back of a donkey in the middle of the action. According to Savary, who served under Desaix, Denon's zealous recording of artifacts and scenes in the midst of battle made him "a constant sub-ject of astonishment" to the soldiers.[10] If he feared anything, he wrote, it was only "loss of time and lack of pencils, paper, and talent."[11] Later, the editor of the *Voyage dans la Basse et la Haupte Égypte* would praise Denon for his courage and the scope of his ambition, which in this instance seemed to rival even Napoleon's: "What does [Denon] do at the height of the action? He observes. What is his weapon? A pencil . . . for it is not with the Mamelukes he is at war, but with oblivion."[12]

There are several accounts of how Denon met Napoleon. In the sim-plest, he met the future first consul and emperor in Paris at the salon of Josephine, after Napoleon had returned from the Italian campaign. A different story has it that Josephine herself visited Denon's studio in 1795, where she fell under the spell of his work and invited him to her salon in the rue de la Victoire. Denon went, along with an "inti-mate friend" who had also been a childhood friend of Josephine's—an enigmatic acquaintance mentioned by several sources but who remains unknown. Bonaparte, who was present at this meeting, disliked Denon's companion and for this reason took an immediate dislike to Denon as well, although Josephine would later bring him around. In yet a third version, Denon attended a ball given by Talleyrand. It was a hot night, and the servants were slow to replenish the drinks. Denon had unwit-tingly taken the last glass of lemonade when he noticed a young offi-cer standing nearby, looking thirsty. This young officer was Napoleon. Denon offered his glass, and they struck up a conversation about Italy, where Denon had worked as part of the French diplomatic legation. Names, however, were not exchanged, and Denon withheld the informa-tion that would identify him as a well-known diplomat and the author of a popular travelogue, *Voyage en Sicilie* (1788). Later, at Josephine's salon, Denon saw Bonaparte again, who had now arrived at the rank of general. Napoleon not only failed to recognize Denon as the man who offered the lemonade at Talleyrand's ball but treated him coldly. Denon continued to attend the salon, where he learned of Napoleon's schemes for Egypt. On one visit, the talk turned to Italy. Denon went on about Sardinia and Corsica; Napoleon remarked that he had heard all this somewhere before. Denon reminded Napoleon of the lemonade, and so, finally, Napoleon recognized him. Thereafter, at Josephine's urging, Napoleon invited Denon to join the Egyptian expedition.[13]

Brooding over the crocodile's death at Qena, Denon mused that Egypt itself, like the unfortunate reptile, would never be the same after the French had finished with it. But the results might not perhaps be as bad for European views of Egypt as they had been for the crocodile. Perceptions of the country would certainly be subtler, and, he believed, Europeans would recognize what Denon thought to be the native goodness of the region and its people. Even the poor crocodile's reputation could be amended. Denon refuted popular theories about its ferocity with stories in which he described crocodiles in idyllic scenes on the Nile, in a sort of Egyptian pastoral where life goes on as though the French had never arrived. In contrast to the vicious predator of popular legend, Denon painted a very different picture of the crocodile—and, by extension, of Egypt itself—as a kind of Enlightenment innocent, attuned to the rhythms and interdependencies of the natural world, peaceful unless provoked. His verbal portrait was a far cry from the chaotic violence of the baby crocodile's capture by Egyptian peasants seeking to make a gift of it to the French troops but wholly consonant with Denon's emphasis, in his writings and sketches, on the essential goodness and innocence of the natural world, as opposed to the corruptions of civilization. He took this inversion of popular beliefs about crocodiles so far as to claim that the animal's tough skin attested not to its strength, but its vulnerability. "They are not as ferocious as they are claimed to be," Denon said, "otherwise they would not be covered with so much impregnable armor." He pointed out that crocodiles are social among themselves, and live in families, crawling in the mud of the Nile's islands and banks in search of a warm spot to sleep, or lazing about with their mouths opened to receive the plovers that arrive to clean their teeth. "We know all the stories" about the crocodiles, Denon wrote, "but we have not grasped this one fact": that even though French soldiers, who are "bold to the point of imprudence," teased the crocodiles, "as for myself, I bathe daily in the Nile; on the very quiet nights when I am bathing the supposed dangers never appear."[14] And although the Nile crocodiles were even rumored to eat corpses from battles between Napoleon's army and the Mamelukes, Denon was skeptical. "If they eat corpses from the battles," he said, "they do not especially enjoy them"—a quip with a barb in it, as if Denon wanted to say that the horror of a crocodile eating a corpse was nothing compared to the horror that produced the corpse in the first place.[15]

As Desaix's troops prepared to leave Qena, an officer took Denon aside and showed him another small crocodile only six inches long. According to a report made later by the officer, even though it had not eaten for four months, the crocodile did not seem to suffer. Remarkably, it had neither gained nor lost weight, nor had it been the least bit pacified by privation. The animal's stubborn resilience called to mind a more general problem for the French in Egypt, for it presented the disconcerting possibility of passive resistance that might at any moment turn to violent revolt. With this possibility at the forefront of his mind, Denon slung his bag of sketchbooks over his shoulder and joined the march to Dendera.

The Expeditionary Force

When Napoleon left for Egypt in 1798, he had four hundred ships and more than thirty thousand troops under his command; by the time he arrived in Alexandria that number had swelled to nearly fifty-five thousand, thanks to reinforcements that had sailed from Italy.[16] Although the expedition's military goal was to disrupt British trade routes to the East, Napoleon's interest in Egypt had an additional, more grandiose dimension: to produce a Paris of the East, where the ancient splendor of the pyramids would complement and sanctify modern French ideas of *liberté* (keystone of the early republic's ideology) and scientific rationalism. Egypt would resurrect, in all its antique glory, once it came under the salutary, modernizing influence of French culture—French political structures, French art, and, perhaps above all, French science.

Napoleon had long nurtured a lively interest in the sciences, especially mathematics. On the first Italian campaign (1796–1797), he often called upon the mathematician Gaspard Monge and Monge's close friend and colleague, the chemist Berthollet, for diversion during the slow periods between fighting, preferring their company to that of others in his retinue.[17] By the close of the campaign, Monge had emerged as a vital player in Napoleon's plans for an Egyptian expedition. In the autumn of 1797, Monge began to collect maps and equipment for the trip, and the following March he traveled around France's conquered territories to oversee the removal of objects of artistic and scientific interest to France, particularly the Vatican's Arabic printing press, which Napoleon specifically asked him to locate and appropriate. Like Monge, Berthollet had proved himself indispensable on Napoleon's lightning romp across northern Italy, where he was especially useful for his knowledge of Italian and for his activity as a trustworthy and reliable commissioner in the removal of valuable and noteworthy artworks to France.[18]

Preparing to depart for Egypt, Napoleon formalized the arrangement he had perfected in northern Italy: scholars were to accompany the expedition, certainly to provide him with suitable opportunities for conversation, but primarily to constitute "a task force capable not only of serving the needs of the army, but of rooting French science and technology in the Valley of the Nile."[19] This force, known as the Commission of Science and Arts, originally comprised 151 members, of whom 84 were scientists and engineers, 10 were doctors, and the remainder were artists and scholars of various kinds. Scientists and technicians made up 58 percent of the commission, which was, like Napoleon himself, remarkably young, with a mean age of twenty-five.[20] Most were engineers, recent graduates of the École Polytechnique or the École des Ponts et Chaussées; the youngest was only fifteen, and a handful were not yet twenty.

Staffing the campaign with engineers, scientists, and other civilian professionals was one important part of the puzzle for Napoleon; positioning the campaign as a specifically scientific undertaking was another. In the spring of 1798, members of the Institut d'Egypte gathered to hear Napoleon's plans for an invasion of Egypt. As he explained the important role that scientists, scholars, and artists would play in his plans to conquer this most antique and fabled of all countries, Napoleon rapped the knuckles of one hand on a copy of the surveyor and traveler Carsten Niebuhr's *Arabian Voyage* of 1762, implicitly comparing his proposed expedition with Niebuhr's legendary trip, as one of the first Europeans to undertake a strictly scientific expedition to the Middle East, sponsored by the king of Denmark.[21]

At least for the more established scientists and artists, the journey to Egypt was to be a great, if unsettling and dangerous, adventure. Napoleon's ship, the *Orient*, became a floating scientific institute, where Napoleon and his favorites—among them, General Caffarelli, Berthollet, and the mathematician Monge—conducted evening discussions on topics selected by Napoleon himself, who also chose the participants. The subjects ranged from the origin of manners to the age of Earth, whether there might be life on other planets, whether the world would end in fire or flood, to how much credence one might place in presentiments that appeared in dreams.[22] Napoleon was often stimulated to historical reflections. "The sight of the kingdom of Minos," wrote Fauvelet de Bourrienne, his boyhood friend and private secretary, "led him to reason on the laws best calculated for the government of nations; and the birthplace of Jupiter suggested to him the necessity of a religion for the mass of mankind."[23]

Troops billeted aboard the *Artemise* kept the republican spirit alive by singing revolutionary songs and hymns, while aboard the *Alceste* the naturalist Saint-Hilaire performed galvanic experiments on various objects, including a captured shark.[24] One evening's entertainment on the *Alceste* involved a bit of orientalizing shipboard theater in which a soldier rescued a young, pretty slave from an elderly Turk and married her. These evenings made the savants feel so at home that Monge wrote to his wife: "It seemed as though, from the moment I arrived, I was in France."[25] In these ways, daily life on Napoleon's expedition combined familiar elements of French intellectual life with the thrill and romance that Europeans had long associated with the exotic East. This pattern would continue throughout the otherwise disastrous French occupation, as Napoleon's savants and artists were charged with, among other things, securing entertainment for the increasingly homesick and demoralized troops. Although, as we shall soon see, these activities would eventually have more than a whiff of desperation about them, the members of the Commission of Science and Arts nevertheless busied themselves by setting up French-style cafés, salons, and even a casino in Cairo, as well as a theater where plays by Voltaire and Moliére were performed.[26]

The library assembled for the expedition sampled Western literature, primarily in French, and shows what was considered most useful for an extended stay in Egypt. In addition to the expected treatises on military subjects like artillery and explosives, as well as accounts of the military campaigns of Turenne, Villars, and Conde, the library included classical works such as Plutarch's *Lives* and Thucydides' *History of the Peloponnesian War*; works by Roman historians Justin, Arrianus, and Tacitus; and the poetry of Homer and Virgil. When, as it sometimes did, Egypt figured in these volumes, it was for the most part as a remote, exotic outpost where writing was said to have originated, and which was home to such fabulous creatures as the phoenix, the winged serpent, and—only a little less fantastically—the crocodile. These works served to domesticate the foreign, to render it familiar. The floating library also included specifically French (and often overtly political) literary material, such as Voltaire's *Henriade*, a series of couplets, rendered in alexandrines, the quintessential French poetic form, about the 1590 Siege of Paris; Fénelon's *Telemaque*, a political parable in which a monarch is persuaded of the value of republicanism; Montesquieu's *Spirit of the Laws*; the fables of La Fontaine; the novels of Voltaire; and Rousseau's *La Nouvelle Heloïse*. Neither were the sciences neglected. Although the selection tilted heavily toward classics of exploration such as Cook's *Voyages*, Barclay's twelve-volume *Geography*, and De la Harpe's twenty-four-volume history of exploration, *Histoire générale des voyages*, the library also included Fontenelle's *Essays on the Plurality of Worlds* and Euler's *Letters to a German Princess*. In addition to the Bible, the religious books included the Koran and the Indian Vedas as well various books about mythology. Goethe's *The Sorrows of Young Werther*, which had particularly captured Napoleon, came along as well; there were entertainments by the encyclopedist Marmontel, the playwright Lesage, and the Abbé Prévost; the poetry of Ossian, Tasso, and Ariosto; and an assortment of forty English novels in translation.[27] Most of these books would have been familiar, at least by reputation, to Napoleon's savants. Altogether the library's contents convey an idea of what it meant to be educated, French, and with imperial designs. It comprised a collection of comforting touchstones for those who might need to remind themselves that Egypt, far from being a completely foreign and exotic place, had been linked to the West for centuries, if not millennia.

If, for some, the expedition's promise of marvelous adventure was realized when the ships left France, for others the journey was a hardship. The trip itself was first of all a mystery. To keep the British in the dark, no more than 40 of the 17,000 soldiers and civilians on the expedition—the highest-ranking officers and Napoleon himself—knew where they were headed when the first fleet of ships left Toulon, nor did they understand much about their mission except that it would involve some damage to British interests.[28] Denon, on the *Junon*, was less comfortable than most of the other senior members of the expedition. By June, he recalled in his *Voyage*, "our provisions were nearly expended; our water so fetid as to be scarcely potable. The useful animals had disappeared, while those that fed on us multiplied one hundred-fold."[29] Even conditions on Napoleon's ship, the *Orient*, left much to be desired. Moreover, the scholars were surprised by the army's procedures and unaccustomed to the rigors and hierarchies of military life. Savants thought they would be treated as a single homogeneous group, but, upon leaving for Egypt, they learned they had been divided into five classes, with salaries that depended on rank. Most civilians, unaccustomed to the military practice of fitting everything and everyone into a hierarchy, disputed their status. Saint-Hilaire observed that not only did *all* the savants claim to be worthy of first-class standing, those who complained loudest were "precisely those least worthy of the distinction."[30] These sorts of petty, divisive squabbles beset the expedition from the outset and certainly did not make the savants beloved by rank-and-file soldiers. To further complicate matters, some generals and admirals did not like being classed with the savants, especially younger ones. When Bonaparte attempted to quell the squabbling by demoting all members of the civilian commission by one rank each, except for members of the Institut National, the solution proved untenable.[31] Rivalry for Napoleon's favor proved so intense on the *Orient* that, to preserve morale, he took his meals in his quarters, accompanied only by his two closest aides and the occasional savant.[32]

Savants in Cairo

On July 25, 1798, the army arrived at Cairo after victories in Malta and again at Imbaba, outside Cairo, at what Napoleon grandly called the Battle of the Pyramids (a name that has stuck, even though the view from Imbaba affords only the barest glimpse of the monuments). Less than a month later, however, the news arrived of the British admiral Nelson's successful surprise attack on the French fleet, assigned by Napoleon to safeguard the coastline in the Bay of Aboukir. After Nelson's siege sank all but two of the French ships, including the *Orient*, effectively trapping the expeditionary force in Egypt, Napoleon had little choice but to turn his attention to inland pursuits. Among these was the establishment the Institut d'Égypte, modeled on its Parisian predecessor and dedicated to the pursuit, by the French in Egypt, of scientific knowledge about the country.

The institute was organized formally in Cairo on August 20–22, 1798. The primary purpose of the fledgling organization was the "advancement and propagation of enlightenment in Egypt," including the "research, study, and publication of [Egypt's] industrial, historical and natural phenomena." It was also responsible for "offer[ing] its opinion on different questions on which it will be consulted by the government" and was to be "the privileged instrument of [Egyptian] national regeneration."[33] The institute opened on August 23, 1798, with twenty-six founding members.[34] Its total membership during the occupation numbered fifty-one men. Like the Institut National in France, the Egyptian analog was divided into sections according to subject: mathematics, physics, political economy, and literature and arts. Twenty-six members were assigned to either of the two scientific sections while the rest were divided, again according to their interests, between the arts and political economy.[35] In keeping with its mission to promote progress and industry—two values whose deep roots in the Enlightenment

Figure 4.1. Monge (*left*) and Berthollet in Egypt

should not obscure the questionable possibility of imposing these values on a subject population during a military occupation—the institute sponsored annual prizes for work that promised to improve Egyptian civilization and industry, at least according to those French savants who were to bestow the award.

In addition to Monge and Berthollet, the institute's most prominent scientists included the mathematicians Fourier and Costaz, the astronomer Nouet, and the geologist and (despite his liberal politics) knight of Malta, Dolomieu, all of whom were well-regarded savants before they left France. An additional three (the naturalist Saint-Hilaire, the engineer Malus, and the young zoologist Savigny) became well-known upon their return, in part for the work they had done in Egypt. Napoleon was in the mathematics section with Fourier, Costaz, Nouet, Malus, and Monge, while Berthollet was listed in the section devoted to the physical sciences along with Saint-Hilaire and the engineer Nicolas-Jacques Conté, inventor of the exceptionally hard lead pencil that still bears his name, who was also the head of the French balloonists' brigade in Egypt.[36] Monge was elected the institute's first president, while Bonaparte served as vice president and Fourier as permanent secretary.

The institute organized a series of two-hour meetings that began at seven in the morning on the first and sixth days of the ten-day interval known on the Revolutionary Calendar as the *décade*.[37] Its day-to-day proceedings followed the form of the ancien régime's academies, which had by this time been replaced by the Institut National: participants served on committees, submitted and read memoirs, assigned referees, and identified experts who might be invited to speak on subjects of interest. All told, the institute held sixty-two meetings, the last of which convened in October 1801.[38]

At the inaugural meeting, Napoleon posed several questions for the consideration of the membership, sending them in a decidedly practical direction largely toward the problems of creating and sustaining French power in Egypt, a direction from which the institute deviated little thereafter. Napoleon's questions ranged from the use of improved ovens for baking the army's bread, to whether it was possible to brew beer without hops in Egypt, to the situation of jurisprudence, education, and civil and criminal law in the country, to suggestions for the best place to erect an observatory, to calling for a study of the crucial nilometer on the isle of Roda, which was traditionally used to forecast the extent of the river's annual flood.[39] Commissions were established to study each of these questions, and over the course of the next five meetings their members, working quickly, reported the findings. These were interspersed with presentations on other locally inspired subjects, notably Monge's memoir, in the second meeting, on the mirages that had tantalized the soldiers on their march from Alexandria to Cairo, and a proposal to create a dictionary of common Arabic terms for everyday use by the French.[40]

Napoleon assigned the task of finding a physical home for the institute to Berthollet, Monge, and General Caffarelli, who secured a site on the outskirts of Cairo where they were to establish facilities for printing as well as physics and chemistry laboratories, a library, and an observatory. While those desiderata sound plain enough, in fact the buildings selected to house the institute were sumptuous, decorated with fine furniture and priceless ceramics seized from the fleeing Mamelukes, as well as archaeological treasures. In the library, there was what amounted to a small museum that included two small obelisks, several sarcophagi, manuscripts in Coptic, Arabic, and Turkish, a fragment of a statue of Ramses found in Memphis, and, most precious of all, the Rosetta Stone with its carved inscription in hieroglyphs, an as-yet unknown Egyptian script, and Greek, which had been discovered by Captain Pierre François Xavier Bouchard in the course of a demolition.[41] According to al-Jabarti, the savants were in effect given an entire neighborhood "and all the houses in it" in the al-Nasriya quarter, including the brand new palace of one Hassan Kachef, "which he founded and built to perfection, having spent upon it fantastic sums of money amounting to more than a hundred thousand dinars. When he had completed plastering and furnishing it, the French came and he fled with the others and left all that it contained, not having enjoyed it for even a whole month."[42] Meetings took place in an adjacent palace, formerly belonging to Ibrahim al-Sinnari, a local landlord, intriguer, and favorite of Murad Bey, in a drawing room that would have otherwise been occupied by the harem.[43] Scientists used a long corridor in yet a third palace to trace a solar meridian. An aquarelle by Conté depicts the courtyard of the institute as a pretty, convivial space, cooled and shaded by potted plants; the astronomer Nouet's sundial overlooked it from high on the far wall. The luxurious environment included terraced gardens, colonnades, and menageries that Saint-Hilaire claimed, in an excited letter to Cuvier, rivaled even the Jardin des Plantes in Paris.[44] Half of the institute's thirty-acre botanical garden was given over to the cultivation of indigenous grains, fruits, and medicinal herbs, while familiar plants brought from Europe, like potatoes, grew in the other half. There was also an enormous prickly sycamore in whose shade the savants liked to gather for informal conversation every evening at dusk.[45]

The library particularly impressed al-Jabarti: "Whoever wishes to look up something in a book asks for whatever volumes he wants and the librarian brings them in to him. Then he thumbs through the pages, looking through the book, and writes. All the while they are quiet and no one disturbs his neighbor." The library was open to anyone, and the savants eagerly responded to expressions of interest among the local people. "When some Muslims would come to look around, they would not prevent them from entering," al-Jabarti observed. "Indeed they would bring them all kinds of printed books in which there were all sorts of illustrations and [maps] of the countries and regions, animals, birds, plants, histories of the ancients, campaigns of the nations, tales of the prophets including pictures of them, of their miracles and wondrous deeds, the events of their respective peoples and such things which baffle the mind." Of the books in the collection, al-Jabarti favored those dealing with science and mathematics, and the French translations of Arabic works. "They have a great interest in the sciences, mainly in mathematics and the knowledge of languages," he reported, and even though he frequently criticized the French for their mistakes in Arabic, he lauded their "great efforts to learn the Arabic language and the colloquial. In this they strive day and night. And they have books especially devoted to all types of languages, their declensions and conjugations as well as their etymologies." Al-Jabarti was

Figure 4.2. Entrance (*top left*), courtyard (*top right*), and central hall or *durqa'a* of the restored house of the Berber Ibrahim Kathuda al-Sinnari, favorite of Mourad Bey, near the mosque of Sayeda Zenab, that the French expropriated to house the Institut d'Égypte. Its narrow street is now named after Monge.

astonished to find a copy of the Koran translated into French, and to meet French orientalists who could recite from the Koran from memory, as well as portraits of all kinds, including some of the Prophet. As portraiture was forbidden by the Koran, for many of the library's visitors these were the first examples they had seen.[46]

From a practical point of view, the most important part of the institute was not the meeting room or the library but the workshops where, at their busiest, three hundred artisans did the nitty-gritty work of keeping Napoleon's occupation in supplies that could not be obtained locally or from abroad because of Nelson's blockade.[47] Overseen by Conté, the nine workshops engaged in practical activities such as smithing, clock making, carpentry, and manufacture of instruments for surveying and astronomy. Conté and his workers also made parts for the printing presses, syringes for the hospital, tools to card and loom wool, excavation equipment, and all sorts of army matériel, from small items like sabers, cannonballs, and gunpowder to monumental objects like the windmills that were installed on the isle of Roda as part of a project to introduce French milling techniques to Egypt. These products were vitally important to the savants after the loss of ships carrying scientific instruments in the Battle of Aboukir Bay, and again after the *Patriote* sank in Alexandria's harbor, taking with it a load of apparatus, from tweezers and pins to microscopes, as well as hundreds of pints of preservative alcohol and a supply of botanical pressing paper.[48] In the institute's laboratories, al-Jabarti noted the skill of the savants as they handled their instruments for chemical experiments, the "wondrous retorts of copper for distillation, and vessels and long-necked bottles made of glass of various forms and shapes, by means of which acidic liquids and solvents are extracted," as well as "instruments for distilling, vaporizing, and extracting liquids and ointments belonging to medicine and sublimated simple salts, the salts extracted from burnt herbs, and so forth." The physical and astronomical instruments were no less interesting, especially for the precision and detail of their construction. "They have telescopes for looking at the stars and measuring their scopes, sizes, heights, conjunctions, and oppositions, and the clepsydras and clocks with gradings and minutes and seconds [are] all of wondrous form and very precious." Despite al-Jabarti's effusiveness, when members of the institute attempted to impress local clerics with feats of scientific derring-do, the clerics sniffed, regarding the experiments as parlor tricks.[49]

While many of the projects were directly related to the needs of the army, some of the most interesting work was inspired by the new experiences and the unfamiliar sights, flora, and fauna of Egypt itself. Savants collected technical, scientific, and even local medical data and made observations about Egypt's natural history and geography, from tables of necrology to drawings of the anatomy of the crocodile. Technical papers were soon produced, such as Monge's on the cause of mirages, which appeared in the first volume of the *Décade égyptienne*, and Berthollet's landmark paper on the formation of natron (sodium carbonate) from the double decomposition of salt and limestone in the lakes and soil in the valleys west of Cairo, which appeared first in the *Memoires sur l'Égypte* and in 1800 in the *Annales de Chimie*.[50] And there were the antiquities—numerous, comparatively well preserved, and tantalizingly old. The study of the monuments of remote antiquity held out the promise of revising the history of Western civilization, shifting attention and authority away from familiar classical sources that scholars had already worked over for centuries.[51]

Napoleon had brought several printing presses to Egypt for communicating both with his troops and with the local population. These could print documents in French, Greek, and Arabic. To run the presses and the publications associated with them, Napoleon brought along, among others, the editor Saint-Jean d'Angély, who reached Malta before succumbing to illness, and Aurel, a journalist and the son of the printer who had published Napoleon's 1793 *Souper de Beaucaire* as well as a Napoleonic propaganda organ, *La France vue de l'Armée d'Italie*.[52] From the Italian campaign, Napoleon had learned the importance of the press to the success of a military campaign. Not only was printing useful for communicating with commanders and troops, it also provided a powerful platform from which to chronicle, position, and publicize diplomatic and military activities for local audiences and for supporters back in France.

The institute was also responsible for a number of other publications. There was the scholarly periodical *La Décade égyptienne*, modeled on the *Décade philosophique*, Napoleon's high-minded journal of the arts in Italy.[53] The Egyptian analog contained summaries of articles and essays by the institute's members and basic information about when each work was to be read before it. Compiled and edited by the ruthless Thermidorean Jean-Lambert Tallien and the physician Desgenettes, each issue of the *Décade* cost one French

franc.[54] The newspaper *Courier d'Égypte* contained general information about all things French in Egypt, including a calendar of public
events, abstracts of the institute's proceedings, news of military activities, festivals, and entertainment, and news from Europe. Edited by
Aurel, the *Courier* ran to 116 weekly issues. All of these publications
served the dual purposes of propagandizing the Egyptian expedition
to readers in France when French ships could get past Nelson's blockade to deliver them and, more important, shoring up the morale of the
troops and civilians marooned with Napoleon in Egypt after the rout
by Nelson at Aboukir.[55] Finally, members of the institute would eventually publish papers on their Egyptian research in traditional scientific periodicals like the *Annales du Muséum d'Histoire Naturelle*, the
Bulletin de la Société Philomatique, and the *Journal des Mines,* as
well as monographs, including Denon's travelogue, the *Voyage en la
Basse et Haute d'Égypte*, which was magnificently printed in 1802 in
two huge elephant-folio volumes, of which one was devoted entirely
to Denon's drawings, including the ones that he did of the two zodiacs
at Dendera.

An Artist at Dendera

In Egypt, Denon made sketches on horseback and on his knees, as
well as standing up, without finishing a single one, as he would have
liked to.[56] While this lack of polish serves as a reminder of the real
pressures under which Denon worked—the time constraints imposed
by army maneuvers, not to mention the constant threat of attack—it
is also likely that, in claiming to prefer rough truth to perfected falsehood, Denon made a virtue out of necessity. If his drawings from Egypt
seemed unpolished, Denon asserted, it was because in a whole year
of sketching en plein air he never once found a table sturdy enough
to permit the use of a ruler.[57] At Thebes and elsewhere he sketched
feverishly, worried that military exigencies would cause the scene "to
escape from me."[58] His unorthodox methods nonetheless bore copi-
ous fruit. More than two hundred sketches—from panoramas of the
desert to lively battle scenes to intricate renditions of ancient Egyptian architectural elements—were immediately and widely admired
on their publication in his *Voyage*. The book proved so popular that
it rapidly went through numerous editions in France as well as being
(rather imperfectly) translated into English only a year after its initial

publication. It would also be extensively serialized in the *Moniteur Universel*, a newspaper that not only was tightly controlled by the government but also had a virtual monopoly on information in France at this time.[59] No other contemporary traveler's account of Egypt was to command such a wide and enthusiastic audience. In the words of a modern historian of Egyptology, "Napoleon conquered Egypt with his bayonets and held it for one short year. But Denon conquered the land of the Pharaohs with his crayon [pencil], and held it permanently."[60]

Denon drew the circular zodiac on his second trip to Dendera; on the first, there was hardly time to get a good look, and none at all to make a proper sketch. He was excited to return but unsure whether his first impressions of the ceiling's importance and uniqueness were truly correct. On the second trip, Denon and the others arrived late, staying overnight in the village. He struck out the next morning with a group of thirty men, heading for the ruins "which this time I possessed in the full plenitude of rest and quiet." Upon reaching the temple, Denon's initial feeling of peace was deepened by relief—the excitement he had felt after his first, hasty view of the site was not unwarranted, as he had feared. "My initial joy [at seeing the ruins] convinced me that my enthusiasm for the great temple had not been an illusion born of novelty," he wrote. His reaction was, it seems, immediate: the two zodiacs at Dendera, the rectangular one on the ceiling of the main court and the circular one in the small temple on the top, "proved in the most positive manner the Egyptians' elevated astronomical knowledge." Indeed, he continued, "after having seen all the other monuments of Egypt, this one seemed to me to be the most perfect in its execution, and constructed at the happiest epoch for the sciences and the arts; everything was carefully done, everything was interesting, even important; everything had to be drawn to have everything one wished to report; nothing had been done without a purpose."[61]

As always when on the march, time was of the essence. Denon did not have unlimited oil for his lamps, and at any rate he was at the mercy of General Desaix's schedule, which could at any time require Denon to abandon his drawing in order to travel to some other area of the Upper Nile, following hard on the heels of the beys (who were proving more intractable than expected). Quickly, Denon made his way to the top of the Dendera temple—not easy in his day, when sand and the detritus of centuries blocked easy passage. Sketchbook in hand, he set to work in the small chapel on the roof. "I began with what had been in a certain sense the object of my trip, the celestial planisphere, which occupied a part of the ceiling of the small room built on the roof of the nave of the great temple." It was hard to get a good, encompassing view of the ceiling: the chamber was dark and uncomfortable, and the ceiling itself was highly and confusingly detailed, covered in hieroglyphics and mysterious symbols, some more familiar than others. Undaunted, Denon was motivated, as he later wrote, by the prospect of being the first to return to France with a sketch of the zodiac. "The thought of taking back to the savants of my country the image of an Egyptian bas-relief of such importance gave me the strength to suffer with patience the stiff neck that I had to get in order to draw it." Crouched on the floor, peering at the walls and ceiling lit dimly by flickering torches, Denon worked on the planisphere while he had light; when he did not have enough to continue, he measured and sketched the various parts of the monument.[62]

Of the vast array of figures on temple surfaces, Denon was most fascinated by the repeated appearances of what he mistook to be Isis but was in fact the deity Hathor, though the two are related. That he so quickly identified the figure as Isis points to her prevalence in the classical sources with which Denon and his readers were familiar, in particular in the *History* of Herodotus. On one side of the zodiac, a large figure of the goddess stood straight up, with "both feet planted on the earth, her arms spread across the sky, and she appeared to occupy all the space in between." Denon also reports that she appears elsewhere holding Horus in her arms, nursing him "like a newborn." Next to the circular zodiac stretched another large female figure "that I believed represented the sky, or the year, touching the same base with her feet and hands, and covering with the curve of her body fourteen globes placed on fourteen boats, distributed over seven bands or zones, separated by innumerable hieroglyphs."

Although Denon often resisted the urge to speculate about the meaning of the antiquities he sketched in Egypt, these affected him so forcefully that for once he threw caution to the winds. The images in the chapel, Denon suggested, formed a kind of physical allegory in which elements were arrayed in a puzzling order that related architecture to the natural world. "It is hard not to speculate about the original purpose of this small structure, which is so well done in its details, decorated with images which are so evidently scientific; it seems that the images on the ceiling are related to the movement of the sky, and those on the walls to that of the earth, to the influences of the air and of

Figure 4.3. Denon's sketch of the circular Dendera zodiac

water." This cosmology of the ruins evoked a standard Enlightenment belief: that myth was born in nature and not, as the Renaissance had it, in allegory.

Denon turned from cosmology to history, to the relationship between, we would now say, church and state in ancient Egypt, and to its resonance with contemporary France, where the authority of the Church and the traditional aristocracy had been destroyed during the early years of the revolution—though Catholicism was beginning to experience a revival that, we shall see, would strongly color the controversies that arose once the zodiacs became known in France. Denon was scandalized by the extent to which ancient Egyptian forms of governance were intermingled with religious belief and activities. He compared the great monuments of Egyptian antiquity, such as the Dendera temple and the pyramids at Giza, to the architectural embodiments of

religious power of the Catholic Church, especially St. Peter's basilica in
Rome. The pyramids were especially provocative. Only "priestly, despotic governments could dare undertake to raise [such monuments],"
Denon declared, "and only a stupidly fanatic people would consent to
undertake their construction." At the same time, the ancient architecture, enlivened by evocative iconography and inscrutable hieroglyphics, excited his imagination. "These chambers consecrated to
an eternal night, this mystery diffused over the cult, obscure as the
temples themselves; these secret initiations, so hard to obtain, and
to which no outsider could ever be admitted, about which we know
only from mystical reports; this government and religion which lost
all its power and all its empire as soon as Cambyses had violated the
sanctuaries, overturned the divinities, and carried off the treasures; all
announce that, as it were, these temples contained the *essence* of all,
that everything emanated from them." Denon's "everything" was the
mystifying authority of the ancient Egyptian state, in particular its way
of being both civil and religious at once—a hybridity that unsettled and
fascinated Denon.[63]

Enlarging on the same subject at Karnak, Denon connected ancient
Egypt's theocracy with the revolutionary Terror, making both guilty of
what, for him, was the supreme sin: hostility to pleasure. "What monot-
ony!" Denon exclaimed, standing before a bas-relief of a particularly
forbidding deity. "What melancholy wisdom! What austere gravity of
manners! I still regard with awe the organization of such a government;
its stupendous remains still excite mingled sensations of respect and
dread." The figure, he wrote, "holds in one hand a hook, in the other
a flail; the former, no doubt, to restrain, the latter to punish." Perhaps
meaning to remind his reader of the plight of the captured crocodile
at Qena, Denon complained, "everything is measured by the law and
enchained by it."[64] It was this smothering civil and religious environment, Denon had explained elsewhere in the *Voyage*, that accounted
for ancient Egyptians' penchant for sometimes sublime but more often
simply unappealing architecture. The architecture, he opined, demonstrated the artists' susceptibility to the corrosive effects of dogma, for
"bend[ing] under the weight of fetters, their soaring genius is pinioned
to the earth." Nowhere could he discern any evidence of delight or
leisure: "in order to destroy pleasure, pleasure is converted into duty;
[there is] not a single circus, not a single theater, not a single edifice
for public recreation; but only temples, mysteries, initiations, priests,
sacrifices; for pleasures, only ceremonies; for luxury, sepulchers." The

irrepressible Denon found a salutary warning in the midst of his disappointment: his readers should be aware that their own happiness was vulnerable to the same forces that bedeviled the ancient Egyptians, forces that were still at large in the world: "Surely whenever [that] evil spirit is abroad in France, it partakes of the gloomy ferocious soul of an Egyptian priest who imagines that he could render us happy by making us sad and miserable like him."[65]

Cairo Revolts

While Denon was in Upper Egypt sorting out the implications of ancient theocratic forms of government, in Cairo and elsewhere Egyptians were learning just what French colonial rule meant. The occupiers were hardly the enlightened liberators they wished to appear. When Napoleon's bloody seizure of Cairo forced the city's residents to evacuate, most people took with them only what they could carry, then to be robbed by bandits beyond the city limits. Fleeing Cairene women, al-Jabarti reported with dismay, did not even have time to veil themselves, while many of those left behind were beaten, raped, and killed by the French.[66] In the provinces, the presence of soldiers in previously tranquil villages frightened the inhabitants into hiding, so that even larger settlements became ghost towns. "The dust accumulated in the market-places, because there was no one to sweep them or splash water, and the shops were abandoned." Thieves and thugs roamed the dusty streets, "relations between people ceased, and all dealings and business came to a standstill. The roads in the city became insecure, not to mention outside it." In the cities and towns, prices for the most elementary of daily necessities rose steeply, while weapons and ammunition became especially dear; al-Jabarti reported high prices for gunpowder and bullets, and even simple clubs were in short supply. Chaos reigned in the countryside as well: "Violence flared . . . people began to kill each other. They stole cattle and plundered fields. They set fire to barns and sought to avenge old hatreds, and blood feuds."[67]

After restoring order, or at least some semblance of it, the occupation became less violent but more expensive, as the French imposed various fines, taxes, and rules. They would, for example, order lamps to be lit all night in Cairo's streets, in the homes and shops. If soldiers found the lamps unlit, "they would nail up the shop or house where this had happened, and would not remove the nails until the owners have made an arrangement and paid whatever they felt like demanding," al-Jabarti reported. Even worse, "[s]ometimes they would deliberately smash the lamp for this purpose." Property owners were required to show deeds of ownership if their property was to be kept out of French hands. But when these deeds were brought forward, they were often deemed so insufficient that further evidence of ownership had to be located in the registers, which necessitated the payment of yet another fee. Finally, as the French sought to remake the city for their own purposes, they indifferently razed buildings and other sites of sacred and secular importance, including mosques, council halls, chapels, and palaces, and erected others in their place. "They pulled down the high places and raised up the low places. They [. . .] wiped out the monuments of scholars and the assembly rooms of sultans and great men and took what works of art were left," al-Jabarti complained. They damaged Cairo's Great Mosque "by removing the *minbar* (pulpit) and the *maqsura* or private prayer area reserved for the Sultan and his entourage."[68] When the barking of the local dogs, which ran freely through the streets as a form of neighborhood protection, unnerved the troops, they responded by shooting them all over the course of two nights, which not only made the streets less secure but also increased the threat of plague as rats proliferated.[69] And the French nearly incited a riot when they tried to chop down an ancient sycamore to which local people ritually affixed their lost teeth and hair, in the hopes of causing them to grow back.[70]

Many aspects of French culture did not graft well onto the Cairene vine. Al-Jabarti complained with disgust that "[t]he streets and houses where the French lived were full of filth, infected earth mixed with bird feathers, the entrails of animals, garbage, the stench of their drinks, the sourness of their alcoholic beverages, their urine and excrement, such that a passerby was obliged to hold his nose." When, hoping to impress the Egyptians with their technical virtuosity, the French sent up a hot air balloon, it came right back down, undermining whatever small credibility the occupiers had managed to accumulate. And when they set up a casino and bordello in Cairo, al-Jabarti disapprovingly, and no doubt accurately, noted that these "places for amusement and licentiousness" paved the way for "all kinds of depravities." The Egyptians were not shy about making their distaste known. Forms of public disapproval became so frequent and distressing that police called upon the public to "desist from meddling in and discussing political

Figure 4.4. The ruins of ancient Egypt envisioned by the French savants. The heroic figure carrying a spear at top is Napoleon routing the Mamelukes. "Liberated" Egyptians, dressed in ancient garb, follow behind in grateful freedom.

matters [and] if a group of wounded or defeated soldiers passed their way, not to mock them or clap, as they habitually did."[71] From windows high above the streets around Cairo's Qala'un maristan, a minaretted thirteenth-century mosque and asylum for the insane, Cairo's inhabitants flung javelins at passing French soldiers.[72]

All this culminated in the Cairo uprising of October 1798, during which the institute was surrounded by a hostile mob, trapping a group of unarmed scholars inside. When grenadiers arrived with muskets and ammunition, they simply dropped off the weapons and abandoned the savants to fend for themselves. Although Monge and Berthollet took leadership roles, Denon observed, "each had his own plan but no one felt obliged to obey." Four members of the Commission of Science and Arts—the engineers Duval and Thevenot, the geographical surveyor Testevuide, and Sulkowski, Bonaparte's aide-de-camp—were killed, and all their equipment was destroyed.[73] The grenadiers' abandonment of the savants was not surprising. From the start, Napoleon's military had resented the presence of civilians on the Egyptian campaign. These career men ridiculed Napoleon's interest in the institute, which they mocked as his "favorite mistress," and they expressed even less respect for the savants. In the streets of Cairo, they called the donkeys "demi-savants," and during periods of fighting the soldiers would contemptuously shout: "Donkeys and scholars to the middle!"[74] Napoleon's emphasis on science and his general favoring of the civilians in his entourage irritated officers, who felt the military aspects of the expedition were getting short shrift.

Napoleon's Flight

"L'oiseau était déniché" was how Kléber put it: *The bird had flown the coop.*[75] On August 22, 1799, Napoleon abruptly quit Egypt, taking with him a few of his favorites, including Berthollet and Monge. Denon went as well, not because he was such a favorite, but because he had a manuscript, found in an ancient Egyptian tomb, that was of great interest to Napoleon. The departure from Egypt took place in the utmost secrecy. A handful of Napoleon's intimates, including Denon, had received official letters, to be opened at a specific time, inviting them to meet at Aboukir Bay. When they arrived on the shore, they found a ship, the *Muiron*, waiting for them, along with Napoleon, who explained his intentions as the ship set sail for France on a moonless night, leaving the unhappy Kléber in charge. In the midst of this secret escape, two uninvited guests were discovered and thrown overboard.[76] The passage back to France was smooth; not a single English ship was sighted. According to Denon, Napoleon passed the time easily, doing mathematics or socializing, apparently unconcerned about the mess he had left behind.[77]

Back in Egypt, the institute's vitality was waning. It had become more bureaucratic, as the forces of institutional power vied with fluctuating levels of energy and curiosity among the membership, who had been badly misled by Napoleon's promise of a glorious French-controlled Egypt. As their appetites for discovery dwindled, the members spent less time formulating and solving new problems and more on the management of a growing organization—maintaining its membership, electing officers, and overseeing the collection and organization of reports, drawings, and artifacts, as well as the numerous books, papers, and other materials, including seed samples, that had been sent to the institute from abroad. Committees were formed for virtually everything, including one for deciding whether the *Courier* or the *Décade* was the right publication to carry official notices of the institute's activities—an argument that, as de Villiers observed to Jollois, became quite heated as Desgenettes, the physician who edited the *Décade*, felt slighted by General Jacques-François Menou, who had taken command after the assassination of Kléber, since Menou contended that the *Décade* was merely "a simple journal."[78] By early July 1799 tensions between the members were running high. The arguments that broke out reflected the problems of an organization sited at the uneasy conjunction of knowledge with colonial power. When, before leaving Egypt, Napoleon had tried to nominate a commission to study an episode of plague that had broken out among the troops, Desgenettes objected on the grounds that the risk of further spreading the disease outweighed the benefits that might accrue from its study in vivo. Napoleon parried viciously, calling Desgenettes a "charlatan" and his medicine "the science of assassins." In response, the uncowed Desgenettes asked Napoleon whether he might like to comment on what he thought about the "swaggering science of conquerors."[79]

Upon returning to France, Napoleon rewarded favorites from the Egyptian campaign. Berthollet was appointed a senator, was made an officer of the Legion of Honor, and obtained funds to maintain a private chemical laboratory on his Arceuil estate just outside Paris. This project soon evolved into the Société d'Arceuil, a group of savants who would turn the sleepy village into a hotbed of experiment and debate.

Monge, too, was made a senator and was ultimately ennobled as the Comte de Peluse. As for Denon, Napoleon was so pleased with the success of the *Voyage* that, three months after the book's initial publication, he promoted Denon to the position of director of museums, where he personally managed the removal of artworks from countries subsequently conquered by Napoleon.

Both Denon and al-Jabarti outlived Napoleon. They died the same year, in 1825. Yet the circumstances of their deaths could not have been more different, or more telling. Approaching eighty in Paris, Vivant Denon divided his time between art auctions, where he accumulated more than fifteen hundred priceless artworks, working on a sweeping illustrated history of art and the pursuit of amorous adventures until he caught a chill while browsing around an art gallery and died the following day, in April. He left behind an enormous personal collection as well as everything he had purloined or otherwise accumulated in his work as director of the Louvre. According to Ulric Richard-Desaix (the general's nephew), among Denon's effects was a cabinet of curiosities set in, of all things (given Denon's enduring dislike of religion), a miniature Gothic cathedral. Eighteen inches high, of gilded brass, and hexagonal in shape, the miniature was adorned with flying buttresses that were each topped with a tiny cross. The reliquary's six compartments contained shards of bones of El Cid and his wife Ximéne, found at Burgos; similar remains of Héloïse and Abelard from their tombs at Paraclet; strands of hair belonging to Agnés Sorel, the mistress of Charles VII, and of Inés de Castro, mistress of Pedro I of Portugal; a clipping of Henry IV's moustache; bones belonging to Moliére and LaFontaine; a fragment of one of Voltaire's teeth; the shroud of the beloved seventeenth-century warrior Turenne, who famously rejected the blandishments of Richelieu (including the offer of his daughter); and, as a memento of Denon's time with him in Egypt, a lock of Desaix's hair. The shrine was adorned with an autograph of Napoleon, a piece of the bloody shirt he was wearing at his death, a lock of his hair, and a leaf from the willow tree under which he was buried on Saint Helena.[80]

Al-Jabarti went on to complete two difficult books in an effort to come to terms with the events of the occupation and its aftermath. The first was a defense of his own involvement with the occupation government, while the second attacked Muhammad Ali, now the ruler of Egypt, for his policies. Although al-Jabarti was among the first members of the nascent *nahda al-arabiyya*, the Arab renaissance, a nineteenth-century movement that opposed popular Sufism and superstition, neither book was at all well received; the second, not surprisingly, was banned by Ali, copies were confiscated, and then, to make his retaliation complete, Ali's henchmen murdered al-Jabarti's son. Al-Jabarti wrote nothing more after that, and in 1825 he died, blind, quiet, and confined to his house, reportedly from grief.[81]

5

One Drawing, Many Words

The Rectangular Zodiacs at Esneh and Dendera

Even as Napoleon's imperial picaresque was concluding in Egypt, ideas about it were taking wing in Europe. Scholars and scientists formerly attached to the expedition busily worked through their experiences on paper, describing Egypt in what would soon become a voluminous literature. These articles, books, and letters were supplemented, in many cases, by images—drawings, paintings, and sketches made by savants during their time in Egypt and reproduced later by engravers and printers. According to the pictorial conventions common among graduates of the École Polytechnique, drawings done on the basis of exact procedures were uniquely credible sources of truth. The techniques of descriptive geometry, developed preeminently by Monge, offered a single point of view that could be presented as a coherent and comprehensive whole, recorded by a witness working to rule. Drawings of this sort were thought to bear a specific kind of verisimilitude, independent of their creator: they were designed to compel belief at a glance.

In this chapter, we will trace the early trajectories of images of the Egyptian zodiacs that were produced by members of the expedition. Some were drawn according to the rule-based strictures of descriptive geometry, whereas the ones produced by Denon followed contemporary canons of artistic representation. His images were the first to circulate in Paris, where they lent a degree of representational credibility to ideas already in circulation about Egyptian astronomy, in particular ones associated with the radically antireligious ideas of Dupuis and his followers.

In March 1799 a nine-member commission under the engineer Girard left Cairo for Upper Egypt, including among its complement two young students of the École Polytechnique, Jean-Baptiste Jollois and René Edouard Devilliers.[1] Despite obstructive efforts on the part of Girard, who apparently disdained antiquarian work, Jollois and Devilliers produced drawings of the monuments and their inscriptions

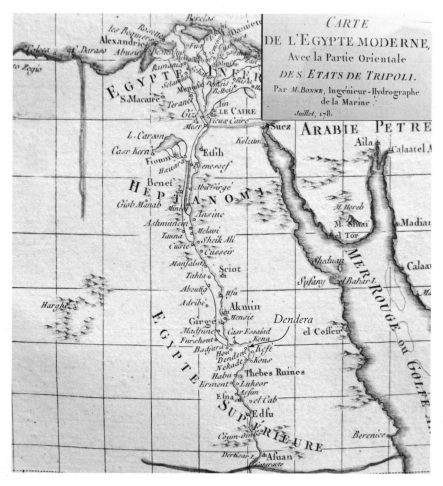

Figure 5.1. Map of Upper Egypt

using the drafting techniques they had been taught by Monge himself.[2] The two of them remained at Thebes to undertake further examination of antiquities after the Girard commission set out for Cairo on August 7. They had encountered Denon along the way there. He himself headed back north on July 4, arriving at Alexandria five or so days later with his sketches of Upper Egypt. (He would remain in Alexandria until sailing for Europe with Napoleon on August 22.) On August 14, Napoleon charged two expeditions with the exploration of Upper Egypt, one under the direction of Costaz, the other led by

Figure 5.2. Jollois (*left*) and Devilliers

Fourier. Both expeditions left on August 20, and together they encountered Jollois and Devilliers at Edfou on September 10. The two young men again met up with the expeditions on September 20, this time at Esneh, to which they had returned on September 12, having first reached it on June 30.[3]

Five weeks before the *Muiron* sailed with Napoleon for France, the young geographer Jean-Baptiste Coraboeuf, who would accompany Costaz's expedition in August, wrote from Alexandria to the astronomer Lalande (who, recall, was also Dupuis' mentor) concerning the ages of the zodiacs that had been found at Esneh and Dendera. Lalande, in France, received the letter sometime that fall or winter.[4] Originally seconded to the astronomer Nouet, Coraboeuf had assisted him in measuring the geographical coordinates of, among other markers, the minarets of Cairo. He accompanied Nouet and other savants on Desaix's campaign through Upper Egypt, arriving eventually at Aswan after Denon's departure, where they inscribed their presence in the temple of Isis on the island of Philae (Figure 5.3). On July 13 Coraboeuf wrote to Lalande about the longitudes and latitudes "determined by citizen Nouet," emphasizing as of particular interest the coordinates of ancient Syene (modern Aswan), which Eratosthenes had placed on the tropic in antiquity. Nouet had measured a difference of over a degree and a half from Eratosthenes's value. An effort had also been made to determine whether the "great pyramid at Memphis" was oriented with respect to

Figure 5.3. The savants' inscription on the temple of Isis at Philae, Year 7 of the French Republic. Because of the Aswan high dam, the temple was dismantled and moved stone by stone to the island of Agilkai about 550 meters from its original position. The move began in 1972 and was completed in 1980.

the cardinal points, but the method used was insufficiently exact, and Coraboeuf hoped for a repeat.

Near the end of his letter to Lalande, Coraboeuf mentioned the zodiacs and speculated about their antiquity. Denon must have shown Coraboeuf some of his drawings, or at least described what he had seen, for at that point Denon's drawings were the only ones that would have been available.[5] Although Denon had been captivated by the great circular at Dendera, returning to it many different times despite the late-spring heat, Coraboeuf did not mention it in his report. He wrote instead about two others—one at Dendera, the other at Esneh. Both were arranged in a pair of parallel rectangular bands.[6] At the time, the colors of the one at Dendera would have been quite striking; traces remain even today (Figure I.3). Coraboeuf's decision to focus on the rectangular zodiacs to the exclusion of the circular is not hard to understand because only the rectangulars seemed to be easy to date by means of astronomy, and he surely knew of Lalande's interest in that.

Like most of his companions on the expedition, Coraboeuf thought that the Egyptian zodiacs were images of the sky as it appeared at the time they were produced. All the zodiacs did indeed seem to float on the temples' ceilings as though they were the sky itself. This dangerous assumption demanded highly accurate reproductions of the images, and it would eventually produce trouble for that and other reasons. Denon (or possibly, perhaps even probably, his engraver) had erred rather badly in depicting one of the Dendera rectangular panels. The upper of the two in the engraving of Denon's drawing has all the zodiacal

Figure 5.4. The great rectangular zodiac at Dendera today, and as drawn by Denon. This and Denon's drawing of the circular one, also at Dendera, were the only printed depictions of the zodiacs until 1817.

figures in the lower panel reversed right to left from the actual object: Figure 5.4, bottom, shows the entire engraving, and the enlargement in Figure 6.1 shows some of the errors. The effect is to make it seem as though the zodiacal images in both panels face in the same direction, to the right. In the panels proper, the figures that appear in the engraving's upper part should face to the left. The point is not a minor one for we shall see later that the directions toward which the figures face were thought to have astronomical significance.[7]

In addition to assuming that the zodiacs were literal drawings of the sky, like many others in his day Coraboeuf was also convinced that the summer solstice was so important for Egyptians that they would have inscribed it in a prominent position. Taking that to be one of the four possible ends of the two Dendera rectangular panels, he simply asserted that "the sun at the solstice is [there] found in the sign of Leo, and care was even taken to indicate that it was approaching Cancer."[8] Coraboeuf wrote nothing more than that, but his reasoning was likely quite simple: he thought the solstice had to be prominently represented by a figure at a terminus of the rectangular. Leo is one of only two zodiacal constellations that are very close to any of the rectangular's four ends, excepting a crablike figure that might represent Cancer, but the latter sits outside the body of the rectangular. Which, he thought, made even Cancer's position significant: if the solstice occurred in Leo when the rectangular was made or designed, then it would certainly have been headed toward Cancer as a result of precession. The only other possibility (besides Leo) for the solstice was Capricorn, which sits at the left in the bottom panel. Coraboeuf did not mention why he rejected that constellation as a site for the solstice. However, Capricorn faces away from the end that it was closest to, with its head pointing toward the same end as Leo's, as though they were marching in sequence, with Leo leading the parade.

Coraboeuf turned next to the rectangular zodiac in the large temple at Esneh. Applying the same logic he had used on the Dendera rectangulars to this one, he concluded that its zodiac begins with the solstice in Virgo.[9] Coraboeuf was sufficiently confident in Denon's reports to assert not only that the "intention of the founders in erecting these monuments was certainly to show the state of the heavens when they built these edifices," but that the zodiacs even provide "striking evidence of the knowledge that the ancient Egyptians had of that astronomical phenomenon, the precession of the equinoxes." Although Coraboeuf did not mention any specific dates in his letter to Lalande, nevertheless the implications were clear: with the solstice in Leo, the Dendera zodiac might date as far back as the middle of the third millennium BCE, and certainly no later than its end; similarly, the solstice in Virgo placed the Esneh zodiac no later than the end of the fifth millennium BCE, perhaps even a thousand or more years before that.[10]

The stylistic differences between Denon's drawings and those done by Jollois and Devilliers several months later highlight the gulf between an artistic sensibility molded, like Denon's, at the court of Louis XV and the method that informed the education of men under the aegis of the École Polytechnique. The ideology and methods of technical instruction had themselves changed significantly during the ancien régime and the early years of the revolution.[11] Central to these trends was Monge, whose emphasis on combining geometry with artisanal skill would eventually be superceded by formal mathematical preparation under the influence of Laplace. However, during the first years of the École Polytechnique, which was initiated in 1794 as the École Centrale des Travaux Publics by Lazare Carnot and by Monge himself, altogether about half of instruction was devoted to descriptive geometry at the school's beginning.[12]

Descriptive geometry had as one of its central aims the control of skill in the production of physical artifacts of obvious military importance, such as gun parts. In 1795 Monge had explained the relationship between skill and truth, claiming that one of the principal goals of the subject was "to deduce from the exact description of bodies all that necessarily follows from their forms and their respective positions. In that sense, it is a method for researching the truth; it offers endless examples of the passage from the known to the unknown."[13] In other words, the technique was less a descriptive than an extrapolative art. Only by rendering an object according to its rules could one attain an appropriate understanding of the object's true character. Any other kind of depiction was likely to mislead, because individual predilections would necessarily enter. The highly personal sketches of Denon could hardly meet the demands of Monge's disciplined "truth," and we can easily observe the differences by comparing Denon's drawings of the temples at Esneh (Figure 5.5) with the ones produced using descriptive techniques by Jollois and Devilliers (Figure 5.6).

For his drawing of the well-preserved temple, Denon had chosen a thoroughly perspectival view, the better to convey not just the object itself, but a sense of its physical and even emotional context. The temple is half-buried in sand between decaying walls; in the distance, a spire of the Arab city rises above it. Here, Muslim Egypt abuts and challenges Egyptian antiquity. In their rendition of the same object, Jollois and Devilliers rip the antique away from its present context and even remove the accumulated sand of centuries in order to display

Figure 5.5. Denon's drawings of the partially buried temple at Esneh that contained the large zodiac (*top*), and of the ruined north temple

Figure 5.6. Jollois and Devilliers' drawing of the large temple at Esneh (*left*), and of the small one north of there

what they could not possibly have seen in its totality: the lower wall of the temple, including designs inscribed on its surface (rather, as they imagined them).[14]

Monge's "truth" had a decidedly odd character about it, since by any reasonable standard Denon's depiction was the more faithful of the two to the temple's contemporary appearance. But that was precisely the point: Monge's techniques, when translated to the ruined monuments of Egypt, did not reproduce current reality, but rather what the savants thought to be the deeper truth of the object's original existence. They aimed, just as Monge had taught, to pass "from the known to the unknown," to seize what they imagined to be the inner, original essence of the monument, cleared of the encumbering detritus of the ages, cleared also of any connection to present-day Egypt. Although many plates in the *Description* do strive to show the monuments covered in sand and in their current state, nevertheless even then the savants often repaired broken edges and straightened crooked lines. Their goal was ever to use the powers of exact science as a sort of time machine, to reincarnate the past glories of Egyptian civilization by returning them to an imagined state of perfection.

Dupuis Vindicated

Lalande was out of town when Coraboeuf's letter reached his mailbox in Paris. Not long thereafter, Grobert, an artillery colonel who was still in charge at Giza (the area around the pyramids), managed to have printed in Paris a description of the region and its history. Grobert had seen Coraboeuf's letter to Lalande and had himself written the astronomer for his views concerning the zodiacs' antiquity, and he included the letter from Coraboeuf in his *Pyramides*, which became quite popular, bringing the issue of the zodiacs to a wide audience. In addition, Grobert included the reply made to Coraboeuf on the absent Lalande's behalf by his assistant, Johann Burckhardt.[15]

Originally from Leipzig, Burckhardt had studied under the astronomer Franz von Zach, who directed the observatory at Gotha. At von Zach's urging, he left for France around 1795 to work at the Paris observatory under Lalande's direction. He took out French citizenship in 1799, succeeded Lalande as director on the latter's death in 1807, and renamed himself Jean-Charles along the way. Burckhardt learned more than astronomy from Lalande. Among other things, he absorbed Lalande's conviction that Dupuis was correct about the origins of that science—and, presumably, correct as well about the celestial and agricultural genesis of religion. As for the zodiacs, Burckhardt wrote, astronomy can give their age "with great precision" because the period of precession was now known to such a high degree of accuracy, provided only that the solstitial points could be pinpointed in the available images.[16] Burckhardt had not seen any depictions of the zodiac at Esneh (at the time of his reply to Coraboeuf, there probably were none in France), but Denon, who was now back in Paris, had shown him his sketch of the rectangulars at Dendera.

The especially great age implied by Coraboeuf for the Esneh zodiac (at least 4000 BCE) resonated with Dupuis' *Origine*. Burckhardt enthusiastically agreed. Indeed, he wrote, "the zodiac of Henné [Esneh] seems to me to dispel any doubts that might remain about that hypothesis [of Dupuis'] which, in an astounding way, rips away the prejudices that had fixed the age of the terrestrial globe."[17] He undoubtedly had in mind biblical prejudices, and in particular the claim that Earth could not be more than about six thousand years old. Had Lalande been in town he would certainly have agreed, for he delighted in anything that put religion in what he took to be its proper place, namely, as a product of human credulity and fear.

The claims made concerning the zodiacs' remote antiquity were hardly confined to the circle of savants. They were discussed not only in Grobert's popular *Pyramides*, printed in 1800, but, also that same year, in several issues of the *Moniteur Universel*. This four-page daily newspaper contained official pronouncements, reports of various kinds, and, often, sections on literature, the sciences, and the arts, including antiquity. Everything in that journal was taken to be, and in principle was, authorized government issue; the *Moniteur* was the mouthpiece of the regime. On August 18, 1800, the paper reported that the zodiacal ceilings dated back five or six thousand years, and, further, that the temples containing them were among the more modern ones in Egypt. Years later, Fourier's good friend Champollion-Figeac, whom we will encounter again, wrote that he possessed a note dated the previous July 22 indicating that Napoleon had himself approved these announcements, in order "to prepare opinion concerning the work of the Commission, with which he appeared to be extremely satisfied."[18]

Print was hardly the only medium used to transmit these startling claims, for they were on many Parisians' lips as well. For instance, Kléber's death in Egypt on June 14, 1800, at the hands of a Syrian determined to avenge the French presence in Muslim lands, resulted in a public oration. On September 23 in Paris, at the Place des Victoires, Egypt's extraordinary antiquity made a most visible appearance. The politician Garat sonorously invoked Kléber's role in revealing ancient Egypt, whose great age "was attested to by monuments before which all the centuries have passed without destroying them, and which, forever standing in the same place, have many times witnessed transformations in the beds of the seas, the forms and chains of the mountains, the order of the celestial bodies."[19]

During these early years of the consulate, before all the savants had returned from Egypt in the fall of 1801, journal and book publishers were essentially unconstrained in printing views concerning the zodiacs and religious matters. "All opinions had free rein," Champollion-Figeac later remarked, and "there were writers who applauded these novelties, others who were alarmed in the interest of sacred chronology."[20] Moreover, the year of Grobert's *Pyramides*, and of the first public announcement of the antique zodiacs in the *Moniteur*, also saw the appearance of Maréchal's decidedly antireligious *Dictionnaire des athées anciens et modernes* (Dictionary of Ancient and Modern Atheists). Maréchal had been imprisoned for several months in 1780 for his *Almanach des Honnetes Hommes* (Almanac of Honest Men). In traditional almanacs, these "honest men" would have been Catholic saints; in Maréchal's they were luminaries of the Enlightenment. Lalande, who knew Maréchal, took over the *Dictionnaire* after the latter's death in 1803, without failing to add several items of his own. Among these was an appreciation of Dupuis, who, Lalande's entry approvingly asserts, "appears to be" an atheist. For atheism, Lalande insisted, was an indication of moral rectitude, an index to the strength of reason. According to Dupuis, the entry asserted, "women, children, the old and the sick, which is to say the most feeble, are the most religious, because reason decreases in proportion to the weakness of the body." (An inveterate womanizer and bachelor, Lalande apparently did not extend to his now-aged self the entry's disdain for the old.) And religion itself, he continued, "is incontestably the greatest plague that has afflicted humanity."[21]

If he had not heard of Dupuis' *Origine* before leaving Germany in 1795, Burckhardt could not have missed it as Lalande's assistant. Lalande's responsibilities included oversight of the Bureau des Longitudes, which had been created in 1795 by the convention with responsibility for the observatories of Paris; direction of the École Militaire; and the timely publication of the official ephemerides in the *Connaissance des Temps*, which was printed two years ahead of the corresponding dates. (The one for Year 13 of the Republic, or 1805, appeared for example in 1803, Year 11.) In the 1803 publication Lalande included news about developments of interest that were connected to astronomy for previous years, including 1800, when he had first heard about the zodiacs. This article extended Lalande's brief report about them, which appeared this same year in his account of astronomical events of note for 1800.[22] He related Coraboeuf's and Burckhardt's comments, concluding that the Esneh zodiac was some seven thousand years old (when the solstice lay in Virgo), which provided "great probability" for Dupuis' hypothesis that had placed the zodiac's origin "fourteen to fifteen thousand years before our time." On the other hand, following Burkhardt, Lalande claimed that the Dendera rectangular was about four thousand years old.[23]

The implications of the zodiacs' dating could hardly have been missed. After all, if the Esneh zodiac had been carved so many millennia ago, before the biblical date of Creation, to say nothing of the Flood, did that not speak to an even older origin for astronomy itself? Its level of sophistication would certainly have taken many hundreds, and even thousands, of years to achieve. Moreover, even the lesser age of the Dendera rectangular shook the foundations of biblical chronology. Sacred chronologers differed somewhat among one another concerning the precise date of the Noachian deluge, though all agreed that it took place sometime between 2500 and 2300 BCE. That date was dangerously close to the very period assigned even to the Dendera rectangular. One could hardly expect the ancient Egyptians to have produced such a sophisticated object in the few centuries after nearly all of humanity had been destroyed by the deity of monotheism.

The Egyptian Zodiacs in Public

Members of the Egyptian expedition began debating questions surrounding the astronomical ceilings virtually the moment they arrived back in Paris in late 1801. Devilliers himself had it in mind to write an account. Early in January of 1802 he had asked Jollois whether, in case the plans for publication of what became the *Description de l'Égypte* (plans just then being discussed in Paris) were to take a long time, would Jollois agree to publish separately their drawings of the zodiacs together with a memoir on them by Devilliers?[24] Their account did appear in print in 1809, in the very first volume of the *Description*. But in 1802 the only widely available remarks about zodiacs beyond those in Grobert's *Pyramides* had been printed in the *Moniteur*, the official mouthpiece of the régime. It is to this account that we now turn, to see how zodiacal antiquity first came to the attention of a broad public in France.

On the twenty-fifth of Pluviose, Year 10 (February 14, 1802), the *Moniteur* printed a letter from a "citizen S.B.," the chemist Samuel Bernard, who had been in Egypt as one of the savants.[25] The letter was addressed to a certain Morand, who was, or had been, a member of the Legislative Body. During the expedition, Napoleon had placed Bernard in charge of coining money in Cairo, using existing Mameluke dies, in order to keep the economy afloat. Bernard later produced an account of Egyptian weights and measures (ancient and modern) for the *Description de l'Égypte*, which was published in 1812.[26] His correspondent was Réné Morand, a doctor from Niort in the west of France and an early enthusiast for the revolution who had been appointed by the Senate to the Legislative Body in January 1800, only to be voted out in 1801. Apparently Morand had often asked Bernard about the results of "our voyagers in Egypt." Bernard reported that Denon had announced a major book (his *Voyage*) and that many other expedition members "newly returned from Egypt" just two months before had brought back "beautiful and numerous drawings, notes and collections."

Morand would hardly have been the only reader of the *Moniteur* eager to learn about the expedition (its military failure having been camouflaged by Napoleon), and Denon's book would soon enjoy a nearly unprecedented publishing success. Bernard immediately established the tone that would characterize most French writing about the savants' work at the time.

> More than thirty people educated in different areas, traveling in a country surrendered, protected by our troops . . . , staying as much as was necessary in each place to omit nothing interesting; parceling out the work in the most suitable ways, informing one another of each one's discoveries and thoughts, and often writing in the very place notes that would not have been so complete or so exact had they been done from memory. Almost every one was taught in draftsmanship, one and the same monument was drawn in all its aspects at once, and a given aspect often by several people.

In a burst of enthusiasm Bernard opined "what confidence must be inspired by narratives and descriptions whose truth will be certified by thirty enlightened witnesses!" How could one possibly doubt the savants' conclusions, soon to be published under government auspices "with the magnificence appropriate to a great nation" (France, that is, not Egypt). How, in particular, could one doubt the conclusions concerning the antiquity of the zodiacs that one especially talented savant had developed, namely, Fourier, who had led one of the two expeditions to Upper Egypt ordered by Napoleon on August 14.

Fourier was to play a significant part in the zodiac controversies, and he was responsible for writing the preface to the monumental *Description de l'Égypte*, to which he also contributed articles on the astronomical monuments, as well as an overview of the savants' results concerning Egypt's science and chronology, including a comparison with the epochs "given by the annals of the Hebrews." The preface and the summary overview appeared in the first volumes of the *Description*, dated 1809 (but issued in 1810), as did his highly abbreviated (and, we shall see, badly misprinted or mistranscribed) remarks on zodiacal chronology. While in Egypt he had already been interested in the zodiacs and either had planned or did read a memoir on the subject before the Institut d'Égypte as early as the end of January 1801. In fact, the month before, on December 1, the *Courier d'Égypte* printed a description of the results of the recently returned expeditions to Upper Egypt. Almost certainly written by Fourier himself (since it bears the hallmarks of his style),[27] the *Courier* report emphasized the zodiacs. "We examined with particular attention the zodiacs sculpted on the ceilings of the temples," Fourier wrote. "They were exactly drawn, and the comparison of them that we made determines epochs far in the past in the history of astronomy and in civil history."[28] We shall turn below to Fourier's interesting career, and to Napoleon's close involvement in it. For now, note that Bernard's letter in the *Moniteur* of 1802 is the only source of detailed evidence concerning Fourier's views on the zodiac during the period shortly after his return to France. Here, in full view of the French reading public, Fourier was seemingly on record concerning the zodiacs' immense antiquity.

Bernard had been on the expedition, and he surely knew Fourier, which likely explains how he came into possession of parts of Fourier's notes on the zodiacs. The iconoclastic excitement that shines through the brief sentences in the Bernard letter is characteristic of a man who, albeit only posthumously, was to become one of the most influential French physicists and mathematicians of the nineteenth century. During his lifetime, however, Fourier's influence was due primarily to his eventual position as secretary of the rebaptized Académie des Sciences during the later restoration. "Until now," Fourier had written in the letter to his friend Berthollet that Bernard had obtained,

the history of men and the sciences and arts had nothing certain and authentic except for very recent epochs, and it was difficult to choose among the chronicles of different peoples; those of Egypt were particularly uncertain. Diogenes Laertius had them reach back to four thousand years before the era of Augustus, and Newton only to a thousand years before Jesus Christ. The discussion of the astronomical monuments which have just been discovered will determine ideas about these different opinions; it justifies the chronology of Herodotus, and it remains certain that the present division of the zodiac, as we know it, was established among the Egyptians about fifteen thousand years before the Christian era, was maintained unchanged, and was transmitted to all other people. This zodiac was evidently nothing other than the primitive calendar of Egypt. When it was established the vernal equinox occupied the sign of Libra; the earth was sown under the sign of Taurus, and the harvest took place in that of Virgo. One cannot avoid recognizing, on inspecting the zodiacs found in the temples, that the figure of Aquarius, crowned with a lotus, was for the Egyptians the astronomical sign of the inundation. It is accordingly natural to place the summer solstice in that constellation, and one thereby determines a position of the [heavenly] sphere, such that the name of each sign becomes, so to say, the natural attribute of the corresponding parts of the year. This confirms perfectly the conjecture of citizen Dupuis.[29]

Fourier had gone beyond even Dupuis' extraordinary claim for the invention of the zodiac. Dupuis, recall, had set the summer solstice at the time of the original zodiac in Capricorn, which produces a date of around 13,000 BCE for its origination. Fourier here claimed 15,000 BCE, which required setting the summer solstice in Aquarius instead, as we can see from Figure 3.4. Where Dupuis had chosen Capricorn because a goat reaches for the heights, as does the sun at summer solstice, Fourier instead chose Aquarius because the water-pourer could represent the inundation, which occurs shortly after solstice. Despite the variation, which was no doubt prompted by Fourier's direct experience of the inundation in Egypt, his scheme accorded fully with the spirit of Dupuis' in attaching agricultural meaning to the zodiacal images, and in vastly amplifying the antiquity of Egypt.

There was more, since Fourier was also convinced that the ages of the temple zodiacs could be coordinated with the periods when the buildings that held them were constructed. "Everything shows," he asserted, "that the buildings which still remain were constructed during the period when the sky was in the state represented." The point was not even debatable because the "reasons for that opinion are so many and so conformable among one another that they naturally exclude all doubt." On that basis the temple at Esneh "reaches back to six thousand years before Jesus Christ; and the beautiful temple at Denderah, the more recent perhaps of those that were consecrated, was, reasonably, constructed more than a thousand years before the siege of Troy."[30]

These were heady words indeed, though years later Fourier's friend Champollion-Figeac insisted that "there exists no writing that could make [Fourier] responsible for the exaggerated opinions that were current in these early times on the subject of Egyptian zodiacs."[31] This would seem to put in doubt Bernard's attribution, or perhaps Champollion-Figeac, tending carefully to his deceased friend's reputation (the remark dates from 1844, fourteen years after Fourier's death), had either forgotten or (perhaps deliberately) overlooked the Bernard piece. Who after all would look up an old newspaper? In any case the *Moniteur* piece was thirdhand: inserted by Morand from a letter to him by Bernard, who was quoting from a letter sent by Fourier to Berthollet. It is worth pausing to sort out the documentary intricacies here, as they suggest that, at least among savants at the center of the *Description de l'Égypte*, making public claims about the age of the Egyptian zodiacs was a matter of some ambivalence. For while Fourier may very well have been unhappy about the Bernard article, it seems unlikely that Morand would have printed the letter without Bernard's permission, or that Bernard, in turn, would have given it without at least Berthollet's assent, especially since it must have been Berthollet who gave Bernard the letter to him from Fourier in the first place—unless it was Fourier himself, who had perhaps kept a copy. For his part, Berthollet may not at this early date have been aware that Fourier had in fact let it be known among the returned savants that he wanted to keep public words about his involvement with zodiacs to a minimum. However, the fact that the astronomical ceilings were to be discussed in an official publication had already been announced in the *Moniteur* in January, the month before the Morand article. Moreover, the announcement continued, Fourier was himself to be in control of the analysis and even the depiction of the zodiacs.

While the older savant dithered over what to say about the zodiacs and when to say it, younger ones hoped for action. In particular, Devilliers was sensitive about this issue since he and Jollois had together done the drawings and were hoping also to write about the zodiacs. On January 12, 1802, he wrote Jollois to report that Fourier had assured him that he himself was unhappy about this early *Moniteur* announcement, that he worried Jollois would be upset, and that he advised Devilliers not to speak too much in public about the announcement. It is possible that Fourier said one thing to Devilliers and another to Morand (for instance, that he would not mind if the letter from Bernard were printed). But in the absence of further information about Morand himself, or other epistolary evidence, it is impossible to know for sure.

Yet we do know what Fourier had himself concluded while in Egypt (in fact while at Esneh itself, where he saw its zodiac on September 8, 1799), that the astronomical depictions reached back thousands of years. The notes that he wrote during the expedition to Upper Egypt are unequivocal. Fourier eventually handed them over to Jollois, whose nephew allowed them to be printed in 1904. "In the second between-colonnation, to the left of the great door of the portico," Fourier observed at Esneh, "we distinguished on the ceiling important sculptures, fairly well preserved, that represent the signs of the zodiac. We will publish a special paper containing the interpretation of these sculptures; it will serve to assign the epoch when the sphere to which this zodiac belongs was established." Going further, Fourier conjectured that the zodiac "goes back four thousand six hundred years before the Christian era" and that the construction of the temple that housed it might be even older. He added that a second zodiac near Esneh (presumably at Dair) might also be a source of confirmation of these observations, "but we haven't examined it this time with enough care. It would be possible to penetrate into the part of the temple of Esneh whose entrance is filled. This research is important, because it's reasonable that we would find other astronomical monuments."[32]

Whether or not Fourier truly wanted to keep his views out of the public eye at this early date, there, in the official newspaper of the consulate, soon to be the organ of the empire, was a letter from him to the most famous French chemist of the day in which Fourier asserted not only that the Dendera zodiac predated Troy by a millennium, and that the one at Esneh was produced long before that, but that Dupuis had in all essentials been proven correct. Burckhardt had written the very same in his letter to Grobert in September 1800, a letter that had become public when Grobert's *Pyramides* was printed later that same year. But Burckhardt was not Fourier, and, unlike the *Moniteur*, the *Pyramides* did not sit each day on Parisian dining tables. Fourier's was a challenge to believers in traditional chronology that could not be ignored—even though Fourier, like Burckhardt before him, had overreached himself. For the zodiacs proper, interpreted in their fashion, certainly did not push back to before 6000 BCE, which was hardly Dupuis' 13,000. Still, even two millennia before the Mosaic date of Creation challenged believers in biblical inerrancy. Now that these zodiacal conjectures bore stamps of approval from men like Fourier and Lalande, whose scientific credentials were impossible to undermine, Dupuis' theories seemed more promising and authoritative than ever. No longer mere iconoclasm, his views were backed by an apparently successful application of mathematical methods to the concrete remains of the human past.

6

The Dawn of the Zodiac
Controversies

The first detailed discussions about the Egyptian zodiacs developed in the years during which Napoleon progressively consolidated his power, ultimately making himself emperor of France. As he shed his adherence to republican and democratic ideals, so these ideals lost currency elsewhere, from the pages of newspapers and articles to the more private realms of academies and salons, making republicanism less legitimate and harder to defend or espouse. Into the resulting ideological gap leapt the Catholic Church, buoyed by a religious revival in the French provinces and by efforts to exact political concessions from Napoleon that met with a degree of success. With the Church looming larger in public and private life, debates ensued about its power and authority, inevitably touching on Mosaic chronology and the dates of Creation and the Flood. The Dendera zodiacs were part of these debates, which took place in progressively more public arenas, from an obscure article attached to a revision of a classical text, to the pages of newspapers in wide circulation like the *Moniteur*, to debates before a newly constituted conservative society at the papal seat in Rome. In keeping with postrevolutionary attitudes of remorseful cautiousness, the past had become fashionable as a source of ideals for the present. Only age reliably conferred wisdom; history was important insofar as it could be used to shore up conservative ideals. At the same time, purely textual interpretations of the zodiacs ceded place to discussions that were inflected by astronomy and mathematics. Even when these disciplines were not skillfully deployed, their presence in the debates lent their users an important and distinct form of author-ity. In the end, what mattered most during these years was political and religious exigency, the uses to which ancient history could be put, the ways in which ideas about history and science could restore and even revitalize religious belief.

Napoleon's Concordat with the Vatican

On June 23, 1796, Napoleon had forced the Vatican under Pope Pius VI to sign the Peace of Bologna, ceding both that city and Ferrara to France. The treaty also required a payment of twenty-one million francs (badly needed by Napoleon for troop support) as well as the transfer to Paris of five hundred manuscripts, a collection of art works embracing a hundred paintings as well as marble busts of Junius and Marcus Brutus, the inclusion of which surely suggested something of Napoleon's ambitions, if these were not already clear enough. As rich as these concessions were, the Directory wanted more. In particular, it wanted to put an end to the papacy's opposition to the republic's Civil Constitution in respect to the clergy. It demanded as well the repudia-tion of anything related to the Inquisition, including Pius VI's efforts to ensure against "excessive familiarity with the Jews" by rescinding the limited rights and freedoms granted to them under the comparatively tolerant Clement XIV.[1] Pius's horrified refusal to agree to the new Civil Constitution kept negotiations in limbo for a year, sorely trying the young general's patience.[2]

At this time a resurgence of religious feeling in France had become a matter of political concern, and Napoleon, who would a year later try to placate Muslims in Egypt, developed a similar plan for Catholics in Italy and France. Early in 1797 General Henri Clarke had advised Napoleon that "France is again become Roman Catholic, and we are perhaps in the point of needing the pope himself, for causing the revo-lution to be seconded among us by the priests, and consequently by the country, which they have found means to govern again." Noting the vulnerability of a polity with multiple, potentially competing loyalties, to church as well as state, Clarke mused, "If one could have annihi-lated the pope three years ago, it would have been the regeneration of Europe. By overthrowing him at the present moment, should we not run the risk of separating for ever from our government a great number of French submissive to the pope, and whom he can rally round him?" The Vatican's negotiations with the Austrian court, once discovered particularly infuriated the Directory, which remained opposed to any accommodation, since, it wrote Napoleon on February 3, "the Romish religion is the one of which the enemies of liberty could, for a long time to come, make the most dangerous use." By February 1797 all of northern Italy was in Napoleon's hands. On February 17 the Vatican formally capitulated with the Peace of Tolentino, which not only reiterated the earlier provisions at Bologna but required a further, huge indemnity, the formal surrender of the cities of Bologna and Ferrara and the entire Romagna region, as well as the transfer to Paris of paintings, sculptures, manuscripts and antiquities. Despite these victories, control of the region remained sufficiently uncertain for the Directory to urge that care be taken to avoid "rekindling the torch of fanaticism in Italy instead of extinguishing it."[3]

A little over a year and a half later, on the 18th of Brumaire (November 9, 1799), Napoleon, recently arrived in Paris from Egypt, acted with others to displace the Directory. Through various maneuvers he was designated first consul among a triad on December 12. Under the new constitution that followed, which dropped elections altogether, this position gave him effectively full power since the three consuls appointed the Senate while the first among them appointed the other two. Napoleon's consolidation of control and abandonment of representative government continued apace over the next two years: the following May he moved into the Tuileries palace; in June he defeated the Austrians altogether and took full command of Italy—Desaix, just back from Egypt, having been killed at Marengo while leading a counter-charge that produced victory over the Austrians under General Michael Melas. A peace treaty signed at Luneville in February 1801 was followed by the signing of a Concordat with the Vatican under the new pope, Pius VII, in July. The French army in Egypt surrendered to the British six weeks later, while the Concordat was publicly proclaimed only the following April, when Napoleon could arrange appropriately impressive pageantry at Notre Dame.

Lingering questions about popular loyalty to the Church, especially in the provinces, gave rise to a peculiar attempt by Napoleon to canvass the state of religious sentiment in France. According to the recollections of Pierre-Marie Desmarest, the director of secret police, toward October 1800 (ten months before the signing of the Concordat) one religious group, the so-called theophilanthropes, caught Napoleon's attention. Formed in 1796, this group held to a deistic vision that combined a conviction in personal immortality with an explicit repudiation of Catholicism as a form of superstition. Napoleon was told that theophilanthropy was based on natural law and aimed to cultivate the love of virtue, that it was entirely moral and social in aim. He was uninterested in such a thing and likely did not believe theophilanthropy was quite so devoid of potentially destabilizing beliefs in the supernatural. As Napoleon saw it, any religion had to tell you "where I come from and where I'm going," and history amply demonstrated how that could easily have political consequences. Shortly after the theophilanthropes came to his attention, Napoleon ordered his survey, which was to examine not only French religions, but "every kind of superstition, prejudice and popular custom having any taint of spiritualism." According to Desmarest, the project, which took several months, uncovered a diverse and widespread religiosity. "The material was truly abundant, ranging from marvels and pilgrimages to sorcerers and card readers! The number of adepts surpassed belief; and that in every class of society, and, in many places, among the bulk of the population." France, Desmarest pointedly remarked, was neither materialistic nor "limited to indifference or a pure deism."[4]

By 1801 Catholicism had revived significantly throughout the country, especially in the provinces, and via the Concordat it acquired recognition as France's majority religion, albeit without also attaining status as the country's official creed per se. The state agreed to pay clerical salaries, the quid pro quo being that the clergy were required to swear allegiance to it, which would soon mean to Napoleon himself. The very first article of the Concordat allowed the free exercise of Catholicism while nevertheless subjecting its "public" worship to whatever police regulations the state chose to put into effect. Cognizant of the state of religious and superstitious belief in the country at large through the survey he had ordered, Napoleon now aimed to enhance his control by altogether erasing potentially disruptive discourse about religion from the public sphere. Just two days after the signing, he ordered that it be made "known to journalists, those who are political as much as to those who are literary, that they must abstain from speaking about anything that might concern religion, its ministers and its diverse creeds."[5] The prohibition of opinions about religion actually bolstered the Church's influence to a limited degree by eliminating the opposition. The theophilanthropes, who were explicitly and vocally anti-Catholic, were for example quickly proscribed.

In May 1802 the Treaty of Amiens was signed between France and Britain. The terms of the treaty were favorable to the French, indeed extraordinarily so. The writer Madame de Staël, an admirer of Napoleon from the days when she thought him a republican, recalled a dinner where she listened with ever-growing astonishment to the British ambassador. "England gave up all her conquests," she wrote in exasperation, "she gave up everything. Without compensation, to a power that she had constantly beaten on the sea." De Staël was exaggerating, but only somewhat. The British did indeed relinquish the territories that they had conquered, save for Trinidad and Ceylon, while France agreed to leave Italy, having evacuated Egypt the previous fall. "And so," de Staël concluded, "the expedition had no result other than to make everyone chatter about Bonaparte."[6] Rumors swirled that Napoleon had himself ordered Kléber assassinated—which was hardly likely, de Staël noted, since he was too far away to "arm a Turk against the life of a French general." Ever more rapidly now did the "institutions of monarchy make headway against the shadow of the republic." Napoleon organized "a praetorian guard; the diamonds of the crown adorned the sword of the first consul." In de Staël's view, "each step of the first consul proclaimed ever more overtly his limitless ambition." During this brief period of peace, Napoleon expanded his political and social control and expanded as well the French navy in preparation for further war with Britain. For their part, rich Britons swarmed into Paris, anxious for the fashions, luxuries, and lively conversation they had sorely missed for over a decade, while literary and scientific exchanges between the countries resumed, at least for a time.

To some among those who were present, the celebration of the Concordat at Notre Dame in April 1802 must have recalled the spectacle that had graced the former Temple of Reason nearly a decade before. Only now, instead of a Goddess of Liberty, the archbishop of Aix (the very man who had read the sermon at the cathedral of Reims on Louis XVI's coronation) represented the Church in its acquiescence to Napoleonic power (Figure I.2). The redoubtable de Staël was again on hand, and she understood the truth behind the ceremony. "The day of the concordat," she wrote, "Bonaparte took himself to the church of Notre Dame in the carriages that had been the king's, with the same coachmen, the same footmen marching next the door; he'd followed court etiquette in the smallest detail; and, though he was first consul of a republic, he worked hard at all of the royal apparatus."[7] Recognizing Napoleon's abandonment of republicanism as a step on his way to absolute power, de Staël did not have long to wait for the next one. In May Napoleon was elected consul for life through a plebiscite in which an absolute majority of eligible French voters, tired of incessant war and deprivation, immolated democracy on the altar of security.

Denon's *Voyage* was translated into English and printed during these eventful months. A tremendous success in both France and England (and in the United States as well), the English translation of the *Voyage* went through no less than five separate printings at five different publishers in London, as well as one each in Dublin and New York, in about a year.[8] The subscribers to the first edition in France (three large, folio volumes of text and one of plates) included Bonaparte, who ordered twenty-six sets bound in vellum and twenty in cloth, for distribution to appropriately important recipients. An elegant binding of the *Voyage* was a status symbol, conferring something of the glories of an ancient civilization and a modern, widening empire upon the owner. Talleyrand ordered one set in each binding. Denon recorded that the "emperors" of "Germany" and Russia each ordered copies in vellum. Sir Sidney Smith in England, who had negotiated the Convention of El-Arish with Kléber for the French surrender in Egypt (ignored thereafter by Admiral George Elphinstone, Viscount Keith, but eventually agreed to under similar terms) ordered a set in cloth. The subscriber list was lengthy and illustrious.[9]

Despite the numerous printings it went through in England, the *Voyage* was not received there with unadulterated praise. The young Lieutenant Colonel Robert Wilson was one vocal critic. He had served in the British expedition to Egypt and wrote a history of it that reached print in 1803 during the Amiens peace, shortly after the appearance of the English *Voyage*. Incensed by what he saw as French boasting in the face of unquestioned defeat, Wilson was especially annoyed by Denon. That "philosopher," he wrote, "proves himself a most obsequious courtier, using that bombast in the relation of the battles he was a spectator of, which has rendered every public French dispatch during the war, with some very few exceptions, ridiculous." Even worse, Denon had slipped away from Egypt with Buonaparte [*sic*], "the general" who had acted "contrary to a sacred promise, that whenever he returned to France, [the rest of the commission of savants] should accompany him." Having put Napoleon in his place by referring to him merely as "the general," and reminding his readers of Napoleon's Corsican origins and duplicity, Wilson was free to knock Denon as well. "Denon," Wilson opined, "endeavours to repay his patron; but perhaps his former associates may not be so obsequious, irritated particularly as they must be at this second march being stolen upon them" by the publication of the *Voyage*. Napoleon had also placed Denon in charge of the artworks and antiquities that had been, and would continue to be, expropriated by the French from conquered territories.[10]

"What a Triumph for Unbelievers!"

In the years following Napoleon's election as consul for life, the signing of the Concordat, and the phenomenally successful publication of Denon's *Voyage*, the first scholarly reactions to the Egyptian zodiacs began to appear in print. The earliest detailed account, by one Ennio Quirino Visconti, appeared as a supplement to the influential second edition of the classical scholar Pierre Larcher's translation of Herodotus's *History*.[11] Visconti's account, framed as a letter to Larcher, deployed familiar methods of humanistic textual interpretation in support of the view that the zodiacs were of relatively recent vintage, a position that allied Visconti to the Church and helped to reconstitute Larcher, who had been tainted in the eyes of the religious Right years before by associations with philosophes.

Long believed to be a prodigy, Visconti had distinguished himself at the age of twelve with an exemplary translation of Euripides's *Hecuba*. His father, Giovanni Battista, had replaced the flamboyant antiquarian Johann Winckelmann as the Vatican's surveyor of antiquities after Winckelmann's murder by a thief in 1768. Although Visconti intended his son for a career in the church, Ennio Quirino had other ideas, particularly after meeting Teresa Doria, whom he married. His father, who had already obtained positions for him at the Vatican, including the office of custodian of the library, obligingly adjusted his paternal expectations. Interceding once more on Ennio Quirino's behalf, he obtained the pope's permission to release his son from the church, whereupon Ennio Quirino took up a position as curator of Prince Sigismondo Chigi's library.

Chigi was by hereditary right perpetual marshal of the Conclave, the gathering of cardinals that takes place to choose a new pope. He had acted in that capacity in the 1774–1775 meeting that chose Pius VI. Although Chigi's link to the papacy would seem to be evidence for the young Visconti's continued close connections with the Vatican, he was not as loyal as his participation in the conclave might suggest. In fact, Chigi was the probable author (anonymously, and just in time for the conclave) of a satirical musical drama that held the Roman curia up to ridicule for corruption and intrigue.[12] Hardly new occupations for the curia, these vices nevertheless seemed particularly noisome in a period when philosophes were busily, and often amusingly, associating religion with deceit and oppression. When his authorship was eventually suspected, Chigi was exiled to Padua. He and Visconti no doubt had many discussions about the Vatican during the times that they spent together there, which perhaps makes less surprising Visconti's actions following the declaration of the Roman Republic early in 1798.

Visconti became one of the new Roman Republic's consuls, having informed the governing French Directory of the contents of the Vatican Museum (the Pio Clementino) which he and his father before him had curated. Not long afterward, he assisted the transfer of artworks to Paris. However, a financial scandal coupled to difficulties that Visconti had with the new regime soon removed him from the consulship. Early in 1799 Rome was briefly recaptured by forces under Ferdinand IV, king of Naples, whereupon Visconti decamped with his family to Paris.

Once there, he was entrusted for a short time with the purloined collections at the Louvre. The first consul soon decided to create a new museum, to be called, modestly, the Musée Napoléon, for the same material. Although Visconti hoped to be chosen as its head, he lost out to Denon, recently returned with Napoleon from Egypt. The first consul often gave with his right hand what he took away with his left. And so in 1803 Visconti occupied a new chair in archaeology at the Institut National. The institute had been reorganized to suppress the now-suspect class of moral and political science, which Napoleon thought to be filled with the talkative and untrustworthy followers of *idéologie*. Visconti joined its newly constituted Third Class, ancient languages and history. His discussion of the zodiacs had appeared the year before he took up this new appointment, in Larcher's *Herodotus*, preceded there by Larcher's vituperative remarks about unbelievers. Given Visconti's behavior scarcely two years earlier in assisting the French occupiers to despoil the Vatican collections, the inclusion of his zodiacal remarks in this volume might seem to be peculiar, but a few facts from Larcher's biography make sense out of this apparent contradiction.

Larcher had become a member of Inscriptions et Belles-Lettres in 1778, having won a prize from it three years before; his new translation of Herodotus appeared eight years later.[13] He was already well-known for having quarreled with Voltaire over the latter's *Philosophie de l'Histoire*, with its bitingly amusing antagonism to all forms of religious belief and practice.[14] Larcher was urged to critique the *Philosophie* by several ecclesiastics who knew that he thought Voltaire's erudition deplorable, to say nothing of the philosophe's moral character. He eventually obliged with a *Supplément*, though reluctantly (he was himself friend to a number of philosophes).[15] Larcher was hardly gentle, and neither was he concise or reticent. Thoroughly enraged, and probably worried as well, given Larcher's evident knowledge of languages and ancient history, Voltaire went on the offensive in his *Défense de mon oncle*.[16] (In this work, as well as in the *Philosophie*'s dedication to Catherine II, Voltaire pretended to be the author's nephew, whom he fashioned as one Abbé Bazin.) He called the *Supplément* a "pedantic libel" filled with erudition, which was no compliment since Voltaire, like d'Alembert (who nevertheless defended Larcher to him), disdained that sort of learning as bookish and inconsequential. When Larcher replied at some length, he discovered that his learning was no match for the philosophe's wit.[17] "Confining myself within the limits of erudition," he tried to answer, "I have been content to display the plagiarisms, the false citations, the badly understood passages and the ignorance of history and chronology."[18] With characteristic mockery and irony, Voltaire thereupon pressed the advantage, discomfiting even his friends by the ferocity of his critique.[19] Though Larcher lost this battle, he nevertheless went on to produce sundry works of erudition, including one in which he argued against the claims of the Dutch scholar Cornelius de Pauw, who thought China to have originated as an Egyptian colony (an idea that at the time had its adherents, including the orientalist and linguist de Guignes).

By 1802 the cultural climate had changed, and so had Larcher. Now seventy-six years old and thoroughly disaffected with the political and religious developments since the revolution, Larcher had become sufficiently devout that he excised from the new edition of his *Herodotus* all of his earlier remarks about biblical chronology, which he now considered to be impious.[20] Yet they had hardly been provocative, at least compared to the philosophes' standard rhetoric against religion. Although the notes in the first edition do indeed run suggestively counter to biblical orthodoxy, Larcher was not explicit about the fact. He asserted, for example, that the Egyptians were likely less ancient than the Ethiopians, who followed the Scythians. Of the latter he noted that they were too advanced to have been submerged in the Deluge, suggesting that the Universal Flood had not perhaps been all that devastating.[21] There is some evidence that by 1795 Larcher regretted his earlier associations with philosophes, including the Baron d'Holbach, and certainly by 1802 he was part of the growing chorus attacking "impiety," or the increasingly secular quality of everyday life in France, including the lives of its intellectuals. Visconti's additions to the revised *Herodotus* fit the old scholar's new attitudes, though Visconti likely shared few of them.

In his new edition of 1802, Larcher assaulted the Egyptian chronology that Volney had developed in his *Voyage*, for it "contradicts the Scriptures." Taking a leaf out of Voltaire's book (having apparently learned something from their dust-up years before), he remarked sarcastically that Volney, whom he insisted on calling Chasseboeuf-Volney or just plain Chasseboeuf, "ought to recollect that he was a candidate for a prize at the Académie des Belles-Lettres on a subject relating to chronology, and that his Memoir was rejected with indignation, and very deservedly. I would recommend to him to study chronology, or rather never to write on any subject which has the smallest relation to it." Further, according to Larcher, the Noachian Deluge occurred in 2328 BCE, appallingly close to the purported age of the Dendera rectangular zodiac. Moreover, Coraboeuf and Burckhardt had absurdly claimed that the Esneh rectangulars were built "217 years before the creation of the world according to Father Petau."[22]

"What a triumph for unbelievers!" Larcher fulminated. But if it was a triumph, it was short-lived, because Visconti, whom Larcher praised as "that celebrated antiquarian, whose knowledge and talents are universally recognized, and whose eye, turned to ancient monuments, easily discerns their different ages," had required only a "glance" at the two zodiacs to judge them recent.[23] This was to be expected because "God laughs at the vain efforts of men." Although Larcher did not personally know Visconti, he was well acquainted with the Arabist Silvestre de Sacy, who, deeply religious himself, had asked Visconti to help Larcher in the matter of the zodiacs.[24] De Sacy was intensely interested in Egyptian matters, having that same year published a commentary on the Rosetta inscription addressed to the chemist Jean-Antoine Chaptal, then minister of the interior (he had succeeded Napoleon's brother Lucien in that capacity), where he claimed to discern proper names in the hieroglyphs. Chaptal had sent de Sacy copies of the inscription, to which Chaptal had access because he established the commission that would produce the *Description de l'Égypte*.

A Philologist Objects

In his letter to Larcher of May 8, 1801, that was printed in the new *Herodotus*, Visconti did not indulge in remarks supporting biblical

Figure 6.1. Denon's drawing and a photo of part of the Dendera rectangular as it exists today: a crab or insectlike symbol on the right, a figure pouring water from both hands usually taken to represent the Nile flood, and Sothis (Sirius), represented by a recumbent cow with a star between her horns. Note that the figures in Denon's sketch face in the wrong direction. For a color image of Figure 6.1, bottom, see Figure I.4. The blue background in the photo represents the sky.

chronology or religion but capitalized instead on his knowledge of Greek texts and history for interpreting the Egyptian zodiacs.[25] He asserted, first of all, that the zodiacal symbols in the Dendera rectangular were disposed in the Greek fashion called *boustrophedon*, after the manner in which a field was plowed using oxen, by turning the ox at the end of one row to begin the next. That fact alone, he implied, suggested an origin no earlier than Alexandrian times. Of course, Visconti had not seen the actual zodiac. His assertion had to rely on secondhand testimony, perhaps a sketch, and possibly a report from a witness. Which raises a basic question of evidence: What had Visconti actually seen? Was it Denon's engraving, or his original drawing?

It is unlikely that he had seen the engraving. First, the print made from it shows no *boustrophedon* arrangement at all because in both of the print's panels the zodiacal figures face the same direction. Ironically enough, the original panels may indeed be interpreted in that fashion because their figures are reversed in direction from one another (Figure 5.4: though not visible in this photo of the rectangular as it exists today, the figures on the panel shown here below the columns have their figures reversed from the panel shown above the columns). Since Visconti wrote to Larcher before Denon's *Voyage* reached print in 1802, either Denon must have shown him an original drawing that did not have the engraving's reversal or else Visconti was relying on verbal portrayals. The completeness of Visconti's account perhaps suggests that he did have a correctly oriented drawing rather than merely a description. If so, this drawing could only have been Denon's since there was, at the time, no other in France. We know that Denon's depiction of the Dendera rectangular had not been engraved as late as January 8, though the circular had just been finished.[26] Engravers could (and did) err, either by omission or by introducing elements that were not in the drawings from which they worked, and we will see in a later chapter just how deceptive this could be for anyone who relied solely on reproductions. Because Visconti knew Denon quite well, he would have had ample opportunity to see Denon's original (and perhaps correct) drawing in the course of their relationship.[27] Jollois and Devilliers would later strongly criticize Denon for the perceived inaccuracies of his depictions, to which he apparently never replied.

Visconti directed his principal argument against Burckhardt's associations. If, as Burckhardt had suggested, the asterism Leo begins the Dendera rectangular and marks a time in remote antiquity when the summer solstice was located in it, then the succeeding constellations should also map to appropriate moments of the year. But, he averred, "the balance [Libra], symbol of the equinox, is in its position, that is to say this sign follows that of the lion [Leo] after the interval of one asterism only; this could not have occurred if the lion were solstitial." Visconti's claim amounts to saying that since Libra would naturally represent an equinox (though Coraboeuf and Burckhardt, as well as

Fourier, thought otherwise), Leo cannot mark a solstice—which is seemingly true since the asterisms are separated by Virgo only, whereas two other constellations should separate the one that contains an equinox from the one that contains a solstice.[28]

Visconti suggested an alternative astronomical correspondence. You could see, he argued, that the annual inundation must have been represented by the figure of Isis on a boat pouring water from two small vases (Figure 5.4, on the far right of the upper panel; see Figure 6.1 for details). This figure occurs in a region that corresponds to the space that should be occupied by Cancer (identifying the latter with the crab-like figure on the extreme right), "and we know that the annual flood of that river [the Nile] arrives at the beginning of summer."[29] To the left of the water-pouring figure, Visconti spied the symbol for the god Sothis (a cow), representing the brightest star in the sky, Sirius the dog star. Most historians and philologists in Visconti's time, and much before then, thought that the ancient Egyptians had associated that star with the onset of the Nile's yearly flood. Since the rise necessarily occurred after the summer solstice, Visconti had accordingly (if implicitly) established a connection between Cancer and the solstice that would push the period of the zodiac at least a millennium and a half later than Burckhardt would have it. But Visconti thought it even later than that because all the zodiacal figures, he noted, resemble those of the Greeks, thereby "proving" that the "opinions of the Greeks were not unknown in Egypt."[30]

Visconti went on to offer a new understanding that bypassed altogether the issues raised by equinoxes and solstices and that firmly moved the Dendera zodiacs into Greco-Roman times. Suppose that the panel end that marks the rectangular's beginning did not signify the summer solstice, but, instead, the start of the year according to Egyptian calendrical practice. The Egyptian civil calendar, which was used by Greek astronomers, eventually contained exactly 365 days, thereby making it quite simple to keep track of time. Over time this calendar would diverge from the seasons because the sun takes about an extra quarter day each year to return to the same place on the ecliptic, for example, to the summer solstice.[31] Much ink had been spilled by Renaissance humanists over Egyptian calendrical matters, most especially by Joseph Scaliger, but for his purposes Visconti needed only one datum. He wrote nothing about the details, remarking only that the beginning of the Egyptian year occurred with the sun in Leo between about 12 and 132 CE.[32]

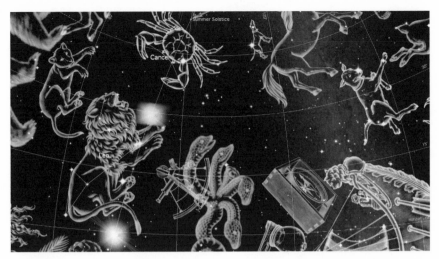

Figure 6.2. The sun enters Leo at the start of the Egyptian calendar year as it rises on August 19, 12 CE (*bottom*), and leaves as it rises on July 20, 139. During this period the summer solstice occurs past Cancer, and not in Leo.

Visconti's dating derived from remarks in the third-century CE grammarian and philosopher Censorinus's peculiar little compendium, *De die natali*. In his twenty-first chapter, Censorinus noted that a hundred years before his time the first day of the Egyptian year fell on the 13th of the calends of August. The first day of the Roman month was called the calend, and the 13th would be the thirteenth day *before* the first, including the calend itself in the count, which gives July 20.

Now, Visconti suggested, suppose that the panels represent an era during which the sun always rose in Leo on the first day of the Egyptian civil year. Starting from Censorinus's date, you could count back Egyptian years of 365 days each to see when this would have occurred. He asserted, correctly, that the entry of the sun into Leo at the start of the Egyptian calendar year would first have taken place around 12 CE, and it would still have been in Leo on that day, but just barely, about 132 CE, a century before Censorinus's time (Figure 6.2).[33] That would place the Dendera rectangular sometime during the first century and a half or so of the Roman Empire—quite recently indeed.

Visconti continued his argument with evidence from inscriptions. He noted that the exterior corner of the portico that contained the rectangular carried one in Greek. Though Denon had not been able

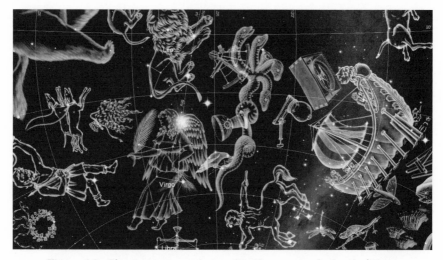

Figure 6.3. The rising sun on August 29, 10 BCE, near the head of Virgo

to copy it, Visconti probably learned of its existence directly from the artist. If we had the inscription, he continued, we might be able to solve the entire question, presumably because it might shed light on the date of the temple's construction (though it would later be argued that the dates of the temple and of the zodiacs within it need not be the same).

Visconti turned next to the Dendera circular, which, he noted, looked like a true planisphere, a projection of the celestial vault onto a plane surface. That impression, which had certainly occurred to the savants in Egypt, would become a matter of great importance years later, when the zodiac arrived in Paris, though at this time Visconti, with his meager knowledge of astronomy, could do little with it. The object's design was nevertheless sufficiently irregular that Virgo, he claimed, could be placed at the starting point of the zodiacal symbols.[34] Given this, he could advance a calendrical interpretation for the circular zodiac as well. It was during the reign of Augustus that the Julian year of 365 and a quarter days was correctly adopted, producing the so-called Alexandrian Calendar.[35] That calendar's year began on August 29, which worked well because throughout the reign of Augustus the sun rose on that day near the head of Virgo (Figure 6.3). Ergo, we are back again to Greco-Roman times.

First Reactions from the Religious Right

Visconti's analysis of the zodiacs came at a particularly opportune time for the religious Right. His critique in Larcher's revised *Herodotus* appeared just two months after the *Moniteur* article that contained Fourier's claim for the zodiacs' remote antiquity (which neither Visconti nor Larcher had seen before their own remarks had reached print). At nearly the same time, Napoleon's laws concerning the role of the Catholic Church in France were officially proclaimed at Notre Dame, on April 18, 1802. These events had a significant impact on the first controversies over the zodiacs, and so we turn now to them.

After a great deal of difficulty, including a not atypical tantrum on Napoleon's part, the original agreement between the republic and the Vatican had finally been hammered out and approved by the National Assembly during the previous summer. It recognized the Church as the majority religion in France while making it subject to whatever "police regulations" the government considered "necessary to public tranquility." On April 8, ten days before his elaborate display at Notre Dame to celebrate the agreement, Napoleon preemptively added a set of "Organic Articles" to the original agreement. These required, among other provisions, that the French government had to approve any papal order before its publication in France and, further, placed stringent requirements on the exercise of papal authority within the republic. Processions were forbidden in areas not exclusively inhabited by Catholics, and even the ways in which priests dressed were now subject to governmental regulation. Although many of these provisions were never fully carried out, their very existence perpetuated and exacerbated the religious Right's sense of being under siege. The zodiacs bubbled to the surface at precisely this moment, when sensitivity to anything that threatened to undermine religious conviction was extremely high.

The first printed reaction from increasingly vocal religious conservatives to the very public claims for zodiacal antiquity came on February 21, 1803, in the form of a highly favorable review of Larcher's *Herodotus*. The review appeared in the widely read daily, the *Journal des Débats*. Founded in 1789 to publish the decrees and debates of the National Assembly, the *Débats* was taken over in 1799 by the brothers Bertin and had become an organ for the religious Right. In the year that had elapsed between the Concordat and the appearance of the *Débats* review, many nonjuring priests had returned to France, emboldened by Napoleon's recent moves to placate believers by abolishing proscriptions against Christian Sunday services, and by his willingness to deal with the Vatican, as evidenced by the Concordat. Royalist sympathizers returned as well, encouraged by Napoleon's abrogation of ceremonies commemorating the execution of Louis XVI.[36]

In keeping with its religious conservatism, *Débats* was particularly bold in furthering an antiphilosophe agenda. On December 10, 1803, for example, it castigated philosophie, which it pilloried as "this impious language that teaches the people to disdain the faith of their fathers, this seditious language that teaches them to revolt against authority, this corrupting language that outrages morality, encourages vice, and removes all impediments to the passions . . . a code of atheism . . . a code of immorality . . . a code of bloody revolt."[37] The first consul would no doubt have been pleased by *Débats*' approval of submission to "authority" (though the authority it had in mind was the Catholic Church rather than the consular government), but the journal turned into outright opposition early in the empire, calling for restoration of the monarchy. Eventually Napoleon had had enough of its constant attacks on philosophie and, more to the point, of its quite apparent approval of the Bourbons. In 1805 Napoleon, now the new emperor, had the publication renamed the *Journal de l'Empire* and kept it under close watch.

During the decade or so after 1797, conservatives revived and developed the old anti-Enlightenment themes that had animated their writing during the years before the revolution. In the process, they solidified a core of social, political, and religious convictions that had not previously been articulated as a unified group. Indeed, conservatives as a self-conscious faction can scarcely be said to have existed before they had coalesced in aggrieved reaction to expressions of late Enlightenment values. Their position was reactionary, but not in the simple sense of clamoring for a return to a situation that had existed before philosophie, as they saw it, had eroded French society. The Right's vision of what had earlier obtained was just that—a vision, some elements of which did match aspects of past reality, but which was built carefully to serve ideological purposes. Although since the 1790s conservatives had been approvingly sounding themes of patriarchy, the family, social hierarchy, obedience to authority, the inherent imperfection of humans, and (above all else) religion, specifically Catholicism, by the early years of the nineteenth century their tone had become increasingly strident and triumphalist. Of the various conservative journals, the *Mercure de France Littéraire et Politique* and *Débats* were particularly vocal. In issue after issue, these journals lovingly elaborated a retrospective utopia, an idyllic past in which wives submitted to husbands, and men to monarchic and churchly authority. This world, they argued, had been lost, destroyed by philosophie, the ideology of a thousand cuts that had disastrously culminated in the bleeding lesion of the revolution.

Because the Right's narrative of the past had to be invented and sustained against counterclaims, the journals were especially concerned to police history. For these antiphilosophes, who insisted on humankind's intrinsically and irremediably sinful nature, only centuries of hard-won experience could provide the learned traditions that might bridle otherwise-ungovernable passions, specifically the revolutionary ones of *liberté, egalité, fraternité*. Where for philosophes the past was a record of humanity's uncertain climb out of the depths of prejudice and superstition, for reactionaries the past provided the sobering wisdom required to convince a chastened people to submit humbly and righteously to the authority of throne and altar. To take just one among a myriad of examples, on May 27, 1802, *Débats* pinpointed the philosophes' "absolute contempt for experience" as their primary mistake. These men, it continued, "regarded the lessons of the past as so many errors; the maxims consecrated by the wisdom of the centuries were only, in their eyes, superannuated stupidities. Their presumption put the most respectable traditions amongst the number of most ridiculous tales." It was long past time for the French to rescue enduring beliefs and traditions before they vanished altogether. "Thus it was in vain," railed the author against the claims of the philosophes, "that humanity has aged. Thus it was in vain that thousands of years amassed our knowledge, enriching the treasure chest of history and furnishing modern generations with resources of instruction."[38]

Nine months later, the *Débats* review of Larcher's newly orthodox *Herodotus*, now thoroughly cleansed of its previous heresies and containing Visconti's rejection of the Dendera zodiacs' great antiquity, rang the same notes of Catholic revivalism. The reviewer relished Larcher's redemptive piety and trumpeted the iniquity of Voltaire—that arch-enemy of religion and (ipso facto) of morality—who had so imperiously and ignorantly attacked Larcher, presented here as a modest and humble scholar.[39] The reviewer quoted Larcher's introductory mea culpa, that "intimately convinced of all the truths taught by the Christian religion, I removed or rewrote all the notes that might wound it." The confession of a redeemed sinner ever surpasses the testimony of those without the purifying experience of the once-fallen, and so the reviewer lauded Larcher for removing "some notes that were little favorable to religion." The reviewer took note as well of Larcher's rejection of the zodiacs' antiquity (without, however, paying any attention to the details of Visconti's analysis, or even mentioning him, perhaps because of what *Débats* would surely have seen as his regrettable performance in Rome a few years earlier).[40]

Accademia di Religione Cattolica and Domenico Testa

Several months before the review appeared in *Débats*, but certainly well after the *Moniteur* article by Bernard that contained Fourier's letter on the age of the zodiacs, the first reaction from the home of the papal seat appeared under the official imprint of the Roman Accademia di Religione Cattolica. Gian Domenico Testa's *Dissertazione sopra Due Zodiaci* (Dissertation on Two Zodiacs) was presented to the academy on July 5, 1802, preceding by ten days the first anniversary of the Concordat's signing, and following by three months its public proclamation in Paris, Testa having been elected to the academy on May 13.[41]

The Accademia di Religione Cattolica was founded in 1800 by a priest named Giovanni Fortunato Zamboni. Zamboni had a publicist's interest in promoting fidelity to dogma all the way to the outer reaches of the Church's influence. His passion was strategic, for he intended to devote the academy to the production and spread of counterarguments to heretical views. "The object of this academy," read the first of the organization's thirty-two bylaws, "will be to promote the study of the Catholic religion to block all current errors, and to save the young, even the secular among them." Preliminary meetings were held throughout 1800 until, on December 4, Pius accorded the academy official recognition. Presided over by a president (the first of whom was Domenico Coppola, titular archbishop of Mira) the organization's hierarchy included four "promoters," a secretary, a custodian and subcustodian, and twelve censors to ensure dogmatic fidelity, particularly since the academy occasionally, though not invariably, printed the "dissertations" that were read before it. The meetings aimed to combat arguments that undercut dogma or that undermined divine action as revealed in scripture. The members heard disquisitions demonstrating, for example, that "it is impossible the world could have been formed by the causal concourse of atoms," or that "miracles are not the effects of occult natural forces."

Among the very first printed disquisitions was Testa's treatise on the zodiacs. The next year he read a second paper, this one on the miraculous eclipse said to have occurred at Christ's death.[42]

Testa had spent nearly his entire life in the church. From the age of eight, when he was admitted to the seminary of Palestrina, Gian Domenico had impressed his teachers. He progressed rapidly in philosophy, entered the church, and became in 1773 at age twenty-six the professor of logic and metaphysics at the Collegio Badinelli in Rome. Testa was particularly interested in sense perception, and in 1776 he published a Latin dissertation entitled *De sensuum usu in perquirenda veritate* (On the Use of the Senses in the Search for Truth). There he argued that human senses, adapted to the necessities of life (presumably by the deity), can be neither imperfect nor misleading. Instead, by virtue of godly wisdom, the senses yield perceptions that correctly match the world—an oblique return to the scholastic doctrine that properly working senses convey the true essences of things, albeit couched in language adapted to more modern ways of talking. According to Testa's admiring biographer, *De sensuum* was well received at the time; four years later Testa wrote more along the same lines. These works display Testa's relentlessly abstract, confrontational style. He would criticize the arguments of others who had themselves made observations or done experiments, without himself having observed or done anything of the kind. He turned later to questions of acoustics, where he argued against physical or mathematical explanations for sonority, asserting instead that true perception of nature occurs essentially without mediation through simple stimulations of the sense organs that are little different whether the sensing organ be the ear or the eye.[43]

In 1785 Testa became secretary to the papal nuncio in Paris, Antonio Dugnani from Milan. There he made a number of friends, including, it seems, the great arch-atheist Lalande, which speaks well either of Testa's diplomatic sagacity or of Lalande's gregariousness, probably both. Having, however, failed to imbue Lalande, the insectivore and connoisseur of porn, with belief in the moral virtues of the supernatural, he turned instead to the astronomer's nephew, who was more susceptible. Testa and the nuncio remained in Paris throughout the early years of the revolution, but the September 1792 massacres of clergy forced them to flee, aided by a woman whose identity Testa never revealed. He returned with Dugnani to Milan, where he embarked on a second round of philosophical inquiry that once again rigorously avoided anything that could not be encountered in a library.

In 1793 Testa took up epistolary arms against the opinions of two naturalists, both also clerics. Alberto Fortis had argued that the fossil fish found on Monte Balco in the region of Verona had been transported there in ancient times from warmer climes during a general cataclysm. Testa argued vehemently otherwise, though he neither visited Monte Balco nor acquainted himself with the region's fossilized fish, which could easily be seen in the collection of Vincenzo Bozza, where they had been used by Fortis. He claimed contra Fortis that the fish were native to the area from a time when it had been warmed by volcanic fires—a time, he asserted, that was hardly remote, reaching no earlier than Greek antiquity.[44] Testa had not the slightest geological evidence for any of this. For proof, he relied entirely on a passage in Theophrastus according to which the Black Sea had once been frozen. Apparently irked by a critical rejoinder to his attack on Fortis, whose anonymous author Testa took to be the naturalist Giovanni Volta, he attacked Volta as well.[45]

Despite his elaborately detailed counterarguments, Testa was not gripped by natural science per se. Like any upstanding member of the academy, he was primarily concerned to counter claims that might open the way to a critique of Mosaic chronology, and perhaps for more than purely doctrinal reasons. Testa accordingly rejected anything like a series of cataclysmic events over long periods of time and, referring to Archbishop Usher, asserted that "the interment of the Bolca fish could have taken place between 2,207 and 1,500 years before the arrival of Christ." That postdated the Flood, put, for example, by the seventeenth-century chronologer Denis Petau at 2329 BCE; Bishop Usher had it a bit earlier, at 2349 BCE. Neither the naturalist Buffon, with his restricted age for Earth of seventy-five thousand years, nor certainly the arch-philosophe Voltaire, who opted for great antiquity and multiple cataclysms in the antireligious diatribe that Larcher had so imprudently critiqued, could move Testa from strict biblical chronology. In a note written in his copy of Testa's argument against Fortis after the time of the quarrels, but during Testa's lifetime, one G. B. Brocchi remarked acidly that "the author is the A[bbé] Testa of Rome, who, being very foolish in natural history, set to writing this nasty book in order to court the papacy, thinking that the Bolca fish might give the lie to Moses where he speaks of the Deluge. He wanted a cardinal's hat."[46]

Testa never became a cardinal, but he certainly did gain entry to the papal court in Rome. The French physicist Jean-Baptiste Biot, who, we shall see, became involved in the zodiac controversies after the Dendera circular's arrival in Paris, visited Rome in March 1825 with his son. Biot was by that time overtly religious and thoroughly in tune with the reactionary climate that characterized the new court of Charles X, the Comte d'Artois, who ascended the throne in May 1825, his brother Louis XVIII having died the previous September. Biot hoped for an audience with the new pope, Leo XII. This pontiff—a member of the "zelanti," fanatics who seethed with anger at modern reforms, such as the emancipation of Jews and freedom of the press—was even more rabidly reactionary than the French court would soon become. Biot and Testa had corresponded about the zodiacs, and when the Frenchman was unable to gain an audience with the pope he turned to Testa for an introduction. Help was quickly forthcoming because it turned out, rather to Biot's surprise, that at the Vatican Testa occupied a high position and was especially esteemed by the new pontiff, who was just then beginning his ferocious battle against every liberal reform that he could think of.[47]

Testa's Assault

"Oh Moses! Oh divinely inspired author!" declaimed the fiery defender of sacred chronology before the pious assembly of the Accademia di Religione Cattolica. "Can it be true that among the ruins of the very country that was the stage for your wonders, and for your glories, we now discover monuments that belie, that contest, that irreparably destroy the history you wove of the world's creation?" One can easily imagine the silence that followed, and the skepticism that perhaps flickered in the minds of at least one or two of the gathered prelates, as they waited to see how Testa would make good on such an extravagant opening. Do not fear, Testa reassured them, for "Moses is accustomed to triumph over his enemies."[48] Since Moses could not be present to defend himself, Testa stepped bravely to the bar and laid out his brief for the youth of the zodiacs about which, he complained, "so much noise is being made everywhere."

Just as Testa never saw the fishes of Monte Bolca, neither had he seen any representations whatsoever of the zodiacs before proclaiming that they, like the mountain fossils, derived from ancient Greece. All that he had to hand were Coraboeuf's and Burckhardt's letters as printed in Grobert's *Pyramides*, and Lalande's comments in his report of 1800 on developments relevant to astronomy for the Bureau des Longitudes.[49] He had probably also at least heard about Fourier's letter to Berthollet that had been printed in the *Moniteur* the previous February, but it contained no further evidence of any sort, just Fourier's claim for the zodiacs' remote antiquity. If, at least in retrospect, these unsupported claims hardly seem reason enough to mount a full-blown counteroffensive, Testa was not deterred. He did not need to see any depiction of the actual objects in order to counter claims for their great age because he knew what had really taken place.

One might suspect, he wrote, that these zodiacs had been ingeniously fabricated to confuse the pious, presumably by the unbelieving French savants. But that would be wrong. The French had not made those deceptively impious objects; the ancient Egyptians themselves had. That antique people, he went on, ambitious to be thought of as the first among nations, had always claimed great age for their history. Indeed, the Greeks had mocked them for it. How better to confound the doubting Greek than to contrive a monument whose inscribed surface testified to Egypt's unparalleled age? A nation so "vain and arrogant" might well have done just that. It is almost certain that this is just what did happen, Testa went on, because the Egyptians could not have known about the zodiac at the early dates that the astronomically minded savants had assigned to them. He was willing to accept, at least provisionally, the calculating savants' claims for what the zodiacs seemed to show, but he thought that the "vain" Egyptians had deliberately concocted the carvings to fool the credulous Greeks into thinking that Egypt's antiquity reached immensely far back.

Testa's reasoning about this, if it can be called that, seems to be based on his claim that the deity Jupiter Amun was "the greatest and most ancient divinity" (originally in its Theban guise as Amun, later as the syncretic Zeus or Jupiter Amun in Greco-Roman times) and was always associated with Aries the Ram. Presumably, therefore, the appearance of Aries in the zodiac attests to the importance of this deity at the time of the zodiac's invention or first use in Egypt. Since even the egregious Dupuis had associated Jupiter Amun with the vernal equinox, and since the zodiac at its first use in Egypt had Jupiter Amun as Aries the Ram, it seemed to follow that the zodiac appeared when the vernal equinox lay in Aries. For that there was indubitable textual support, at least in Testa's opinion. The early fifth-century CE Latin writer and philosopher Macrobius, in his *Saturnalia*, remarked that Egyptians associated Aries with the sun. According to his account, a ram lies for the six months before the spring equinox on its left side, and for the six following months on its right.[50] That clinched the argument, for,

Macrobius continued, the ram behaves just like the sun, which, before the spring equinox, "traverses the right hemisphere and then, for the rest of the year, the left hemisphere."[51]

Testa turned next to calendrical issues. The ancient Egyptians, it was known, had originally used a 360-day calendar. But how could they ever have done so, if their knowledge of the zodiac had reached back to the vast millennia that the savants had suggested? Such a calendar would have come rapidly into conflict with the agricultural year since the return of the sun to the vernal equinox—the solar year—takes about 365 and a quarter days. A learned people, a people so knowledgeable about the zodiac, would almost at once have adopted a different system, one that connected their calendar more directly to the stars. If the introduction of the zodiac in Egypt reached back millennia before Moses, how could it be that the Egyptians were, as he thought, still using a 360-day calendar in his days? Testa of course admitted that the Egyptians did eventually add five extra days to their year, which would certainly have helped maintain the concord between calendar and seasons, at least for a time. Though some, he noted, would carry the introduction of these additional five days back to 1325 BCE, and so to near the time of Moses, this was still vastly later than the savants' date for the origin of the zodiac in Egypt. "To have a zodiac," he sarcastically remarked, "to have it for centuries, and yet to ignore how many days are in the year, that's a contradiction that cannot be supposed for the Egyptians, whose genius, whose studiousness, whose astronomical science are celebrated with such great praise."[52]

For extra measure, Testa threw in some remarks about the "famous Sothic cycle." This concerns the significance for the ancient Egyptians of the dog-star Sirius, which, we have seen, Visconti had also brought to bear on the question of dating. The issue of the cycle recurred several times during the zodiac controversies and can even today raise questions, albeit not of the sort that engaged religious firebrands of the early nineteenth century. It has always been confusing and requires explanation.

The most important event of the year in ancient Egypt was the rise, flood, and subsidence of the Nile, upon which the rhythm of everyday life depended. One notable astronomical event had a striking connection to that supremely significant occurrence. Sirius disappears from the night sky when the sun, in its yearly path around the ecliptic, is close enough to it that the star cannot be seen. Some seventy-two days after Sirius's first disappearance, the star will again be visible for a short time before the light of the rising sun obliterates it. Each night thereafter Sirius will be visible for longer and longer periods before sunrise. That day of first appearance, known to Greek and later astronomers as the star's heliacal rising, was thought to have been associated by the Egyptians with the forthcoming flood of the Nile, which occurs each year after the summer solstice. The association posed a problem on which Testa now seized because he did not think that Sirius's heliacal rise maintained a constant relationship to the summer solstice throughout Egyptian history.

To understand Testa's reasoning, begin with the 365-day calendar, which was certainly in use by the time of Ptolemaic Egypt. Within a century after its invention, that calendar would already have been out of synchrony with the seasons by about a month. If the solar year were exactly a quarter day longer than the calendar year, and if Sirius's heliacal rise always occurred the same number of days from the summer solstice, then the heliacal rise and the 365-day calendar would also diverge from one another at the same rate that the solstice diverged from the calendar. In consequence of that difference, both the solstice and the rise would come back into coincidence with the calendar after 1,460 years. That period constituted the so-called Sothic cycle, about whose existence, Testa wrote, the ancient Egyptians so proudly "boasted." However, this "famous" cycle scarcely testifies to the Egyptians' profound astronomical knowledge because it is incorrect. It is wrong by 36 full years, Testa claimed. For it ignores the difference between the tropical year and the sidereal year, the latter being the number of days that it takes for the sun to return to the same star. The tropical year is independent of the stars, for it measures only the number of days for the sun to return to a specific point in its apparent motion about Earth.

A Sothic cycle of 1,460 years assumes that the star Sirius will be in approximately the same place with respect to the sun when the sun again reaches the summer solstice, thereby keeping the number of days between the two events (heliacal rise and solstice) very nearly constant. That, however, cannot possibly be so, Testa argued, because precession moves the stars themselves around the ecliptic, resulting, he thought, in 36 extra years beyond the 1,460 for the return to coincidence between Sirius's rise and the 365-day calendar. The ancient Egyptians seem not to have known about this defect, in which case they must have been altogether ignorant of the "precession of the equinoxes." Yet, he continued, how could this possibly be so if they had known so much about astronomy to have constructed the temple zodiacs so many millennia ago?[53]

Since the issue of Sirius's rise would continue to haunt the zodiac controversies in later years, it is important to understand what does occur. Let us look first at the relation between the solstice and the heliacal rising of a star that sits pretty nearly on the ecliptic itself. For an ecliptic star choose Regulus, a very bright one on the left foreleg of the constellation Leo, and for the horizon choose ancient Thebes, present-day Luxor. There, in 3100 BCE, Regulus rose heliacally on July 22. This was three days after the sun reached the summer solstice, making Regulus's rise a good marker of the coming flood. Move ahead some 2,100 years, to 1000 BCE. In that year Regulus rose heliacally near the beginning of August, while the summer solstice occurred near July's start, so that there was now an entire month between its rise and the solstice, which would make the star's rise completely useless for signaling the flood.

Sirius, however, does not lie on the ecliptic. Precisely for that reason, and uniquely at the latitude of Luxor, matters are quite different for it than for an ecliptic star like Regulus. At Luxor in the year 2500 BCE, Sirius rose heliacally toward mid-July, just before the solstice. Moving ahead, to 1000 BCE, we find that Sirius still rose heliacally in about mid-July, while the solstice occurred nearly a week and a half earlier. So the date of the solstice has moved from shortly after the heliacal rising to ten or so days before it—not a large shift in comparison to the one for Regulus. Consequently throughout the entire course of Egyptian dynastic history, the heliacal rise of Sirius in fact remained a reasonably good marker for the summer solstice, and so for the flood, despite precession. In effect, Luxor's latitude combines with the position of Sirius off the ecliptic to compensate for the effect of precession for several millennia. This is where Testa stumbled. He thought that precession would affect the connection between heliacal rise and solstice in the same way for Sirius as for a star on the ecliptic. Because it does not, an Egyptian value for the Sothic cycle would have worked quite well throughout most of pharaonic history. Figure 6.4 illustrates the situation.

Testa's mistake was entirely understandable; even able astronomers could easily make it, but it would certainly undercut his claims in the eyes of those who knew otherwise—as, we shall see, Fourier certainly did. But, for Testa, the argument from precession made it "very certain that the zodiacs of Dendera and Esneh do not have the antiquity

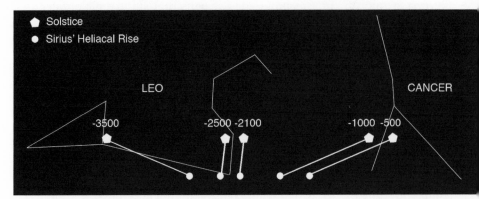

Figure 6.4. The sun at the heliacal rise of Sirius and at the solstice from 3500 to 500 BCE. As the years pass, the line that connects the solstice to the heliacal rise rotates clockwise. Since the sun's annual motion is in the direction from Cancer toward Leo, a rise takes place before the solstice in a given year if it occurs further to the right. From 3500 BCE until about 2100, the rise either precedes the solstice or is approximately coincident with it. The solstice thereafter moves slowly past the rise, until by 500 BCE the rise follows the solstice by about two weeks.

attributed to them."[54] Moreover, he asked, how could it be that the precession of the equinoxes was not known until the time of Hipparchus, if, as the savants asserted, the zodiacs depicted the summer solstice first in Virgo (at Esneh) and then in Leo (at Dendera), whereas in Hipparchus's time it was in Cancer? While Testa's question seems to beg the point, since the savants asserted that precession itself was known before the second century BCE, Testa had an answer to that as well. If the zodiacal indications of precession were inscribed so visibly on these temples, how could Hipparchus in Alexandria not have known about it from the start? It is hardly likely that he would later have been thought to have discovered something whose effect was well enough known to have been carved in stone, as it were. Testa concluded that the zodiacs and the temples themselves must all have been constructed after, not before, Hipparchus's time.

To bolster his argument, Testa adduced more evidence, again from astronomical sources. The Dendera zodiac depicts Libra, but Testa asserted that in the remote eras of their history the Egyptians had not known that constellation, whose location had in those times been occupied by the claws of Scorpio. The Alexandrian Greeks were the first to have inserted Libra—"an incontestable fact admitted everywhere in general," he noted. Dupuis had argued against this late origin for Libra, but his reasoning, Testa was certain, failed in the face of countervailing evidence. For Eratosthenes himself had not mentioned Libra in his *Catasterisms*, and neither did it appear in Aratus's poem or in Hipparchus's commentary on Aratus. The temple zodiacs must therefore date from Greco-Roman times. And so, Testa mocked, "after so many convulsions and so many cries, the mountain of work gave birth to a mouse."[55]

But a slight problem remained. If indeed the Dendera zodiac dated to such a late period, why then did it seem to show the summer solstice in Leo, instead of where it would then have occurred, namely, in Cancer or Gemini? Well, Testa explained, in fact the Dendera zodiac does not show the solstice at all because the solstice per se was not important for the ancient Egyptians. Rather, it depicts the position of the sun among the stars at a different but particularly significant moment in the past. The Egyptians, Testa averred, had their own version of sacred chronology: they thought that the world had begun when the heliacal rise of Sirius coincided with the entry of the sun into the constellation Leo. This connection between Sirius's rise and the sun in Leo was, he went on, constantly maintained for religious reasons. Its importance was due to the connection between the rise and the Nile flood: "since the inundation of the Nile," he averred, "was and is the source of well-being for the Egyptians, it is not surprising that they regarded the canicular, that is Sirius, as the most beneficent of all the stars, and that they joined its birth to that of their year."[56] This is what the zodiac depicts, which therefore has nothing to do with the solstice.

It does, however, have to do with the Nile flood, for "due to the influence of Leo," the Egyptians thought, "the Nile arrived more abundantly, and in consequence with greater fertility." This, Testa thought, was why the Egyptians placed such great emphasis on the connection between Sirius's heliacal rise and the sun's position in Leo. In which case at some point in their history the rise and the flood must both have taken place when the sun was actually located there. Now, the solstice was in the muzzle of Leo, he noted, around 1322 BCE, which would therefore have marked the proper moment for the Egyptians to have established the connection. At that time, Testa assumed—correctly in fact—Sirius did also rise heliacally around the time that the sun reached the solstice, and so did then signal the flood.[57] But by the time of Hipparchus over a millennium later, Leo's muzzle had moved well away from the solstice (in fact all the way through Cancer and to the verge of Gemini), and so any connection between the solstice, on the one hand, and both the position of the sun in Leo and the heliacal rise of Sirius, on the other, would have been lost. Consequently the link between the flood and the rise would also have been broken, and that, according to Testa, had been the original reason for the Egyptians to have insisted on the importance of Sirius. He was certainly correct that the solstice would over time move out of Leo, breaking any connection between Leo and the flood. But Testa was altogether wrong that the link between Sirius's heliacal rise and the flood would also have been broken. He had again incorrectly assumed that the heliacal rise would always occur when the sun reached the same position among the stars, namely, Leo's muzzle, moving right along with precession. It does not: the link that broke connected the sun in Leo to the solstice, not the solstice (and so the flood) to the rise of Sirius, which persisted until well into Alexandrian times.[58]

As for the savants' claim that the Esneh zodiac had the solstice in Virgo, Testa replied that there was no evidence for the association. But even if the sun were depicted in Virgo, he continued, *even then* it did not signify a solstice. No—it signified instead the month of September in the years following the adoption by the Egyptians of a new 365-and-a-quarter-calendar under Julius Caesar in 46 BCE. For in Egypt, he asserted (citing Macrobius),[59] it was in September that the new system was adopted, and in those days the sun rose in Virgo during that month, as indeed it certainly did. Moreover, Caesar's victory at the battle of Actium also took place in September. What more could one ask?

For Testa, science could and should be co-opted and redeployed in the service of religion. Indeed, there was no better way to prove the truth of a religious claim than by recourse to science. "You see," he ended, "that since arguments like these [of the savants] fight our sainted religion, the study of the natural sciences, which has always been useful for its defense, has today become absolutely necessary." Testa saw himself as turning the weaponry of science back upon the savants themselves, just as he had previously done in arguing about the fossils of Monte Balco, or, even before, about the ability of the senses to apprehend the essential natures of things.

The appearance of Testa's critique during the very months that the Concordat with the Vatican came into full effect marks a significant moment. For a period thereafter the publication of anything that might cause religious controversy (particularly by upsetting the right) was enjoined.[60] During the last days of the consulate and the first years of the empire, conservatives became extraordinarily vocal, while expressions of Jacobin sentiment were strongly repressed. Within several years the situation had in some respects reversed, as Napoleon and his functionaries began to detect as great, or even greater, threats to the empire's legitimacy from the monarchist, and religious, Right. Before then, from 1802 to 1806, debates over the zodiacs were strongly affected by censorship, as indeed were other expressions of views that might produce religious controversy. Control of the press was primarily the responsibility of a wily and politically resilient man, Joseph Fouché, whose career led him from Oratorian to Jacobin to head of the General Directory of Police. As we shall see, his complicated efforts to manage the printed word had an important impact on the course of the zodiac controversy.

7

Ancient Skies, Censored

> For me it matters little whether the world be old or young, I gave
> the zodiac an age that I believe it really has.
>
> —Charles Dupuis, 1806

Joseph Fouché, Director of the Ministry of Police

Though Joseph Fouché's long political career was marked by frequent shifts of loyalty, to one allegiance he remained constant: regardless of how events unfolded, he always found ways to profit by them. Trained originally by the Oratorians (a secular order) to teach mathematics and physics, by 1791 Fouché had become involved in the period's political ferment in Nantes, his original home. He had been transferred there by the Oratorians in the vain hope that he would stay out of politics. In his previous post at Arras, Fouché had met a young lawyer by the name of Robespierre, to whom he may have lent money to attend the Estates-General. For a short time Fouché had also become involved with Robespierre's sister, Charlotte, and the collapse of that relationship permanently estranged the two men.[1]

The first signs of Fouché's political flexibility, if not as yet his judgment, appeared at Nantes, where he agitated against slavery, going so far as to write about it to the Girondist Brissot. His outspokenness did not go over well locally since many of the bourgeois in town had interests in the slave-holding colony of Saint-Domingue. In a sudden reversal that showed his typical elasticity, Fouché wrote a second letter to Brissot, countermanding the first. His about-face earned Brissot's disdain, but it dampened local antagonism to such an extent that, after marrying the daughter of a prominent Nantes official, Fouché was sent by the city to the new National Convention in the fall of 1792. Both the monarchy and the Oratorian order had by then been abolished.

In Paris, Fouché initially did little, but his career and prospects changed abruptly with the trial of the king. Though expected by his bourgeois constituents to vote for the unhappy monarch's pardon,

Fouché opted for death. He had perceived that power was rapidly shifting in the direction of the radical and action-oriented Montagne faction, led by the Jacobins, from the more moderate Gironde. Fouché soon associated with the Hébertists in supporting measures against the church, which made him more radical even than Robespierre, who, we have already seen, eventually guillotined the Hébertist leaders. Until then, the increasingly agile Fouché remained in a strong position. In March 1793 he was sent to Nantes as a "representative on mission." Nantes was thought to be near falling to the Vendée rebellion, but by June the city seemed no longer to be in danger, and Fouché went on to Troyes to obtain troops.

While there, he put down a rebellion in Nevers, a move that would assuredly have made him popularly disliked, except that along the way he acted as a sort of Jacobin Robin Hood. He liberated silver and cash from local churches, urged the poor against the rich, and promised to split the proceeds between the deprived and the government. His anticlerical actions in Nevers had in fact served as something of a model for Chaumette (the same man who had used Dupuis to justify saving the icons on the façade of Notre Dame), who visited his mother there on September 18. Four days later Fouché was in the local cathedral, declaiming against the clerics and challenging "ministers of religion" either to "get married, adopt a child, or house and feed an elderly person"—or else suffer the loss of their pensions.[2]

From Nevers, Fouché went at the end of October to Lyon, which had been recaptured following a rebellion on October 9, and where Collot d'Herbois had already been sent by the convention. The two representatives had been ordered to eliminate all resistance. Unlike d'Herbois, who despite his flintiness had not thus far carried out his charge, Fouché took the convention's command seriously, to say the least. Although he was at first comparatively moderate, he soon took some of the most vicious and extreme actions of the entire Terror. From December 12 through April 5, Fouché and d'Herbois presided over the massacre of 1,682 people, including women and, it has been said, even children. Among those executed was the father of André-Marie Ampère, who would become the founder of electrodynamics in France. The young André-Marie did not speak for an entire year thereafter.

The ferocious exterminations in Lyon were in part responsible for the fall of the Hébertists at the hands of Robespierre and Danton. Even before their demise in April and May 1794, Fouché had once again changed direction and begun to moderate his actions, thinking, it seems that Danton, less fanatic than Robespierre, would emerge on top. But Danton himself fell to the guillotine on April 5—the very day that the executions in Lyon ceased. Robespierre had defended Danton until very nearly the end but had ultimately agreed to his friend's arrest, pressed by the even more extreme Billaud-Varenne. After Danton's fall, the lawyer from Arras took full control, though Danton had correctly predicted that he, too, would soon fall under the blade. Summoned to Paris, Fouché was forbidden by Robespierre to defend his actions in Lyon before the convention as a now-disgraced Hébertist. He was expelled from the Jacobins—something that would stand him in good stead when he had later to efface his earlier connections. However, the growing reaction against the Terror's extremism, stimulated in no small part by pervasive fear that anyone might be next, sapped Robespierre's influence. Fouché took advantage of the situation by forging contacts between others increasingly opposed to the self-proclaimed Apostle of Virtue, leading to Robespierre's downfall on July 27 in the Thermidorean reaction. Eventually those associated with Jacobinism, including some former members of the Committee on Public Safety, were jailed or otherwise punished. For once, Fouché's political intuition failed him. Caught in the reaction, on August 9 he was denounced and sentenced to exile in Guiana, a tropical hell with its moist heat and fevers.

Though his sentence was commuted, for the next few years Fouché disconnected himself from politics, leading an underground existence that may have included a stint as a pig farmer. During this time, he also wormed his way into military supply lines and developed contacts in the Parisian underworld that would later prove most useful. Then, in the turmoil of 1797, after initially supporting a royalist conspiracy that seemed on the verge of success, Fouché shifted his allegiance to one of the directors, Barras, sensing that he would prevail. Pleased by Fouché's service, Barras persuaded his fellow directors to make him ambassador to the Cisalpine Republic, which Napoleon had carved out of northern Italy. After Fouché surprisingly proved too liberal in his policies for the Directory's comfort, he was transferred to the new Batavian Republic in Holland, where he successfully countered anti-French activities. A new member of the Directory, the Abbé Sièyes, fearing revivified Jacobinism as well as the continued plots of the royalists, took advantage of Fouché's unique abilities and contacts and had him moved back to Paris to head the Ministry of General Police. On July 26, 1797, Fouché dispensed with Jacobins, shuttering the club of which he had been such a dedicated member not so many years before.

Napoleon continued to find work for the ever useful Fouché. He eventually appointed him to head a newly configured General Ministry of Police for France.[3] Ever the opportunist and intriguer, Fouché laid out an elaborate plan for the ministry that would have given him powers potentially rivaling the first consul's own. Napoleon accordingly limited the structure, placing local police commissioners under the control of the Ministry of the Interior rather than, as Fouché wanted, his Ministry of Police. Interior at the time was headed by Chaptal (whom Napoleon in his inimitable way later alienated by seducing Chaptal's mistress practically in front of him). Moreover, Paris was the only municipality to have its own, special Prefect of Police, headed by one Dubois, who, like Fouché, had direct access to the first consul.

Censorship Instituted

Fouché's censorship began modestly but quickly expanded. On November 22, 1799, he issued an order forbidding the police to send journalists any news "bulletins or notes." Two weeks later, theaters were informed that no plays could be staged that "might become a subject for dissent." To that end all dramatic works concerning the revolution had to be submitted to the ministry for examination. Then, on January 17, 1800, several weeks before the first decrees establishing the General Ministry of Police and the Paris Prefecture were issued, the press was subjected to full-blown censorship. Only thirteen among the sixty-three active political journals survived the suppression, and the creation of new journals was proscribed altogether. Public opposition to this draconian action was virtually nil, as the populace preferred order and security to even the miniscule risk of chaos represented by a free, or at least freer, press. On April 5 Napoleon ordered Fouché to ensure the virtue of the remaining journals, making certain that what appeared in their pages was properly moral and patriotic.[4]

The institution of full-fledged censorship accorded well with the sentiments of the apostles of religious revival, men like Bonald, whom we have already met, and Sabatier de Castres—or, rather, it did so when censorship silenced the disturbing voices of philosophie. Trained like Fouché by the Oratory, and originally a social moderate, Bonald turned counterrevolutionary, emigrated in 1791 and wrote a *Théorie du pouvoir (Theory of Power)* in which he argued for a "science of society" grounded in hierarchy, authority, and the organic nature of the social order, whose best guarantor was apparently the Catholic

Church. Sabatier, for his part, was initially tonsured in the church but had fled the institution's embrace, eventually reaching Paris in 1766, where he became part of the circle around the philosophe Helvetius. He reverted, however, to the defense of religion and emigrated early in the revolution. His 1810 *Apologie de Spinosa et du Spinosisme*, which nearly turned Spinoza into a Christian advocate, captured the religious conservatism of the moment. "A people of philosophes or of men without religion cannot exist," he opined, "[it] cannot even be imagined; whereas a religious, even a superstitious, people will sustain an empire and make it flourish." Such a community, he said, "is very easy to govern. It is bridled, saddled, shod, and a child can lead it." Religious belief was not merely a prophylactic against crime or a source of models for virtuous behavior but could be directed to serve political ends. "Superstition is itself a powerful bridle in the hands of a government that knows how to direct it towards public utility."[5] Napoleon himself held rather similar views, albeit without Bonald's sanctimony.

One of Napoleon's savants played a starring role in the regime's efforts to render inconvenient ideas powerless. To further ensure propriety in the fourth estate, Napoleon ordered his librarian, Ripault—the same man who had run the library of the Institut d'Égypte in Cairo—to scrutinize newspapers, brochures, books, in short any publication, for questionable views. Between five and six in the morning, after having made his inspection, Ripault was to report on what the day's publications contained in respect to "religion, philosophy, and political opinions."[6] Trained as an antiquarian, Ripault was a close friend of Devilliers, had accompanied the Costaz expedition (his name was inscribed at Philae—see Figure 5.3), and had assisted Fourier in working on Kléber's plan for a volume describing the expedition's findings, a plan that, back in France, evolved into the *Description de l'Égypte*.[7] Devilliers, Ripault, and Fourier had furthermore traveled to Egypt together on the *Franklin*.

Ripault was the author of the 1800 report produced by the arts commission for the decamped Napoleon on the antiquities of upper Egypt—ironically so, in view of the use to which Napoleon would put him back in Paris. For in it he had written that the zodiacs at Dendera "represent the state of the heavens at the distance of four thousand eight hundred years from the time when we behold them."[8] The report reflected the general opinion among the savants still in Egypt concerning the likely date of the Dendera rectangular zodiac, though not of the one at Esneh, which was thought to be much older. Ripault shared that

Figure 7.1. Ripault (*left*) and Napoleon in Egypt

view as well, for he was certain that there had been in Egypt a "generation of monuments" so ancient that they reached "to an epoch anterior to that which the Christians assign to the creation of the world"—a remark guaranteed to exercise believers whether in France, Great Britain, or Italy.[9]

With the signing of the Concordat in July 1801, and especially following its formal proclamation in April 1802, many nonjuring priests returned to France. This did not altogether please their juring brethren, the constitutional clergy, who had sworn oaths to the state and who, inspired in part by traditional Gallican sentiment, were fearful of future domination by Rome. One nearly contemporary account, entitled *Mémoires d'une femme de qualité sur le Consulat et l'Empire* (*Memoirs of a Woman of Quality under the Consulate and the Empire*), describes the situation. It was ostensibly written by a certain Olympe, Comtesse D . . . , but was in fact the work of an astonishingly productive literary hack and memoir fabricator, Etienne-Léon Lamothe-Langon. He had frequented Parisian literary salons between 1807 and 1811, after which he served as Napoleon's subprefect in Tolouse for two years, then in Livorno, and, during the Hundred Days, in the Aude. It is possible that these so-called *Mémoires* were in some respects at least based on the diary of Zoë Talon, Comtesse du Cayla, later a partisan of the aristocratic Ultras and an erotic favorite of Louis XVIII.[10] Lamothe-Langon's oeuvre requires a cautious approach, to say the least. His 1829 *Histoire de l'Inquisition en France* (*History of the Inquisition in France*), for instance, which described murderous fourteenth-century witch trials, proved in 1972 to be a complete fabrication.[11] He was nevertheless present in Paris at the time that concerns us, spent many hours at the salons, and may have had access to Talon's diary. His perceptions and descriptions of the period's events must be taken with a grain of salt, yet they do reflect at least his own experiences, and perhaps those of Talon.

Contemptuous of the clergy, Lamothe-Langon reserved a special animus for the returned priests of 1802, whose eagerness to ingratiate themselves with the first consul he saw as base opportunism. "We could not admire enough, we other royalists, however much we tried to emulate them, the alacrity with which the priests forgot the Bourbons, on whose behalf, the day before, they had addressed their prayers to God in the canon of the mass." He continued, "I well saw the truth of a remark by Barrère de Vieussac: 'the abbés,' he claimed, 'are like cats, loyal to the house and not to the master; they always caress whoever brings them the best food, and would support even more those who will let them climb their backs.'"[12] De Vieussac was hardly an impartial observer, since he had once advocated the complete extermination of the proroyalist Vendéen rebels of the revolutionary years. He was scarcely alone in these views. Hippolyte Carnot, the son of the former director Lazare, reflected in his 1837 obituary of that epitome of juring priests, the Abbé Grégoire, constitutional bishop of Blois, that the "emigré clergy, no less senseless than the nobility in its counter-revolutionary projects and hopes, did not offer models of Christian resignation to foreigners. Called back to France by the Concordat of 1801, when Napoleon worked to copy the *ancien régime*, they rushed fanatically down that path of reaction and used all their means to efface every vestige of religious and political reform."[13] Grégoire was so opposed to the Concordat that he resigned his bishopric on October 8, 1801.

Neither Ripault nor Fouché was happy with the growth of the religious Right, to which Napoleon willfully turned a blind eye despite the contradictions that religious sentiment presented to his followers from the old days, who were increasingly excluded from the corridors of power. Ripault, proving overly liberal for Napoleon, was replaced in

1804 by an elderly Italian, the Abbé Carlo Denina. Though no doubt less liberal than Ripault, Denina was hardly a rabid antagonist of eighteenth-century philosophie.[14] Nor was Ripault alone in falling afoul of the first consul's desire to mollify the religious Right.

On September 13, 1802, the entire Ministry of General Police was suspended, and its functions were placed under the control of the Ministry of Justice, then directed by Claude Régnier, with the policing functions turned over to Pierre-François Réal. Although no records provide unimpeachable evidence concerning the precise reasons for the suppression, Napoleon never did trust Fouché, and with good reason, not least because of Fouché's Jacobin background. Fouché had moreover worked subtly, if unsuccessfully, against the move to make Napoleon first consul for life. He had posed as a confidant of the consuls to the Senate and insinuated that Napoleon wanted only to augment his power, that he would actually be embarrassed by an offer of a lifetime consulateship.[15] This intrigue had certainly angered Napoleon. Vehemently opposed to any collapse of distinction between church and state, Fouché let his opposition to the Concordat become well-known, and Napoleon's brother, Lucien, took particular umbrage at the resistance.[16] Writing as a rejected and self-justifying ex-minister years later, Fouché admitted his antagonism in his *Mémoires*. "I did not share the opinion," he wrote, "that we ought to come to a concordat with the court of Rome. . . . I represented that it was a great political error to introduce into the bosom of the state, where the principles of the Revolution had prevailed, a foreign domination, capable of giving trouble. . . . it was reviving in the state that mixture of the spiritual and the temporal which was at once absurd and fatal."[17] Fouché had also undertaken moves that he attributed to orders of Napoleon but that were instead his own interpretations of them, designed to minimize as best he could the power of repatriated clerics. For example, on February 13 he had written an illuminating letter to Portalis, director of the Department of Religion, who had been one of the negotiators and signatories of the Concordat, and who subsequently worked to structure the Civil Code, or, as it was soon known, the Code Napoléon. "In virtue of the orders of the first consul," Fouché directed, "I've made known to different authorities that the resigned bishops who are permitted to return to France must not stay in Paris but go to the communes I indicate to them. I will follow this rule in respect to all the former bishops whom the first consul will in general permit to return to France. This is purely a police affair, and I'm only giving you notice of it to be persuaded that your correspondence will contain nothing contrary to the measures of public order that have been entrusted to me."[18]

In resisting the Concordat, Fouché once again proved himself a shrewd interpreter of the prevailing political wind, if not of the immediate consequences for his position. The Concordat had fostered a great deal of discomfort and anger among republicans, especially in the army. According to the head of Fouché's secret police, Desmarest, the agreement "produced in France, not merely a shock, but a serious crisis, I would say even a threatening one, though it was hardly sensed by the public. Its focus was in the high military. Most of the principal chiefs of the army, collected together in Paris, loudly expressed their discontent with the act, whether in vexation with an institution which they had fought, or whether because they saw there the first step of General Bonaparte in separating from their ranks and elevating himself, without them, to other destinies, or whether, finally, because of jealousy over rival ambitions."[19] These dangerous reactions would play out months later in Napoleon's conviction that one general in particular, Moreau, aimed to replace him. Anything involving religion or indeed supernatural belief had a strong potential to undercut Napoleon's increasingly dynastic aims and would certainly have raised red flags if bruited about in public.

The censorship had, we have seen, been instituted on January 17, 1800, while the Ministry of General Police was suspended on September 13, 1802. Seven months before the suspension, on February 14, the letter containing Fourier's claims for the antiquity of the zodiacs had appeared in the *Moniteur*. Given the prevailing attitude of the army's upper echelons and of literati toward the Concordat, this letter, with its disdainful repudiation of scriptural chronology, appeared at a critical juncture. If, as Fourier wrote, "until now the history of men and the sciences and arts had nothing certain and authentic except for very recent epochs," then clearly the Bible, that record of superstition and priestly oppression, had nothing whatsoever to say about human history. These were dangerous views to promulgate at a time when Napoleon was still only first consul (though three months later he would become consul for life). Remarks of that sort could easily overheat the military, fanning the flames of its antagonism toward the Concordat.

Napoleon had taken care to ensure a degree of synergy at the very top of the power structure in respect to censorship. Of the two men most directly involved in its working, one was closely associated with expedition members who were convinced of the great age of the zodiacs, while the other was strongly opposed to the Concordat and had reasons to fear the influence of the religious Right. Ripault was to inform Napoleon each day about what had appeared in the papers, while Fouché had ultimate responsibility for seeing that objectionable material was kept out. Anyone who read Fourier's *Moniteur* letter would at once have recognized that it was destined to generate intense opposition on the right, as indeed it did. The letter could hardly have been printed without the knowledge of at least Fouché and possibly of Ripault as well—and if Ripault had followed strictly his brief to tell Napoleon of matters that might raise religious issues, then the first consul would also have learned about Fourier's claims early one morning. Clever intriguer that he was, Fouché would no doubt have seen that the letter amounted to a direct assault on religious sensibilities. It may well be that the printing of Fourier's zodiacal dating was a deliberate ploy on Fouché's, and possibly Ripault's, part to undercut the growing influence of the Right, or, more subtly, perhaps to give the impression that, despite the Concordat, the government had not succumbed to pressure from Rome.

Fouché eventually lost his position, whether because of his opposition to the Concordat, his attempt to deflect Napoleon's acquisition of life status as consul, or more likely for both reasons, at least for a time. Ripault would also lose his job two years later because of excessive liberality. Fouché, however, did not suffer overly much. Napoleon arranged for him to be appointed a senator and to be granted the rent of the former crown lands of Aix, amounting to between twenty and twenty-five thousand livres per year—as well as allowing him to keep half, or 1.2 million livres, of the surplus he had accumulated in secret police funds. Neither did Fouché lose power for long. He kept in close contact with the first consul, providing him with tidbits from the secret agents whom he had previously hired and who continued to supply him with information. Eventually, the pressures of circumstance led Napoleon to reinstate him. The adroit Fouché had proved himself too useful to do without.

The Press and the Religious Right

Throughout the years between 1800 and 1810, a considerable number of right-wing tracts and essays reached print. Among the first, and certainly the most influential, was the four-part *Génie du christianisme* (*Genius of Christianity*) by the Breton author, Chateaubriand. Son of the taciturn former ship's captain and ongoing slaveholder, the Comte de Combourg, and an exceedingly pious mother, the young François-René eventually decided for the military instead of the church. He was commissioned a second lieutenant and by 1788 had made captain. Visiting Paris in that year, Chateaubriand, whose father had died in 1786, met several of the principal poets of the day. Like them, he sympathized with the early aims of the revolution, but by 1791 he had become disillusioned and had left for America. A year later he returned, married under family pressure, and joined the royalist resistance, only to be wounded at Thionville that year, when he was left for dead on the battlefield. He was thereupon brought to England, alone. Making his way to London, Chateaubriand scrounged for work as a French tutor, absorbed English literature, had at least one love affair, and, most important, became intensely religious circa 1798.

According to his later testimony, Chateaubriand's decision to embrace religion was occasioned by a pointed letter from his sister, Julie de Farcy, whose piety rivaled her mother's. To be sure, the Terror had deeply affected the family. Of Chateaubriand's four siblings, his brother had been executed in the last months of the Terror, while his mother and both of his sisters had been imprisoned. Shortly before her death in May 1798, Chateaubriand's mother apparently told her daughter Julie how wounded she had been by the critique of Christianity that Chateaubriand had articulated in his *Essai sur les révolutions*, published in 1797. That letter, he recounted, reached him only in 1798, after Julie, too, had died. As a result, he declared, "I wept and I believed." Perhaps, but Chateaubriand had been at work on the *Génie* long before Julie's death. His motivation for writing it likely had as much to do with a desire to make a literary splash at a time when religion was making a decided comeback as with a sentimental conversion romantically inspired by his mother's and sister's deaths.[20]

Chateaubriand had never been antagonistic to religion per se, but rather to the hierarchical form it had taken under Catholicism. By 1799 he had changed his mind about that as well, for the *Génie* is nothing if not an argument for the organic connection between society and church. In it he sounded themes that would soon be echoed by the emerging Romantic movement. He waxed eloquent about the beauties of ruined churches, linking decayed monuments to nature, and both to nostalgia for an idyllic medieval polity saturated with belief. Morality, he opined, always forms the very basis of society, and morality could not exist if the world were solely material. The moral universe had to emerge from "a more stable world than that." Unfortunately, he complained, "some philosophers thought that religion was invented to sustain morality; they did not perceive that they had taken the effect for the cause. It's not religion that flows from morality, but morality that is born from religion . . . for it is certain that, when men lose the idea of God, they are thrown into every crime despite laws and executioners."[21]

Chateaubriand's conviction that without belief in supernatural guidance society would fall inevitably into depravity and crime was (and remains to this day) a long-standing theme of the religious Right—one that conservatives continually insisted was proven by the Terror. Critics of the Enlightenment had long argued that materialist preoccupations would spawn full-blown atheism, which necessarily produces social disintegration. Chateaubriand went further, blaming the evils of materialism on science. Abstract science, and especially mathematics, was not merely dangerous, it was also useless. It tempted people to laud scientists at the expense of the real heroes, those who labored in the fields and on shop floors. "We falsely attribute to our sciences," he insisted, "what belongs to the natural progress of society." For "the great discoveries scarcely ever produce the effect we anticipate. The perfection of agriculture, in England, is less the result of some scientific experiments than that of patient work and the industry of the farmer obliged ceaselessly to worry the ungrateful soil."[22] The gnarled toilers of the earth, simple believers in revealed truth, were the authentic foundations of Chateaubriand's organically united social world.

Chateaubriand's insistent dislike for what he saw as the desiccated constructions of theoretical science—the science, as he saw it, of *systems*—was hardly novel, since Rousseau held much the same opinion, though he had not married it to a veneration for Catholicism. But Chateaubriand wrote at a time when the partisans of the religious Right felt increasingly assertive, and his eloquence lent force to their movement. Passionately articulate, with a rhetorical flourish that appealed to the sensibility of many young Frenchmen, and not only to them, Chateaubriand extended his critique to any assertions that challenged the biblical account of creation. Nobody, he insisted, could deny the truths of Mosaic cosmogony. Genesis alone provides a true account of the world's origin—true because the Mosaic narrative is "natural" and "magnificent," as well as "easy to conceive and most in accord with man's reason."[23]

The *Génie* first appeared in 1802, with a dedication to the First Consul. Chateaubriand may have already known about claims for the immense antiquity of the Egyptian zodiacs; indeed, he could hardly have missed them given the publication of Fourier's letter in the *Moniteur* on February 14. Whether or not he had, Chateaubriand was certainly well aware of Dupuis' *Origine*, and in the *Génie* he castigated all attempts to attack scriptural chronology using astronomy. Those who try to do so "seek in vain in the history of the firmament . . . proofs of the world's antiquity and of the errors of Scripture. Thus, the heavens, which narrate the glory of the Most High to all men, and whose language is heard by all peoples, say nothing to the unbeliever. Happily it's not the stars that are silent; it's the atheists who are deaf."[24] The pretenders to zodiacal antiquity must accordingly all have been deaf unbelievers.

The *Génie*'s tremendous public success, coming just in time to support the reconciliatory intent of the Concordat, led to Chateaubriand's appointment by Talleyrand as attaché to Rome—rather to the author's disappointment, who had hoped for a position with true governmental influence. In Rome he caused sufficient trouble to be recalled, whereupon he was sent to the Swiss canton of Valais. This too did not last long, for he resigned on learning of the duc d'Enghien's assassination by Napoleon's agents.[25] Returning to Paris, Chateaubriand began writing for the *Mercure*, one of the two principal organs of the religious Right; it had been in the hands of Louis de Fontanes since early in the Consulate. A poet, author, and something of a libertine in his youth, Fontanes had supported moderate monarchic change early in the revolution and had left Paris for Lyon to marry the wealthy Chantal Cathelin. He and his wife managed to flee ahead of Fouché's massacres. After Thermidor they returned to Paris, only to leave France again in 1797, decamping for England after the failed conservative coup of Fructidor.

While on the other side of the Channel, Fontanes became close friends with Chateaubriand. Returning to France in 1798, Fontanes decided that the future lay with Bonaparte. This decision was no doubt helped along by his having taken Napoleon's outspoken sister, Elisa Bacciochi, as his mistress, who had introduced him to her brothers Lucien and Napoleon. Monarchist in outlook and religious in sentiment, Fontanes capitalized on his connections to the Bonapartes to turn the *Mercure* into a major organ for conservative propaganda. In 1802 he was elected to the Legislative Corps, then in 1803 to the Académie Française, and a year later he took over the legislature's presidency. He had Chateaubriand removed from the list of proscribed émigrés, and on his friend's return to Paris Fontanes introduced him to literary and political society.

Figure 7.2. Fontanes (*left*), and Chateaubriand

Chateaubriand began also to write for the *Mercure*. From the outset the journal pushed the envelope of acceptability, and Fouché made certain that Napoleon knew about it. His report to the emperor on October 6 noted that "diverse representations had been made to the *Mercure de France* concerning the bad spirit that habitually dominates this sheet, on the stubbornness of the editors in attacking the principles at the base of the government and in embittering spirits with dangerous discussions."[26]

Chateaubriand and Fontanes had charge of the *Mercure* (which Chateaubriand eventually bought from Fontanes for twenty thousand francs, only to have it expropriated by Napoleon—who however paid him the full price), while the *Journal des Débats* was in the hands of Joseph Fiévée from 1804. Fiévée, a royalist, had emerged after the fall of the Directory to literary fame and to become a clandestine informant to Napoleon himself. He nurtured the hope, as did many others who shared his monarchist views, that the first consul would facilitate a restoration, if not of the Bourbons, then at least of the influence and power of property and religion, the two not being altogether different in the minds of conservatives. To that end he warned Napoleon about the comparative dangers of Jacobinism, a refrain he harped on in his *Journal*. Fiévée tried also to insist on the introduction of a formal structure for censorship, in contrast to Fouché's more limited approach, which was to deal only with politically dangerous material. This more rigorous and extensive censorship, naturally, would not extend to Fiévée's highly vocal and deeply conservative journal. This irritated Fouché. *Débats*, he reported to Napoleon on October 19, "perseveres audaciously in preaching intolerance and the proscription of all the men of the revolution. The good articles in it were inserted by order." Fouché was himself among the "men of the revolution" that the authors no doubt had in mind.

Napoleon's self-transformation into an emperor on December 2, 1804, eventually altered his tolerance for the scribblings of the *Mercure* and *Débats*, but not only for reactionaries. He had decided the year before becoming emperor that the idéologues were dangerous for their republicanism. Moreover, with unusual prescience he had little respect for what was evolving into the incipient social sciences. In fact the very name "idéologue," with its pejorative whiff of the unthinking partisan of system, was given them by Napoleon himself. They themselves preferred to be called *idéologiste*, suggesting a scientific analyst of social matters. Napoleon had accordingly abolished the Third Class of the Institut National in 1803, which had been the idéologues' stronghold, with its members scattered to other areas. He nevertheless retained some affection, if not respect, for these partisans of *idéologie*, who occupied significant positions and continued for the most part to publish with comparative freedom.

Censorship during the Consulate and empire was usually imposed only on books, plays, and newspaper articles that advocated or seemed to advocate open opposition to the regime. Napoleon, whose own literary talents were hardly insignificant, respected literary and artistic work and continued to hope that under his administration the arts would flourish. But his liberality always stopped at two points: he would brook no overt challenge to his power—always subject to doubts concerning his legitimacy as a Corsican product of revolutionary times—and he had no respect whatsoever for the intellectual talents of women, whose rights were if anything even further abridged by the Code Napoléon. Mme de Staël especially annoyed him, though even her critical works were not enjoined until the very last one, *De l'Allemagne* (*On Germany*), in 1810 went too far in its Romantic admiration for contemporary German culture, in particular what she convinced herself was its love of liberty.[27]

The ceremony that transformed the first consul into France's emperor was tightly controlled and choreographed down to the least detail. Pope Pius VII was brought to Paris for the occasion, and Notre Dame was restored to its prerevolutionary glory. Although the pope did not crown the new emperor (that would have signified submission to Rome) nevertheless the ceremony was replete with religious symbolism. It began with a Latin hymn, the millennium-old *Veni, Creator Spiritus*—a hymn to the triune deity of the Athanasian creed—sung the moment that Napoleon and Josephine, now the imperial consort, together entered the sanctuary. Mme de Staël, a Protestant and moderate believer herself, wrote in revulsion that "ecclesiastics of all ranks never let pass an opportunity to praise Bonaparte in their fashion, that is to say by calling him the envoy of God, the instrument of his decrees, the representative of Providence on earth. The same priests no doubt later preached a different doctrine. . . . The catechism received in all the churches during the reign of Bonaparte threatened eternal punishment to whomever would not love or defend the dynasty of Napoleon." She concluded with a remark that seems particularly prescient: "In truth, only those nations have sincere piety where the doctrine of the church has nothing to do with political dogmas, where priests exercise no stately power, where, finally, one may love God and the Christian religion with all one's soul, without thereby losing and above all without obtaining the slightest earthly advantage."[28] One may doubt whether any form of superstitious sanctimony deserves to be called "pious," but alliances between church and secular power have inevitably proved to be unstable.

The public binding between church and empire as expressed in the Concordat was not, in the end, especially durable. It did not spring from any deeply held conviction on Napoleon's part but was primarily a matter of political convenience. Within a year, even Napoleon had become uncomfortable with the Right's overt monarchism, and especially with its stridently antiphilosophe tone. His own legitimacy was after all a final product of the revolution. More to the point, the continual harangues from reactionaries raised the specter of factionalism, which led Napoleon to pay heed to the newly restored Fouché's warning about counterrevolution. Angered by the *Journal des Débats'* monarchist gloss, Napoleon had the name changed to *Journal de l'Empire* a year after becoming emperor, despite attempts by Fiévée to placate him—and Fiévée was now responsible for ensuring that it did not again cross the line.

Chastized for Atheism, Lalande Redates the Zodiac

Near the time that Napoleon shackled *Débats*, he also became irritated with Lalande's increasingly visible atheism. This had reached truly magnificent proportions with the publication of Lalande's continuation of Maréchal's *Dictionnaire des athées* (*Dictionary of Atheists*) in 1803, the last year of the Consulate. There the new emperor could have read that "The true atheist is not to be found among those bloody and hypocritical heroes who, in order to open the way to conquest, advertise themselves as the protectors of religions to the nations that they propose to dominate, and amuse themselves, in the bosom of their families, with human credulity."[29] Napoleon was no doubt not a "true atheist" by these standards; he was only a "bloody and hypocritical" conqueror. Little wonder that the day after Christmas in 1805, Napoleon wrote a letter specifically about Lalande to the minister of the interior, at that time Jean-Baptiste de Champagny. The letter was read to the assembled classes of the institute two days later, with Lalande present. Humiliatingly, he was not only denounced for his atheism but threatened vaguely with punishment for undermining public morale.

"It's with sadness," wrote the emperor to Champagny,

> that I learn a member of the Institut, celebrated for his knowledge, but fallen today into infancy, has not the sagacity to hold his tongue and looks to speak, sometimes through statements unworthy of his former reputation and of the body to which he belongs, sometimes in professing atheism, a principle destructive of all social organization, which takes from man all his hopes and consolations. My intention is that you gather the president and the secretary of the Institut and that you charge them with making known to that illustrious body, which I have the honor to be a member of, that it must send word to M. Lalande, in the name of the body, that he is not to print anything further, and not to darken in his old age what he did in youth to obtain the esteem of the savants, and if these fraternal invitations are insufficient, I would be obliged to remember that my first obligation is to prevent the poisoning of the people's morale. Because atheism is the destroyer of all morality, if not among individuals, at least among nations.[30]

The president of the institute, the botanist René Desfontaines, reported that poor Lalande, "present at the meeting, prays the Institut to receive his declaration that he will conform entirely to the intentions of His Majesty."

Napoleon's extraordinary and peremptory order apparently had its effect, perhaps more than "His Majesty" anticipated, since the aged and publicly chastened Lalande expired a mere eighteen months later. Possibly Lalande had previously irritated Napoleon, since the newly crowned emperor had almost certainly read the report of his secret police for July 18, 1804. There he could read Lalande's purportedly sarcastic remark concerning a new creation of which Napoleon was particularly proud, his Légion d'Honneur, that among the "12,000 newly discovered stars, not one is fixed." Two days later the ever-vigilant Fouché, who likely shared Lalande's views about the Légion, added a palliative note to the effect that Lalande had not been responsible for the remark since he had been at Nevers for some time, and that "since the sarcasm in question had a scientific color, it was natural to attribute it to the chief clerk of astronomy."[31] Late the previous year (1803) had seen the appearance of Lalande's final views about zodiacs, which had been prompted by the publication of Testa's screed and Visconti's remarks in Larcher's *Herodotus* in 1802; the remarks were printed in two separate publications near the same time.[32]

The old astronomer and insect gourmand may have been averse to keeping his religious views private, but he was neither so foolish nor so self-abandoning as to continue loudly to oppose religion in the face of certain punishment. It is scarcely surprising to find that he had also changed his mind about zodiacal antiquity, though he barely moderated his disdain for believers. All he said about Testa, whom he knew (and who may well have sent him his dissertation), was that he "undertook to prove that [the zodiac] does not go back more than 300 years before the common era."[33] He gave Larcher a bit more attention. In contrast to Testa, Lalande continued, Larcher had produced "a sortie against unbelievers who carry one of the Egyptian zodiacs back 6,000 years . . . and his only reason is that this would go back 217 years before the creation of the world according to Petau, as if the creation were susceptible of calculation." So much for Larcher.

With Visconti Lalande was more careful. He did not challenge Visconti's claim that the architecture of the Dendera temple seemed to be Greek in style. Instead, he first revised his earlier agreement with Burckhardt concerning the Dendera rectangular's age. "For myself," he now wrote with unaccustomed reticence, "I remarked from the engraving by M. Denon, that Cancer is figured in the two lines, at the head of the descending signs and at the end of the ascending ones; which proves that the solstice was near the middle of Cancer, and that goes back 3,000 years." Note Lalande's claim concerning the position of Cancer in respect to the "ascending" and "descending" signs; we will return to it shortly. Burckhardt had asserted four thousand, not three thousand, years, and Lalande had in 1800 repeated the claim—a date uncomfortably close to the biblical deluge (circa 2500–2300 BCE) for the ancient Egyptians to have evolved such a sophisticated civilization after the nearly total annihilation of humanity. At the time, however, Lalande had not seen Denon's drawing, which he now examined in the light of Visconti's remarks. Eudoxus, Lalande continued, had described a sphere "according to an ancient tradition that came from Egypt or India." And so it would have been "altogether natural" for the Dendera zodiac to have been produced in Grecian times but to have reflected a much older tradition. This was a move that others would later make as well, albeit embellished in various ways.

The most striking thing about Lalande's final published words on the zodiacs is their circumspection. He did not recall Dupuis' claims, as he had in 1800, and neither did he say anything more about Esneh beyond sarcastically noting Larcher's reliance for critique on the purported date of creation. Considering Lalande's usual vigor, and the fact that just the year before he had gone on record as a devoted atheist with an entire panoply of critical remarks about believers and those who exploited the credulous, it is unlikely that his newfound caution arose entirely from a disinterested weighing of the evidence. In fact, there is unmistakable evidence that Lalande could not have changed his mind on the basis of any new or old data at all.

On December 4 Fourier permitted the astronomer to see the expedition's own drawings of the zodiacs provided that he "made no use of them before him [Fourier]," according to a note that Champollion-Figeac possessed in Lalande's hand. "I saw them on the fifth with the cortège of the emperor," Lalande had written, which inclined him "to change the calculation of M. Burckhardt."[34] This would seem to indicate that he had changed his mind on seeing the expedition's drawings, which were more exact than Denon's. That cannot be what happened. The differences between Denon's and the expedition's depictions of the Dendera rectangular, and there are several, are utterly beside the point for Lalande's new reasoning, which moved the Dendera rectangular's date forward a millennium, well out of the biblical danger zone. If this same drawing had been available to Lalande at the outset, nothing would have been different. His belated change of mind more than likely reflects his perception of the dangers in transgressing religious sensibilities.

The climate was nevertheless beginning to shift. At first, the change was subtle; from the outside, nothing much appeared to have altered. For instance, despite the emperor's warning to Fiévée concerning the tone of the newly named *Journal de l'Empire*, its publication continued apace with him at the helm. But then, in 1807, the emperor replaced Fiévée altogether, and the journal ceased to blame philosophie for all the world's ills. At very nearly the same time, Chateaubriand pushed the *Mercure* too far, having written an article that likened Napoleon to Nero. That astonishingly ill-advised comparison resulted in his removal from the journal, followed by the *Mercure*'s amalgamation with its arch-enemy, the *Décade philosophique*, thereby neutralizing two opposing and mutually opposed irritants at once. The combined journal, which kept the *Mercure*'s title but not much else, leaned more toward philosophie than to the religious Right.[35]

Sometime after 1805, then, the atmosphere had begun to change, though Lalande did not live long enough to see much of it, and in any case Napoleon would not have tolerated overt expressions of atheism. In 1806, just as it was becoming comparatively safe to return to discussions with religious consequences, Dupuis entered the fray, publishing two pieces on the zodiac during that year. Although Napoleon was surely aware of Dupuis' views, which he would have known at least since the Egypt expedition, his reaction, if he had one, was not significant enough to merit recording. Dupuis had long been well-known to the first consul, since the two men had for years been entangled in a series of political involvements. After Brumaire, Dupuis had been elected to the Legislative Corps for Seine-et-Oise, had become the legislature's president, and had then joined the Senate. Shortly thereafter Napoleon awarded him one of the very first Légions d'Honneur—not a prize that the first consul would bestow on a scholar whose views he thought to be pernicious.

Napoleon did nevertheless keep tabs on Dupuis' atheism in various ways. At some point during the Consulate, a deputation from the Institut National visited the first consul. Although the date of the visit was not recorded, it likely took place between August 2, 1802, when Napoleon became consul for life, and the spring of 1804, when his transmutation into emperor began. The most probable date would be about the time that Napoleon decided to eliminate the institute's Third Class containing the highly vocal minority of intensely antireligious idéologues; this occurred formally on January 23, 1803.[36] Taking explicit notice of Dupuis' presence among the members of the deputation, Napoleon said to him, "Is it not true, Monsieur Dupuis, that Jesus Christ never existed?" To which Dupuis replied, "Sire, that is my opinion." In view of the fact that the Concordat would be celebrated at Notre Dame the following April, Napoleon's question was anything but innocuous. Neither was this the last time that Napoleon brought up Dupuis' views in front of others. One evening Dupuis was invited to dine with the first consul and a group that included the liberal constitutional prelate, the Abbé Grégoire. Though scarcely a friend of the nonjuring émigrés who were now being let back into the country, Grégoire was no unbeliever in fundamental Christian doctrines. Napoleon noticed Dupuis and Grégoire, seated next to one another, engaged in friendly conversation. Apparently surprised to see the two on such affable terms, the first consul asked Grégoire, bishop of Blois, how it could be that Dupuis' opinions didn't "raise a wall of separation between him and the author of the *Origine des Cultes.*"[37] No less forthright in Napoleon's presence than Dupuis, Grégoire answered, "That might have been the case, had not Dupuis and I had a religion in common; that's the religion of the Republic." A provocative reply, since the astute bishop could scarcely have been unaware of the first consul's dynastic intentions.

Dupuis' Final Words

Dupuis was supposed to accompany the Egypt expedition but had not done so. His interest in its results was, for obvious reasons, extremely keen, and there is little doubt that he had seen the various reports on the zodiacs almost immediately. Yet he published nothing about them until 1806, nearly four years after Visconti's and Testa's efforts to counter claims for their great age—and so to counter as well the connection that had been made between them and Dupuis' theory in the *Moniteur* letter by Fourier. One of the two pieces that he printed that year discussed the chronology and mythology of the zodiac. Although quite long and replete with elaborate cross-cultural comparisons with India and China, it was essentially a reprise of arguments in Dupuis' *Origine.*[38] The other is a fourth the length and directly concerns the points at issue, being more in the way of a polemical reply to Visconti and Testa.[39] It could have been written reasonably quickly once the critiques had appeared, and, given the issue's importance for Dupuis, probably was, at least in draft form. Yet neither was printed until several years later.

Dupuis would have been quite aware that Napoleon was paying attention to his work, given their encounters, and he would also have known just how sensitive Napoleon was at the time to public statements that might trouble religious waters. As a member of the Institut National, Dupuis would surely have been there when the letter castigating Lalande for atheism had been read. However, by 1806 the climate had softened, primarily because the Right's continual harangues were becoming dangerous in these early years of the newly founded dynasty, whose suspect legitimacy had its seat, if anywhere, in the consequences of the revolution. We have seen before that Dupuis knew how to detect which way the wind was blowing, even if his performance when in office was at best lackluster. Delaying publication until more propitious times would have been entirely consistent with his behavior during and before the revolutionary years.

Dupuis tended always to visualize representations of the heavens, for he was not comfortable with the aridity of astronomical computation. Recall how he had had his friend Fortin construct a precession globe that could be manipulated to test his theory. He envisioned something similar here, in reacting to the several claims concerning the Egyptian zodiacs. He discussed only the rectangular zodiac at Dendera, ignoring the circular, not least because only Visconti had as yet (1806) commented on the latter, which was much harder than the rectangular to interpret.

Working from the Denon drawing (there was no other available to him), Dupuis insisted that the Dendera rectangular had to be viewed in a very specific way. Take your celestial globe, Dupuis instructed, and adjust it so that the north pole sits at 26 degrees above the horizon, which will place you at the latitude of Dendera. Next to the globe place your copy of Denon's drawing, and orient it as in Figure 7.3, with the feet of the sky goddess at the bottom. Now, he continued, imagine yourself to be at the celestial globe's center looking outward with your face turned toward the north. The local meridian runs north-south along the globe. If you now turn your celestial globe so that the tail of the constellation Capricorn lies on the meridian, then it will divide the zodiacal constellations into two sets in just the same way that they are divided in the figure, with Aquarius to your right and Sagittarius to your left (remember, you are at globe center—see Figures 7.3 and 7.4). To Dupuis' way of thinking, then, the two columns of the Dendera rectangular must be oriented with the sky goddess's feet at the bottom in order properly to grasp their designer's intention.

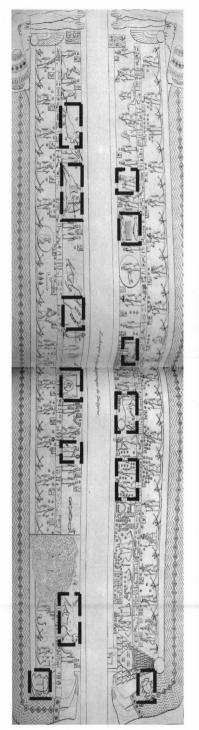

Figure 7.3. Dupuis' understanding of the Dendera rectangular, based on Denon's drawing (Figure 5.4). The outlined regions in the "left" column contain, from the bottom, Cancer, Leo, Virgo, Libra, Scorpio, Sagittarius, and Capricorn. The "right" column regions contain (again from the bottom) Cancer, Gemini, Taurus, Aries (just above the fold), Pisces, and Aquarius. Note the unequal spacing between the signs. According to Dupuis, the left panel designates the western, and the right the eastern, segments of the ecliptic above the horizon at the time and place where the zodiac was carved.

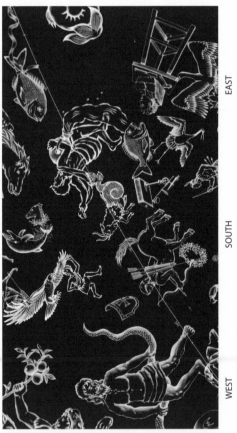

Figure 7.4. A representation of what Dupuis envisioned. The dotted line diagonally across the center traces the ecliptic.

Dating requires knowing the position of a solstice or an equinox with respect to a constellation. To that end, Lalande had asserted that Cancer is depicted "at the head of the descending signs and at the end of the ascending ones." Which, he had concluded, "proves that the [summer] solstice was in the middle of Cancer."[40] In itself this proves nothing of the kind because twice a day this configuration always occurs no matter what the year might be. Lalande had, however, further assumed that the monument was designed to represent the daily course of the sun, and he had decided that the goddess's feet, where Cancer is depicted, represent the sun at noon, with the succeeding signs in either panel then being in rise or set.

Dupuis felt that Lalande had "rather supposed than proved" his claim. He, Dupuis, would instead "suppose nothing." He would leave "what's represented at the top of the monument, at the highest point." And "what's represented at the feet of the figures, will remain lower." Like Lalande, he assumed that the monument does represent the daily course of the sun at some time. But unlike Lalande, Dupuis insisted that the goddess's feet must correspond to the sun's lowest diurnal position below the horizon, and the goddess's head to noon (not depicted, but at top in Figure 7.3). He himself would now "seek indices that can lead us to fix the place of the summer solstices, and that of the equinoxes, to discover and not to suppose them, as most of our astronomers have done." To do that, he would resort to what was known about "the science of the hieroglyphic language." Even though "we are little advanced in our knowledge of it," nevertheless Dupuis felt confident that "after thirty years of work, we know enough to be guided in that research, and to resolve a problem that astronomy alone cannot."[41]

These were brave words indeed for 1806. While recent efforts to use the Rosetta Stone for deciphering hieroglyphs seemed to have borne some fruit, the script remained a mystery. In 1802 Johann Akerblad, a Swede, had published a seventy-page letter to De Sacy on the Rosetta inscription. De Sacy had been a member of the Académie des Inscriptions, and then of the Institut National from its inception in 1795, and was certainly known by Dupuis. In 1799 he had also published a treatise on "universal principles" of grammar. Akerblad, a diplomat by trade, knew Coptic, the language spoken by Egyptian Christians that had long been thought the final version of ancient Egyptian proper. The Coptic script itself, though Greek for the most part, also employed seven symbols from Egyptian demotic, a very late script that had not been widely used for monumental inscriptions. By comparing the demotic with the Greek in the light of Coptic, Akerblad and De Sacy had identified a number of words, including personal pronouns. Both had assumed that, like Coptic, the demotic was alphabetical through and through, which it is not. Neither, however, had pushed ahead with hieroglyphs proper, and both continued to embrace the long-held assumption that the hieroglyphs were a symbolic aracana and could not be understood otherwise. Dupuis actually did little with Akerblad's and De Sacy's work. He used it instead as a sort of rhetorical flourish, to support claims he intended to make based directly on his own understanding of the symbols on the Dendera rectangular.

Dupuis decided that the zodiac actually represents the state of the sky on the very day, indeed at the very hour, of its depiction. Since the head of the goddess represents the zenith, and her feet the region far below the horizon, the two panels in between must depict the constellations from the diametrically opposite points east and west. Halfway down from the zenith—at the fold in Denon's depiction (Figure 7.3)—Dupuis spied the horizon. The figures from that line on the right-hand panel accordingly depicted the constellations that rose from the eastern horizon to the zenith, and those on the left depicted the constellations that were headed to the west, all at a particular moment. But at what hour, and on what day?

Significant moments of the day were always sunrise and sunset, so Dupuis looked for an emblem of the sun near the horizon. He found it just above the fold near the symbol for Aries in the image of a child sitting on a lotus leaf (Figure 7.5). In his *De Iside*, Plutarch had written that the Egyptians always depicted a child on an "aquatic plant to represent the sun rising from the breast of the waters."[42] And so the rectangular zodiac must surely depict the moment of sunrise. But on what day of the year? The lotus-seated child appears near the eastern horizon along with Aries the Ram, while Capricorn rules the zenith, in which case the day had to be one on which the sun rose in Aries. At this point Dupuis again reached back to his deep knowledge of classical texts. In his *Stromata* (*Miscellanies*), Titus Flavius Clemens, better known as Clemens Alexandrinus, an early Greek theologian, had written that the spring equinox in Egypt was symbolized by the sparrowhawk. That bird appears near Aries seated on the head of a dog, who is drawn back-to-back with a goat. According to Virgil in the *Georgics*, the goat was said to fix the spring equinox by the sun's rise, while the dog (Canis Major) did so by the sun's set.[43] And so, Dupuis concluded, the Dendera rectangular must represent six in the morning on the day of a spring equinox in Aries![44]

Figure 7.5. The sun next to Aries—the lotus-sitting child at lower right, with the sparrow-hawk atop a dog backed by a goat at upper left

Dupuis had dispatched the questions of the day and hour, but one last mystery remained, the most important one of all: What year? This too could be answered with appropriate attention to symbolic meaning. With Aries in the east at the spring equinox in the middle of the rectangular zodiac, the winter solstice would have to be above the horizon on the meridian, putting it at the top of the zodiacal column. That could be used to fix the year, so Dupuis looked to see whether an appropriate symbol could be found there. At the top of the column, near Capricorn, Dupuis spied "a man who walks having only one leg, or whose legs are so united that they make but one." That symbol, he asserted, signified the winter solstice according to Horapollo.[45] Here Dupuis was misled by Denon's drawing. Denon had indeed depicted the united-leg figure that Dupuis claimed, if we except the likelihood that even in Denon's drawing the figure's protruding chest suggests a female rather than a male. However, the actual symbol clearly shows legs parted, not united (Figure 7.6). This is particularly ironic because Dupuis had manipulated Horapollo's words to fit the incorrect Denon figure. Horapollo had in fact written that a figure with "two feet together and standing means the course of the sun at the winter solstice."[46] These words describe the actual symbol but not Denon's drawing, so it seems that Dupuis had unnecessarily massaged the original text to fit an incorrect depiction, when a more accurate drawing, which he had not seen, would have worked perfectly well. That raises questions about just how reliable the stupefyingly lengthy correspondences of his *Origine* were—it is likely that no one has ever read entirely through them all—but, even

Figure 7.6. Denon's single-legged figure (*left, at upper right*) and the same figure by Jollois and Devilliers

after Jollois and Devilliers' more accurate drawing became available, the mismatch between Dupuis' and Horapollo's words remained unremarked, at least in print.

With this Dupuis had the winter solstice in Capricorn, and so the summer solstice had to be in Cancer, near which he naturally found an appropriate symbol—the "pyramid" of sunlight, for that tallest of all ancient buildings obviously symbolizes a moment when the sun reaches its highest point in the sky (Figure 7.3, lower right). And if the solstices must be in Capricorn and Cancer, then the year that corresponds to the Dendera rectangular must lie somewhere between 2548 and 388 BCE. If, as Dupuis thought likely, two important astronomical designators called the colures[47] ran through the middles of the constellations, the date would be about 1468 BCE —long before the period insisted on by Visconti, whose stylistic arguments Dupuis rejected, primarily on the grounds that he thought the Greeks had borrowed their architecture from Egypt.

One might think that Dupuis had now set to the side his own earlier links between the zodiacal constellations and the agricultural seasons since he did not use them in 1806. However, he was not aiming to establish the epoch of the zodiac's original invention, for which he continued to adhere to his argument based on agricultural and natural periods.

The question here was, instead, the epoch of the temple zodiacs, not the epoch of the "primitive" sphere when the seasonal connections had first been established. Dupuis had even suggested in his lengthy *Origine* a possible alternative interpretation of the associations, years before the Egyptian zodiacs were seen by the French savants. He had based his primary argument in the *Origine* on the constellation that *rises* with the sun during a given season at the latitudes of Egypt. But suppose that the Egyptians' had not looked to the eastern horizon just before the rising sun. Suppose instead that they had looked to the western horizon, to spy which constellation *sets* with the rising sun. Refiguring the seasonal links on that hypothesis would shorten Egyptian chronology by many thousands of years, in fact placing the origin of the zodiac only "two or three thousand years before the common era." Dupuis had not thought the possibility reasonable at the time of the *Origine*; neither did he mention it in his 1806 discussion of the Esneh zodiac, which he dated to about 4000 BCE. And he remained convinced that the original, "primitive" sphere dated back over twelve millennia.[48] His alternative associations would nevertheless prove quite useful a few years later in the hands of Fourier and Jollois and Devilliers when they produced a revised dating for the temple zodiacs.

Dupuis likely nourished a desire to reply to Larcher as well as to Visconti and saw this as a fine opportunity to do so. By reframing the antiquity of the carved zodiac as a matter of some small, unthreatening span of time rather than millennia, Dupuis was subtly reminding the reader that Visconti's remarks had by contrast appeared in Larcher's *Herodotus* following "a violent sortie [by Larcher] against unbelievers who make the world a little older than he does in his chronology." Larcher's aim, Dupuis knew, was to "forearm the public against the trickery of professed unbelievers," among whom Dupuis certainly figured. He would also have been aware that the climate was still not favorable to the overt embracing of positions that would enrage believers. Although a date of 1468 CE placed the zodiac long before Greco-Roman times, nevertheless it did not push close even to the Flood, much less to a time before creation itself. "For me," Dupuis judiciously (if misleadingly) remarked, "it matters little whether the world be old or young, I gave the zodiac an age that I believe it really has."[49]

These were Dupuis' last printed words on this or any other subject, since he died three years later, in 1809. The year after his death, the first officially sanctioned publications of results from the expedition were at last available to the public, results that again raised the politically unstable issue of time. Not long thereafter, time itself confronted the zodiacs when the naturalist and developer of the new science of comparative anatomy, Georges Cuvier, published a discussion of them in his influential four-volume *Recherches sur les ossemens fossiles de quadrupeds* (*Researches on the Fossil Bones of Quadrupeds*). For Cuvier and indeed other naturalists of the day, the question was not the date of creation, which few among them thought to be recent, but rather what the zodiacs had to say about the span and events of human history, for that was the central point at issue between the savants and those who opposed them. Cuvier's *Recherches* appeared in 1812, two years after the first volumes of the *Description de l'Égypte*. Those volumes contained Fourier's influential *préface*, which discussed the aims and results of the expedition, together with Jollois and Devilliers' circumspect analysis of the zodiac. It contained as well the efforts of another member of the expedition, the engineer and geographer Edmé Jomard, to find traces of lost knowledge in the remnant stones of Egyptian antiquity. We turn first to Fourier, his colleagues on the expedition, and to Cuvier as they grappled with the perplexing, and politically dangerous, zodiacs.

8

Egypt Captured in Ink and Porcelain

Jean Joseph Fourier

When the remnants of the Egyptian campaign returned in the early fall of 1801, they found a France effectively controlled by Napoleon as first consul. Many changes in the political and social order were already under way, changes that soon affected Fourier. He had returned at first to his teaching duties at the École Polytechnique, but not for long. On February 7 the first consul wrote a letter to Berthollet asking him to inform Fourier that he, Napoleon, wished Fourier to take the newly vacated position of prefect of the Department of Isère at Grenoble, 350 miles southeast of Paris. The prefecture had not done well under the recently deceased Ricard de Séalt, whereas Fourier's organizational abilities had been amply demonstrated in Egypt. Berthollet reached Fourier via Monge, and the new prefect was in place by mid-April.[1]

Jean Joseph Fourier was the ninth of twelve children born to his father's second wife. When Joseph's mother died in 1777 his father, no doubt overwhelmed by the thought of having to care for so many children, abandoned the two youngest to the local Foundling Hospital before expiring himself the next year. The ten-year-old Joseph nevertheless did well, first at a local preparatory school in Auxerre and then, in 1780, at the regional École Royale Militaire, which had been placed under the direction of the Benedictines. Here his interest in mathematics grew, though he excelled particularly in rhetoric. Never in the best of health, the young Joseph was ill from late 1784 until the following November, but success in his studies sent him to the Collège Montaigu in Paris, after which he returned to Auxerre to teach mathematics—and to enter the church, as he began preparations to take his vows at the Benedictine Abbey of St. Benoit-sur-Loire.

At first indifferent to the developing political situation, Fourier was primarily interested in the latest developments in mathematics, physics, and astronomy. He had already written a short paper on algebra that he had rather brazenly asked his former teacher, Bonard, to forward to eminent mathematicians of the day, anxious to learn their opinion of it. Then, on November 2, 1789, news arrived that the Constituent Assembly had expropriated church property. Fourier was to have taken his vows three days later, but he never did; whether because events had made it too difficult or because he refused to do so remains unclear. A month later he visited Paris, where he read his paper on algebra to the Académie Royale des Sciences, returning thereafter to Auxerre as assistant to Bonard at the École Militaire. The Benedictines were permitted for a time to continue their direction of the school despite the fact that clerical orders were formally suppressed on February 13, 1790.

Although two men were murdered during a mob riot in August 1792, Auxerre was spared the guillotine and the excesses of revolutionary tribunals. Nevertheless, a militant Jacobin club arose in the town, indeed among the most militant ones of all, the Society of the Friends of the Republic, or Patriotic Society, founded by Michel Lepelletier. Immensely rich, and originally president of the Paris parliament, Lepelletier became increasingly radical and voted for the king's death, having earlier proposed the general substitution of beheading for the rope in an excess of concern for the pain of the condemned. The evening following the king's execution, Lepelletier was assassinated at the Palais Royale under somewhat mysterious circumstances.[2] His Patriotic Society continued after his demise, and it had a definite Hébertist cast, putting it dangerously to the left of Robespierre.

In the spring of 1793, when the country was riven by defeats in Belgium and rebellion in the Vendée, Fourier was "offered" a place on Auxerre's Committee of Surveillance. This was the first step down a treacherous path. He was sent in June to a nearby town, Avallon, to draft men into the military to counter the Vendéens, and thereafter to other departments as well. By July the twenty-five year-old Fourier had developed something of a taste for Jacobin sentiments at a time when the Republic was in extreme danger. However, by September, when the Committees of Surveillance had became a part of the apparatus of Terror, Fourier tried to withdraw from the one in Auxerre but was reproached for lack of zeal and forced to remain in place. In October he was sent to another town to enforce the draft but was deflected to collect horses for the war. And here, like so many of his contemporaries, Fourier was compromised by revolutionary chaos.

Along the way Fourier had stopped in Orléans, where he became involved in a dispute. Local bourgeois, supporters of the Girondins, were at loggerheads with small-tradesmen and artisans, all Jacobins. Aggravated by bread shortages, inflation, and the draft, the situation became acute in mid-March, when the representatives on mission were verbally assaulted, and one among them, Léonard Bourdon, was wounded in the town center. On March 18 the convention voted Orléans in rebellion. D'Herbois, who six months later would preside with Fouché over the massacres at Lyon, reached Orléans in the company of an ex-Benedictine by the name of Laplanche, and the town fell into the hands of the Jacobins. After Marat's assassination in July, the thirteen men who had wounded Bourdon were guillotined. Laplanche returned to Orléans on September 1 and in an orgy of foul language, publicly disgorged abuse upon the city's administration.

Although arrests followed Laplanche's diatribe, he himself soon took a strong dislike to the local sansculottes, perhaps for personal reasons involving his father-in-law, or perhaps because he was calmed by the submissive attitude of the town's merchants. When Laplanche returned on October 11, the radical Patriotic Society did not greet him, and two days later he began to suppress them, ordering arrests, at which point Fourier arrived on the scene.

Despite Laplanche's newfound antagonism to the Jacobins, Fourier went before the Patriotic Society to defend three of the sansculotte leaders, convinced, he later wrote, that they were innocent of the charges against them. He was as a direct result denounced to the person in charge of the representatives, Pierre-Louis Ichon, himself a constitutional priest who had earlier taught theology at the Collège de Condom and who had voted for the king's execution.[3] The denunciation reached the Committee on Public Safety, effectively compromising Ichon, who had sent Fourier on mission. On October 29 the convention revoked Fourier's appointment and forbade him to exercise similar responsibilities in future.

The queen had been executed barely two weeks before, and the Gironde was to be obliterated two days later. Ichon, understandably if inexcusably concerned with his own neck, came close to ordering Fourier's arrest and guillotining. Fourier, however, managed to return to Auxerre, where the Patriotic Society, as well as Nicolas Maure, a spice merchant and fierce Jacobin who became the deputy of the department at the convention, protected him for a time.[4] The following June (1794), Fourier's situation had seemingly improved locally to such an extent that he became president of the Auxerre Revolutionary Committee. Shortly thereafter Fourier visited Paris, where he met Robespierre at the Jacobin Society.

This meeting with the Apostle of Virtue did not, it seems, go well, because Fourier was arrested on his return to Auxerre on the orders of the Committee of Public Safety. The reason was his support of the Orléans sansculottes, whom Robespierre had himself criticized following the fall of the Hébertists, and who were considered to be dangerously out of control. Locals in Auxerre soon interceded, and Fourier was released, only to be reimprisoned eight days later, on July 17. For the next ten days he surely anticipated the guillotine's blade. But on July 27, the ninth of Thermidor, Robespierre fell, and Fourier was released fifteen days later.

Fourier's brush with death resulted from his association with Jacobins, though its proximate cause was likely his honest conviction that the Orléans leaders were not guilty of Laplanche's accusations. Though he was certainly no Fouché, Fourier had nevertheless headed the Revolutionary Committee in Auxerre, and there is little doubt that he had been involved in detentions in town. His situation changed markedly, at least for a time, when he was nominated as a student to the École Normale in Paris, which ran from December 21 to the following May 30. There he met, among others, Monge. But his Jacobin associations in Auxerre returned to haunt him.

By mid-March Fourier learned of rumors against him—that he was a drunkard, an embezzler, and a supporter of tyranny. These reached the National Convention itself on March 20, though they were apparently never brought formally before it. "It is only too true," the accusation read, "that Balme and Fourier, pupils of the department of Yonne, have long professed the atrocious principles and infernal maxims of the tyrants. Nevertheless they prepare to become teachers of our children. Is it not to vomit their poison in the bosom of innocence."[5] Two months later Balme and Fourier found themselves formally listed as responsible for "tyranny." Maure intervened one last time on Fourier's behalf to prevent the suspension of his stipend at the École Normale, though he was himself soon implicated in the Romme conspiracy and died by his own hand.

Nevertheless, in early June, Fourier was arrested for terrorism by one Bayard, who, when the concierge at Fourier's residence voiced the hope to see him back soon, encouragingly replied, "Come and get him yourself—in two pieces." Fourier was released a few days later and then rearrested. From prison Fourier wrote in his defense that "no one in the commune of Auxerre was condemned to death or judged by the revolutionary tribunal at Paris; that no revolutionary tax was established of any kind whatsoever, that the property of those detained was never confiscated, that no cultivator, artisan, or merchant was arrested, that in what concerns me personally I believe that I introduced a moderation which I did not find in my adversaries, that far from having shared the revolutionary madness of many men I regarded it with horror." And finally, "I have done nothing arbitrarily and nothing that does not emanate from a law."[6] Fourier at worst followed "lawful" orders, a refrain that the events of the twentieth and twenty-first centuries have made all too familiar. And yet it seems that he had done his best even while inflated with Jacobin enthusiasm to act with justice and compassion, at least to the extent that anyone caught up in the chaos of Terror and rebellion can be said to have done so.

Fourier's eventual release remains unexplained, due perhaps to the rapidly changing political situation as the winds blew leftward once again, and as the young Napoleon was brought in by Barras to disperse royalist sympathizers in the streets of Paris. By September or October, Fourier was back in the city and had become a faculty member of the new École Polytechnique, then to join his fellow *polytechniciens* Monge and Berthollet on the Egypt expedition three years later. When he returned to France in the fall of 1801 with the remaining savants, Fourier rapidly became involved in plans to publish the expedition's scholarly results, not least concerning the zodiacs of Esneh and Dendera.

Organizing the *Description de l'Égypte*

During the months between Fourier's return to Paris from Egypt and his departure as prefect for Grenoble, plans were set in motion to compile and print the expedition's results. In late September the two inseparable friends, Jollois and Devilliers, had themselves left Egypt on a rickety Italian ship appropriately named the *Amico Sincero* (Sincere Friend).

Devilliers arrived first in Paris on January 7, while Jollois returned home for a time to Bourgogne. From Paris Devilliers wrote Jollois that Fourier had arrived just afterward, that "two or three diligences" filled with members of the Egypt Commission had also reached the city, and that discussions about collecting the results of the expedition had already begun. Monge and Berthollet, he continued, had their own plan in mind and wanted Devilliers to take charge of editing the expedition's history, which he did not want to do. The first consul had already charged Fourier with developing a plan to put the expedition's results together, regardless of what Monge and Berthollet might have had in mind. Fourier told Devilliers that on the tenth he would meet with Monge, Berthollet, and Laplace to "show them the inconvenience of the form they wanted to give the work," and to learn "whether they hold strongly to their ideas." He would tell Devilliers the results on the twenty-first.[7]

Fourier also brought up the matter of the zodiacs. In fact, Devilliers recounted, he posed a question "that, as it were, I solicited: in case the work of the Commission goes on for a long time [Fourier asked], would you [Devilliers and Jollois] want to publish separately the drawings of the zodiac, with my [i.e., Fourier's] memoir? (This would appeal greatly to him, it seemed to me.) I didn't want to give him the high sign [le dire tout haut] for fear of signaling the dismemberment of the work, but would this be agreeable to you? I didn't answer." As for the zodiac drawings themselves, there was the matter of Denon, whom Devilliers met on the eighth. "His large zodiac is not yet engraved," Devilliers wrote, referring presumably to the rectangular at Dendera since Denon did not draw the one at Esneh. But the Dendera circular was engraved, and "absolutely just as we saw it together at Dendera, that is to say very small, and as appeared to me, very incorrect. He didn't increase the scale of his drawings, but he put several on the same sheet. The text and plates are large in quarto."[8]

Denon's drawings were already eliciting scornful comments from the engineers. On the twelfth one of them (Fèvre, another young member of the expedition) wrote Jollois that he had spoken with De Prony, a senior member of the École Polytechnique, "of what we did in Egypt. I didn't forget to let him know about your work, above all on the zodiacs. He doesn't doubt that we'll provide drawings that will convey Egyptian architecture much better than the pretty pictures of Denon, and that all the works united will form a magnificent production."[9]

By the twelfth the project was afoot, with plans that called for a "historical" part by Fourier, descriptions of monuments with drawings, and diverse memoirs by members of the commission. Devilliers wrote to Jollois that a supervisory board had now been formed to take charge of the project, consisting of Fourier, the architect Legrand, the astronomer Delambre, the naturalist Cuvier, and "an [unspecified] antiquarian." In the meantime Fourier had become worried that Devilliers and Jollois might have taken umbrage at the report in the *Moniteur* that he was to draw and discuss the zodiacs, and so Devilliers had "calmed the disquiet he seemed to have that this would hurt you."[10] Shortly thereafter Jollois arrived in Paris and went to work with Devilliers.

The plan about which Devilliers had written did not last long. In February a formal decree establishing a publication project to be paid for by the government was issued, including funds for the contributors and editors. The chemist Chaptal, having just replaced Napoleon's brother Lucien as minister of the interior, had overall responsibility, with a committee consisting of Berthollet as president, and members Lancret (who had announced the Rosetta Stone's discovery), Conté, Monge, Fourier, Costaz, Girard, and Desgenettes reporting to him.[11] The first meeting took place on June 7. Conté was appointed the editor-in-chief, but he died of an aneurysm in 1805, following the deaths of his brother and his wife, after which Lancret took over, only to expire himself in 1807. Jomard, who saw the world almost entirely through measurement and mapping, became editor thereafter until the project ended in 1827.[12]

The direction of the enterprise became a full-time, and very expensive, affair. Those who had no other occupation were salaried, and the contributors were paid retainers. As of June 1, 1809 there were thirty-six people on the payroll, and by that date the project had spent 1,454,818 francs. More than three thousand reams of paper were estimated to be necessary for the publication, and the production of the copper plates was parceled out among four printers. Late in 1806 Napoleon demanded that sheets be printed off as soon as the plates were ready, and he further required that the work be ready by 1809. The volumes in fact appeared seriatim with dates from 1809 through 1824. The first volumes of the antiquities, three in all, which contained respectively memoirs (Fourier's *préface* to the entire work was separately printed), descriptions, and plates, actually became available only in 1810 despite the title page date of 1809, no doubt formally to fulfill Napoleon's demand. This first issue did not contain any plates of the Dendera zodiacs, but only the one drawn by Jollois and Devilliers at Esneh. The Dendera plates first became publicly obtainable in the fourth volume, which appeared in 1817. Each of the engraved

Figure 8.1. Jomard (*left*) and Fourier in Egypt

plates contained the names of the original drafters at the lower left, and the engraver's name in the middle or lower right. In one case (the Dendera circular) the original drawing was itself also engraved.

During the seven years that the early volumes of the *Description* were in preparation, the first wave of a craze for Egyptian antiquity swept through the country. Historians have aptly named the phenomenon "Egyptomania." Stimulated especially by the tremendous success of Denon's *Voyage*, wallpapers, interiors, furniture, buildings, even jewelry exhibited unalloyed enthusiasm for Egyptian motifs. The temple of Dendera itself was a special favorite. It was even fashioned as a pendulum clock by St. Montcloux et Cie in 1807. The clock sold "without movement" for 1,200 francs; with, for 1,500.[13]

Despite the military failure of his expedition, Napoleon vigorously exploited Egyptian imagery as an emblem of his empire. In 1804 Denon, now at the Louvre, urged the Sèvres porcelain works to produce a set based on Egyptian motifs. His drawings were to be featured on the set's seventy-two plates, including one that would bear the circular zodiac. Most magnificent of all, the set would include porcelain models of the temples of Dendera (Figure 8.2), Edfou, and Philae, the two-storied colossi of Memnon, and even the colonnade of rams leading to the temple at Luxor, ancient Thebes. These were extremely hard to fabricate. They were in the end not based on Denon's own drawings, but on the

Figure I.1. The festival of the Supreme Being.

Figure I.2. Portion of the Coronation of Napoleon at Notre Dame.

Figure I.3. The rectangular zodiac at Dendera.

Figure I.4. Section of the Dendera rectangular.

Figure 1.5. The *Théatre du Gymnase*.

Figure I.6. The Franco-Tuscan Expedition in Egypt.

Figure I.7. One of the many empty cartouches at Dendera.

Figure I.8. Cleopatra VII in the guise of an Egyptian goddess (Mut) at Dendera.

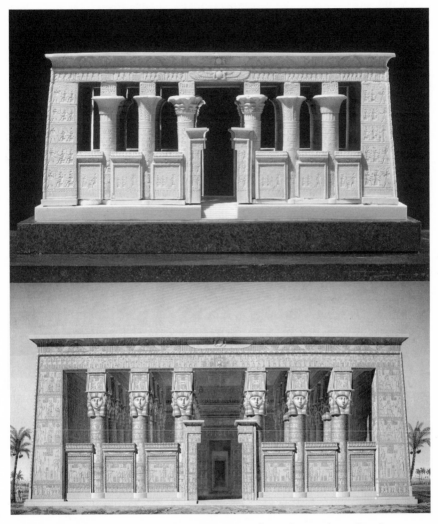

Figure 8.2. Dendera in Sèvres porcelain (*top*), after Jollois and Devilliers'
reconstruction of it (*bottom*). Note the missing faces on the Sèvres model, which
seems to have imported elements of Jollois and Devilliers' rendition of the temple
at Esneh (Figure 5.6).

ones that Jollois and Devilliers had produced for the *Description*. The
designs were produced by Lepère, an architect and himself a member
of the Egypt expedition. Denon ordered the Sèvres company to fabricate all these difficult works rapidly, or else, he warned in an exercise
of transmitted imperial power, Sèvres would "no longer be the premier

manufacturer of Europe." Napoleon insisted that the result must be "of the first, not the second, order" and that it had better be done quickly. The plates proper appeared swiftly (forty-four were ready by the end of 1805), but the difficult porcelain models were not available until July 1, 1808. One complete set was initially produced, which Napoleon sent to Tsar Alexander as a gift; it is today on exhibit at the Kuskowo ceramics museum in Moscow. Josephine ordered a second set but decided that she did not like it. During the restoration that set was given by Louis XVIII to the Duke of Wellington, and it can be seen today at his residence, Apsley House, in London.[14]

The Prefect in Grenoble

Sent reluctantly as prefect to Isère, Fourier settled in at Grenoble. Though unhappy with the posting, he took his responsibilities seriously and, among other actions, was able to effect the long-desired drainage of the swamps of Bourgoin, which required technical acumen and a good deal of political savoir faire.[15] Constantly suffering from the cold, attacked by rheumatism, and afraid that he would be utterly forgotten in Paris, the center of scientific and scholarly work, Fourier took every opportunity that he could to return to the city. Indeed, between 1804 and 1813 he spent nearly three years altogether, though at different times, in the city.[16] Yet Grenoble itself was hardly an intellectual wasteland since it had a significant Egyptological presence. For a short time there was Sonnini de Manoncourt, who had been named by Fourcroy to head the principal college of the nearby town of Vienne. Sonnini was a naturalist who had traveled in South America and had worked under Buffon describing bird species. He had also journeyed to Egypt and Greece and had published in 1799 a three-volume *Voyage dans la haute et basse Égypte* (*Voyage in Upper and Lower Egypt*) with a separate volume of plates. Sonnini had visited Dendera, and he drew many of the figures that he found there, including a rather fanciful one depicting an overarching sky goddess similar to the one that embraces the rectangular zodiac (Figure 8.3), though he sketched neither that nor the circular one on the ceiling of the small roof temple, which he likely had not seen. A member of Grenoble's arts and sciences society, Sonnini gave talks at its meetings about his travels in Egypt. Tired of administrative work, he left in 1807 after only two years. Fourier no doubt sympathized. Grenoble also had the Abbé Gattel, a linguist and partisan of *idéologie*, who was interested in the origins of writing, about which he had written a memoir in 1801.

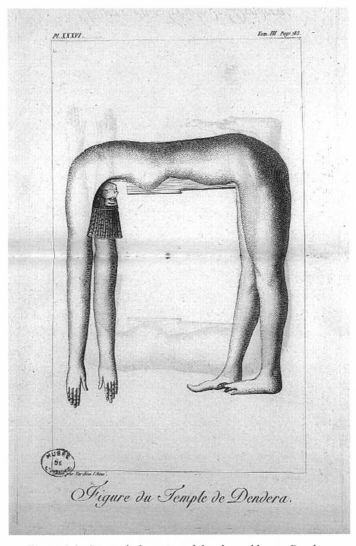

Figure 8.3. Sonnini's depiction of the sky goddess at Dendera

Above all, Grenoble had the two brothers Jacques-Joseph Champollion-Figeac (hereafter Champollion-Figeac) and Jean-François Champollion (hereafter just Champollion). Champollion-Figeac was supposed to accompany Napoleon's expedition, but he literally missed the boat at Toulon. Starting that year (1798) he worked in Grenoble for a cousin, a merchant, having left his younger brother behind in Figeac in the care of his father, who was often away on business, and his ill mother. He brought Champollion to Grenoble in March 1801 and took charge of the boy's education, eventually placing him in the private institution of Abbé Dussert, where he developed a special interest in Hebrew.

Fourier made visits to the local schools, and on one of them he met Champollion for the first time, shortly after arriving in town. He invited the boy to visit him at the prefecture, where he explained the significance of the Egyptian zodiac to the no doubt awed youngster, which indicates how much the issue was on Fourier's mind for months after his return from Egypt. The boy thereafter visited him at the prefecture, where soirées were held at which art, archaeology, literature, and recent discoveries were discussed. Fourier soon referred to Champollion as "a fiery colt demanding a triple ration."[17] A close relationship also developed between Champollion-Figeac and Fourier, who entrusted Champollion-Figeac with the conservation of inscriptions on the walls of an old bishopric undergoing demolition.

The reluctant but hardworking prefect repaired frequently to an isolated mansion, Beauregard, a few miles outside town to immerse himself in the two projects that gripped his interest and that, he hoped, would bring him the scholarly reputation that he had ardently sought since boyhood. And, perhaps, bring him back to Paris as well. The first of these to reach fruition did ultimately ensure Fourier's reputation as one of the most influential mathematician-physicists of the nineteenth century, though not during his own lifetime. There, in the Grenoble countryside, Fourier developed a new way of thinking about the conduction of heat. Like most French savants of the period, Fourier was strongly aware of the powerful example set by Laplace, who was by this time the most influential figure in the Parisian community of mathematicians and *physiciens*. He and his friend, the chemist Berthollet, gathered together a group at Berthollet's country house in Arcueil, just outside Paris, to develop and to make known their ways of approaching natural phenomena. New theories were discussed, physical and chemical experiments were carried out, and memoirs were presented that were eventually published in three volumes.[18] Among the several frequenters of Arcueil, one could find the young Biot, himself an early graduate of the École Polytechnique, whom we have already encountered as, much later, a friend of Domenico Testa's, and who would figure prominently in the revived controversies over the zodiac after the Dendera circular's arrival in France in 1821. Fourier would certainly have known about these meetings and may have visited Arcueil, which would have made even harder to bear his normal distance from the city.

CHAMPOLLION-FIGEAC

(JACQUES.-JOSEPH)

Né le 5 Octobre 1778 — † 6 Mai 1867.

Figure 8.4. Fourier's friend, Jacques-Joseph Champollion-Figeac

Heat formed a major subject of interest to the frequenters of Arcueil, especially to Laplace himself, and in 1804 Biot had produced a short piece that provided experimental results and that briefly stated a "mathematical law of the propagation of heat" for the case of a long, thin bar. This brief result was based on the standard physical image of the time among Parisian physical savants, namely, that thermal phenomena, like all inorganic (and probably organic) processes, were governed by forces exerted on one another and on ordinary matter by the particles of a special substance, in this case the substance of heat, namely, *caloric*. To this way of thinking, the appropriate method to analyze the flow of heat would be to divide a body up into discrete parts and then to make assumptions about the heat that each part sends to its neighbor, or, perhaps, to the body's other parts not far distant. Working that way raises considerable analytical difficulties, and Biot was certainly not able to overcome them. Neither at first was Fourier, who was by far the better mathematician.

Fourier had trouble acquiring the latest work from Paris, and so he might not for a short time have been aware of Biot's attempt at a theory of heat flow, though it is quite certain that he knew about it soon enough. He may in fact have heard something directly from Biot himself, who had taken Fourier's course in mathematics at the École Polytechnique, since Fourier's first draft paper mentions receiving information from Biot about the latter's experiments. Fourier's own first analysis, done sometime between 1804 and 1805, also worked with slices of the heated substance, which posed difficulties of the sort that Biot had likely encountered. Even in this earliest effort, however, Fourier developed equations that he would soon put on a firmer foundation, as well as a mathematical technique, later eponymously named "Fourier series," that, if not wholly unprecedented, was nevertheless unusual and potentially controversial.[19]

In any event Fourier continued working on the problem, and in so doing he did generate a novel way of thinking about the flow of heat. He abandoned the formulation in terms of discrete slices of material and instead introduced a fundamentally new conception, that of *flux*. Instead of considering parts interacting with one another, Fourier evolved the concept of a quantity's flow across a surface, the *flux* being the measure of the flow per unit area. That way of working avoided the several difficulties that he (and perhaps Biot) had previously encountered and permitted him to generate equations for thermal conduction without the sorts of approximations and special assumptions that would otherwise be necessary and that, five years later, Laplace himself did deploy in an effort to arrive thereby at the very equations that Fourier had produced.

During these years, Fourier undertook extensive experimental tests at Beauregard to confirm his results. Then, in 1807, a new draft containing his novel concepts, as well as the full set of equations, his mathematical techniques, and his experimental results, was ready to go. He surely hoped that this would securely establish his reputation among the scientific savants of Paris. Journeying once again to the city, Fourier read an abstract of the new paper to the Institute National's First Class on December 21. As was customary, a commission was empaneled to report on the memoir, in this case consisting of Laplace, Monge (Fourier's colleague from Egypt), and the mathematicians Lagrange and Lacroix. Whatever hopes Fourier entertained of making a major splash were soon dashed because no formal report appeared, and the only reaction was a mostly descriptive account by a young mathematical protégé of Laplace's, Siméon-Denis Poisson, which was printed in an important journal, the *Bulletin de la Société Philomatique*. Poisson hinted that there were "delicate questions in the theory of heat" that required further investigation, but he wrote nothing further in praise or in criticism. The memoir had sunk like a lead balloon, altogether ruining Fourier's hopes for the moment.

All the more reason for the continuing efforts that Fourier poured into his work on what would in the end be titled the *Préface historique* to the *Description de l'Égypte*. Not that he abandoned his theory of heat, far from it. But Jomard was pressing Fourier for results, since the absence of the *préface* was holding up the entire publication. Here, Fourier knew quite well, he had to make an impression, but not only among savants. For the *préface* would be widely read, if not explicitly commented on in print, representing as it was certain to do the official account of what Napoleon's expedition to Egypt had accomplished. Not the least of its readers would be the emperor himself.

Fourier worked and reworked the *préface*, drawing on the help of the Champollion brothers and investigating the minutest details, down to whether or not the words he had used conformed to the latest specifications of the Académie.[20] By early 1809 it was nearly ready; he finished at last on June 28. The following September, Fourier was again in Paris. He had three copies of the *préface* produced at the imperial printer on large sheets in an impressively sized font. One of them was bound in green for the emperor, to whom, it seems likely, Fourier delivered it in person.

Over the next several months Napoleon's interior minister, now Bachasson, Comte de Montalivet, tried repeatedly to pry the *préface* out of the emperor's hands, but to no avail. Finally he succeeded, and the day at last arrived that the *préface* was to be returned to Fourier, along with Napoleon's emendations. Fourier received it directly from the emperor. Champollion-Figeac much later reminisced that Fourier thereby received "warnings from which he profited, and which under Fourier's pen produced the second draft of the discourse."[21] The result finally appeared in print in 1810, albeit bearing the requisite date of 1809.

Fourier's *préface* began with a lament for the present state of a once-storied civilization. He justified the French invasion of Egypt by invoking the tyranny of the Mamelukes and, especially, the "injuries" to French commerce due to the "beys." The failure of the Ottoman central authority to overcome these problems was the reason for the expedition. "But the hero who directed it," Fourier gushed,

> did not limit his goal to punishing the oppressors of our commerce; he raised the project of the conquest to a new height and grandeur, and imprinted upon it the character of his own genius. He appreciated the influence that the event must have on the Orient's commerce, on Europe's relations with the interior of Africa, on navigation of the Mediterranean, and on the fate of Asia. He proposed to abolish the tyranny of the Mamelukes, to open constant communication between the Mediterranean and the Arabian gulf, to form institutions of commerce, to offer the Orient the useful example of European industry, finally to make sweeter the condition of its inhabitants, and to procure for them all the advantages of a perfected civilization.[22]

Fourier's rationalization was often repeated in subsequent decades to justify conquest and empire by other countries. No one bothered to ask the "liberated" Egyptians whether they were willing to pay the price of thousands dead and a polity destroyed to rid themselves of their widely hated tyrants, or whether they wished wholesale to exchange their ways for the benefits of "perfected [French] civilization." Popular uprisings and chaos during the occupation indicated just how eager the inhabitants were to embrace Western ways and control.

Given Fourier's encomia to the pure motives of French imperialism and, especially, to the "genius" Napoleon, what precisely had the emperor changed in the draft *préface*? And how do we know?

Champollion-Figeac had in his possession one of the three copies that Fourier had had printed, in fact the very one bound in vellum that he had delivered to Napoleon. It contains the edits that the emperor ordered, as well as changes that Fourier himself added. Altogether there are twenty-two changes or additions. The ones that Napoleon required all pertain either to his self-presentation as emperor or as hero of the expedition, to justification of the expedition despite its failure, or else to remarks that might offend the Ottomans, whose favor Napoleon was courting in order to counterbalance the British. So, for example, where Fourier had originally written that Egypt could be controlled by "a European power," Napoleon altered this to read by "a European power which, tightly united with the Porte [viz the Ottoman emperor]."[23] Where Fourier had written that "the name of Bonaparte reigns," Napoleon had it changed to read "the immortal name of Bonaparte reigns."

Not all of the changes that Fourier entered were due to the emperor's orders. One in particular is of special interest to us because it concerns the zodiacs and tells us much about Fourier's unsettled views of the matter in 1809. Champollion-Figeac was often with Fourier while he was working on Egyptian matters. He recalled years later that Fourier "devoted an infinity of time and an infinity of calculations to his researches [on Egypt], but he never made them public." These elaborate calculations have apparently disappeared: "his manuscripts weren't preserved, and I wouldn't be able to say what became of them, because their content remains a mystery for science, which hasn't profited from them; these manuscripts are the property of the State; they are deposited at the royal library, but I've never seen them."[24] An intriguing point: were Fourier's computations perhaps removed from the Royal Library by someone who feared and hated their inflammatory religious potential? We will probably never know, but it seems likely that Champollion-Figeac had reasons to be suspicious.

We do know what Fourier was willing to write about zodiacs in 1809, and it is quite different from his opinions while in Egypt or shortly after his return to France. Seven years as prefect, seven years dealing with the pressures and political complexities of administration, and several months dealing with Napoleon's demands for changes had left their mark. In Egypt and in Paris, Fourier had declared his belief that the zodiac at Esneh reached back six thousand years before Christ, and the one at Dendera to a thousand years before the Trojan War. In the draft *préface* that he had given Napoleon, Fourier had written nothing at all about the zodiacs. Certainly, the *préface* had insisted, ancient Egypt had originated the arts and the sciences. Indeed, Fourier had written, "priests trained at the schools of Egypt had observed the skies of Chaldea [Mesopotamia]," and by them "the fundamental truths of geometry and astronomy were discovered." But when had this taken place? The version presented to the emperor remained mute about that.

Sometime after Fourier received the draft back from Napoleon he entered an addition. It followed an emendation to a passage that had originally remarked that "circumstances [for scientific work] were not always favorable, and, in the midst of the many events of war, we were sometimes stopped by truly insurmountable obstacles." The emended passage placed the onus for failure on the unfortunate weakness of the savants themselves, who had "succumbed" to the fatigues, the contagious maladies, the seditions, indeed the "nearly certain perils to which their imprudent zeal had exposed them." These words reek of Napoleon's consistent efforts to shift blame for failures onto his subordinates and were almost certainly, if not written, then at least ordered by him. But the addition that immediately followed the emended passage does not bear the impress of Napoleon's orders. On the contrary, it seems to be a cri de coeur from a Fourier too oppressed by the emperor to be entirely forthright about a potentially dangerous matter, but too torn by his convictions to say nothing at all.[25]

"The consequences that result from the attentive study of the monuments," Fourier added in manuscript, "will never permit the history of Egypt to be confined within the limits of a restricted chronology that was not followed in the first centuries of the Christian era."[26] This alone was a brave remark, challenging as it did the views of Catholic orthodoxy. It would have been even braver five or so years earlier, when censorship of opinions that could raise the ire of the religious Right was particularly acute. By 1809, however, the climate had considerably altered. Relations between the emperor and the Vatican had begun to deteriorate in 1805, when Pius VII did not attend the ceremony at which Napoleon took the Crown of Lombardy, and had become even more strained when Pius refused to annul the marriage of Napoleon's brother Jérome to the American Elizabeth Patterson (which the emperor later arranged to have done by Parisian prelates). Matters deteriorated further in 1806 following the French occupation of Ancona, with Napoleon becoming increasingly angry at what he saw as papal recalcitrance. Then, on February 2, 1808, the French occupied Rome itself under General Miollis. Pius VII protested, castigating the empire for its utter indifference to religion. He excommunicated

Napoleon, following which, in early July 1809, the pope was arrested and exported to Savona, north of Genoa. These events were initially kept quiet; no notice of them appeared in the *Moniteur*, though news soon spread nevertheless. Matters were further exacerbated by Napoleon's desire to divorce Josephine in order to marry someone who could produce an heir, to which Josephine tearfully agreed on the evening of November 30. That too was unacceptable to the Vatican.

Fourier surely heard about these ruptures between the emperor and the church since he had been in Paris from early September 1809, in which case his slight challenge to religious orthodoxy in the *préface* hardly seems overly brave. It seems even less so in that it was immediately followed by words that seemed to take it all back. Although churchly chronology cannot fit Egyptian history, he wrote, nevertheless the monuments of that ancient country "are no less contrary to the view of those who found an exaggerated antiquity of the Egyptian nation on conjectures, and who do not at all distinguish the true historical epochs from calculations that serve to regulate the calendar."[27] Still, we should not too quickly attribute Fourier's reticence to fear of controversy. During the "infinities" of time that he had devoted to matters Egyptian, he had no doubt learned a great deal about ancient calendrical practices and had perhaps decided in consequence that the zodiacs could not after all be interpreted quite so easily as he and others with him in Egypt had originally assumed.

Fourier's preoccupation with zodiacal chronology had begun early in his years at Grenoble. Sometime in 1805 he had asked Champollion-Figeac about an inscription in Greek on the circumvallation just south of the Dendera temple that Denon had reproduced. Champollion-Figeac replied in a subsequently published letter that began with an encomium to Fourier: "you have attached your name to the temple of Dendera; you have sought the explanation of its mysterious ornaments even deep into its sanctuary."[28] Unfortunately, the Greek inscription was poorly done, reflecting "proof of the inexcusable negligence of the Greeks in the engraving of their inscriptions." The passage in question offered a consecration to the "Emperor Caesar" and to the gods honored in the temple. It had been discussed recently in the *Magasin Encyclopédique*, and Champollion-Figeac demurred concerning the views advanced there as well as the translation provided to Denon and printed by him. The point turned on an issue of dating, and Champollion-Figeac argued that the inscription, properly scanned, indicated that there was a dedication of the Dendera temple to Isis when the Emperor Augustus was thirty-one years old, on the eighteenth day of the Egyptian month of Thoth.[29] This did not necessarily imply that the temple had been constructed at the time of this dedication—quite the contrary, because, Champollion-Figeac continued, "it suffices to stop at the sensible difference that one sees between the manner of the Greek inscription and that of the symbolic figures which adorn the interior and exterior of the temple. . . . This disparity strikes eyes trained in analyzing the productions of Art and in the comparison of monuments."[30] This was precisely what Fourier was looking for, and no doubt why he had asked Champollion-Figeac to examine the issue.

Fourier's *préface* mightily impressed Fontanes, now master of the new Imperial University, who was sent an early copy by Champollion-Figeac. "I don't doubt Monsieur," Fontanes wrote in fulsome praise to Fourier, "but that the work in preparation on Egypt will be worthy of the savants who execute it, and of the monarch who orders it; but, in awaiting it, I dare say to you that your preliminary discourse is in itself a beautiful monument. You write with the grace of Athens and the wisdom of Egypt. Your style is all elegant and grave, and it's a long time since I have read anything so good and so solid."[31] Fourier may have received other praises, but if so they were mostly conveyed privately, because when reviews of the *préface* did appear they were almost all matter-of-fact reports of its contents. Given the printed *préface*'s adulation of the emperor and its marked tone of French triumphalism, any real critique would have been politically unwise, to say the least. Champollion-Figeac later recalled that some said of the *préface* that it was "nearly aphoristic, abundant in assertions, and sentences admirably expressed, devoid of any demonstration," though Fourier's manuscript notes, which seem no longer to exist, would have shown otherwise.[32]

Fourier Imagines Ancient Egypt

The printing of the first three volumes of the antiquities part of the *Description* was under way in 1809 and available the next year.[33] The first volume contained memoirs on the monuments, arts, sciences, and geography of ancient Egypt. It concluded with a twenty-one-page account by Fourier entitled "Researches on the Sciences and Government of Egypt." This was the most detailed piece that he ever published on the questions of ancient Egyptian astronomy. The same volume contained an essay by Jollois and Devilliers on the "Astronomical Bas-reliefs of the Egyptians," as well as another by the Orientalist Rémi Raige (who had also been on the Egypt expedition) on the zodiacs, both of which offered many pertinent remarks on the subject, as we will shortly see.[34] The second volume provided Jollois and Devilliers' descriptions of the astronomical monuments. The final volume available in 1810 contained the plates, including their depictions of the zodiacs at Esneh (see Figure 8.6), but, again, not the ones at Dendera, which were not printed until 1817, during the Restoration, in the fourth volume of antiquities. Two volumes each (text and plates) on Egypt's "modern condition" and on its "natural history" were also available in 1810.

Unlike his *préface*, Fourier's words on Egypt's astronomical monuments were not presented for Napoleon's personal approval, and it is here that we find his final, public opinions on zodiacal chronology.[35] His essay divides into four "articles," all very obviously in summary form, and not infrequently repetitive. Neither Jollois and Devilliers nor Remi Raige saw the memoir before the volume reached print, though Fourier did briefly tell the former two what his basic assumptions were.

Fourier's analysis is not transparent because the printed memoir, which is all that we have despite the "infinity" of time that he devoted to the issue, is heavily abbreviated. The memoir's fragmentary character betrays the pressure Fourier was under to finish up, and especially to revise the *préface* once he had it back from Napoleon. That left little time to perfect the astronomical arguments, which were after all intended as a summary to be followed eventually by a fully developed account—an account that, in the event, never appeared, not least because of political events surrounding the Restoration.[36] We can nevertheless extract Fourier's main claims. First, the original zodiac was produced about 2500 BCE, for it was then, he asserted, that the correspondences between the seasons, the setting sun, and the rising and setting asterisms worked out well.[37] This was much different from Dupuis' prior claim of 13,000 BCE, but it connected, we will see, to Fourier's interpretation of the rectangular zodiacs. For these he did not give dates, but rather limits, claiming that the zodiac at Esneh preceded 2100 BCE, while the rectangular at Dendera was built sometime thereafter.

Champollion-Figeac had provided Fourier with many classical references, and Fourier became convinced, like many before him, that the Egyptian calendar was born in connection with Sirius's heliacal rise. Taking the importance of the rise as a fact, Fourier based his arguments on the claim that, in about 2200 or 2100, the rise "passed from the sign [*sic*] of Leo into that of Cancer."[38] It was then, he asserted, that the heliacal rise of Sirius occurred "at the point of division that separates

Leo from Cancer, and it [thereafter] advanced more and more into this last constellation."[39] To understand Fourier's reasoning, we need briefly to examine how the rise moves among the constellations throughout the millennia-long history of ancient Egypt.

To calculate the displacement of the rise over time requires deciding how far below the horizon the sun sits when Sirius appears just before sunrise after its many weeks of complete invisibility. The distance between the sun and the star at that moment is termed the *arcus visionis*. If Earth had no atmosphere, the star would appear on the horizon instants before the upper limb of the sun. This was known as the star's *true* or (in terminology Biot used at the time) *cosmical rise*. However, the atmosphere intervenes, refracting the starlight; the *arcus visionis* is also affected by dust (which scatters light) and the temperature change with height (which alters the air's density and so its refractive index). These factors, which are difficult to pin down, determine how far below the horizon the sun must sit in order for the star to be visible above it. Taking them into account yields the star's *heliacal rise*. A reasonable estimate of the *arcus visionis* for the heliacal rise at Luxor is today taken to be eight degrees or thereabouts.[40]

Figure 8.5 illustrates the effect of three different values for the *arcus* on the position of the sun at heliacal rise with respect to the constellations as seen from Luxor; the values are for zero degrees (thereby computing cosmical rise), eight, and twelve degrees. For twelve degrees, the rise clearly remains entirely within the confines of the constellation Leo throughout the period of ancient Egypt. At eight degrees, the rise still remains within Leo, but it begins to move out of it and into the separation region between Leo and Cancer by about BCE. And for a value of zero degrees, the rise does cross from Leo to Cancer in about 2100 BCE.

Fourier maintained that, before about 2100 BCE, the rise lay in Leo, and that it passed thereafter into Cancer. For this to hold, he must have computed a true rising, one that ignored the effect of the atmosphere. Fourier was a superb mathematician, but he was not an astronomer. In calculating the heliacal rise, he apparently did not think to adjust for the effect of refraction, or at least not significantly, whereas anyone who—like Biot—had measured stellar positions would have taken care to do so. In 1823, when Biot published his own analysis, he castigated Fourier for asserting that the rise moved out of Leo into Cancer, remarking sarcastically that he, Biot, would "ignore what could have deceived such an able geometer as M. Fourier in so simple a calculation."[41] The implied denigration of Fourier's talent as an "able geometer" was certainly

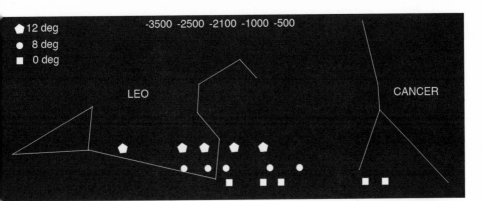

Figure 8.5. Locations of the sun at Sirius's heliacal rise at five dates from 3500 BCE through 500 BCE for three values of the *arcus visionis* at the latitude of Luxor, ancient Thebes.

intentional, and we shall see below how tense the relations between Biot, on the one hand, and Fourier and his colleagues, on the other, had by then become for reasons other than ancient astronomy. Biot's own claim was clearly based on an *arcus visionis* of about twelve degrees, which is what he had years before asserted to be proper for a heliacal rise.[42]

Given this, we can unpack Fourier's claims concerning the zodiacs. Having assumed that the Egyptians recognized the rise as a signal for the supreme event of their year—the Nile flood—Fourier conjectured that a terminus of each rectangular zodiac should represent that event. He further assumed that the heliacal rise corresponded to the end of the Egyptian year, which did indeed occur around that time.[43] Consequently the end of a rectangular would naturally represent the rise. But which terminus marked the beginning of a rectangular, and which one marked its end? The orientations of the zodiacs with respect to their respective temples provided an answer. All three rectangulars (the two at and near Esneh and the one at Dendera) were perpendicular to the temples' entry doorways. Moreover, the figures faced in opposite directions between the two panels that composed each of them. One panel might therefore represent figures *exiting* the temple, while the other would then represent figures *entering* it. Consider, for example, the large rectangular at Esneh (Figure 8.6, top image). Here the temple entrance is to the left, so that the figures in the top half of the panel seem to be marching into the temple, whereas the ones on the bottom

Figure 8.6. Jollois and Devilliers' drawings at Esneh: (*top*) the zodiac on the ceiling of the large temple's portico, and (*next*) an enlargement showing (*from the right*) Scorpio, Libra, and finally Virgo, all facing left. (The lionlike figure just in front of Virgo is not Leo, which is depicted in the upper part of the rectangular just to the left of Cancer.) The bottom two figures, which formed one continuous band but were separately drawn, are from the portico of the small temple just north of Esneh.Virgo would have been to the extreme right, in the ruined portion, preceded by Scorpio and Libra.

half seem to be marching out of it. Fourier therefore took Leo (toward the left in the top half) to be at the *end* of the procession, while Virgo (the leading zodiacal figure in the bottom half) had to be at its *front*.

The Dendera rectangular had to be interpreted in the same way as the ones at Esneh. Figure 8.7 shows its two panels, with the temple entrance to the right of the image. In the upper panel, the zodiacal figures that represent the signs all face to the left, as though they were headed into the temple, whereas in the lower panel they face to the right, as though they were headed out of the temple. Leo marches out first in the lower panel and so leads the procession, while Cancer brings up the rear.

The directions the figures faced were accordingly critical for Fourier, who had only Denon's and Jollois and Devilliers' depictions to work from. We have already seen that in Denon's drawing, which Dupuis had used, the zodiacal figures in the upper panel point in the wrong

Figure 8.7. Jollois and Devilliers' drawing of the Dendera rectangular, with figures correctly oriented

direction. Though that error would not have affected Dupuis' interpretation, it certainly was significant for Fourier. He pointed it out in his careful descriptions of the ceilings in the second volume of *Antiquities*, which was printed in 1818—nine years after his memoir on Egyptian astronomy.[44] And yet, despite their different interpretations, Fourier shared with Dupuis an underlying assumption that reflects their common belief that the zodiacal panels must be literally figurative of the heavens as the Egyptians perceived them in remote antiquity. For Fourier, like Dupuis, insisted on granting the spatial position of the zodiacal ceiling within its building temporal significance. Time and space configured one another for both of them.

What now of the constellations that lead the figures into the temple? They too had a natural interpretation. Since the rise marked the end of the year and signaled the forthcoming flood, the immediately following constellation—and so the one that first *enters* the temple—had to coincide with the period when the Nile had fully covered the land. It would be the constellation that is fully traversed after the inundation commenced (Jollois and Devilliers, we will see, adopted a similar interpretation). Fourier accordingly decided that the Egyptian zodiacs began with the first constellation that was completely passed through after Sirius's rise.[45] With this understanding, his claims for the zodiacs' dates fall into place given his computations, now unfortunately lost, of the locations of the rise at the latitude of Dendera throughout Egyptian history (Figure 8.5, with an *arcus visionis* of about zero degrees). The zodiacs at Esneh must have been designed to represent the state of the heavens no later than about 2500 BCE, when the rise still took place unambiguously in Leo. At Dendera the rise occurred in Cancer, which could not have taken place before about 2100 BCE, when, on Fourier's calculation, it was beginning to pass from Leo into Cancer.

Fourier had also developed a theory concerning the origins of the fixed *signs*, which he also attributed to the Egyptians.[46] As he examined all the zodiacs and thought through their possible astronomical meanings, Fourier decided that the astronomers of Thebes had originated the signs around 2500 BCE, though the carved zodiacs represented the corresponding asterisms and not their fixed correlates. This original system, like the one that the Greeks deployed over a millennium and a half later, divided the ecliptic into twelve zones, each thirty degrees in length. However, Fourier conjectured, the ancient Egyptians had set the summer solstice at the dividing point between their signs for Cancer and Leo, whereas in the later Greek system the solstice marked the division between the signs for Gemini and Cancer. They did so because, he wrote, a solstice or equinox was assigned to the first sign that was completely passed through after its occurrence. Consequently the original Egyptian system was offset by one sign from the later Greek scheme.

So much for the evidence provided by the temple zodiacs proper, but was there more? Was there independent corroboration? Though he did not mention it, Fourier could hardly have been unaware of support for his claims of a kind that would particularly appeal to the material sensibilities of the savants. An article by the expedition's chief engineer, Pierre Girard, on the hydrology of the Nile region appeared in the same volume as Fourier's piece on the zodiacs. It contained a lengthy discussion of the Nile bed's increasing elevation over the centuries based on a comparison between the contemporary height of the flood at its maximum and its height in antiquity, derived by Girard from an inscription on a nilometer at Elephantine.

The nilometer consists of horizontal marks, each about a cubit in length, that are inscribed on a wall at equal vertical intervals.[47] When the Nile rose, its waters gradually moved up the wall, successively covering the cubits. In antiquity the device was used to record the flood's level, but by 1799 at peak inundation the waters reached 2.413 meters beyond the highest mark. An inscription carved about 0.31 meter above this highest cubit referred to the reign of the Roman emperor Septimus Severus. According to Girard, the inscription "seemed to have as its purpose the recollection of an inundation that rose many palms [hands] above the twenty-fourth cubit." Assuming, as Girard did, that the rate of bed rise had been constant over the centuries since that time, then, since Severus reigned from 193 to 211 CE, the bed at Elephantine must have risen about 0.132 meter every century since (taking the beginning of the reign as the date of the inscription). A second nilometer on the isle of Roda in the Nile at Cairo (in which Napoleon had taken a particular interest) consisted of a tall marble column. Indications on it led, Girard asserted, to a different rate of bed rise, namely 0.120 meter per century. Taking the mean between the two rates produced 0.126 meter. Assuming further that the lowest mark represented the height of the flood when the Elephantine nilometer was first created yielded a time interval between 1799 and the original moment of 4,760 years. The Egyptians had accordingly begun tabulating the Nile rise around 2960 BCE. That epoch, Girard continued, marks "a revolution which, changing the mores of the first inhabitants of Egypt and giving them the means for an agricultural life, led them to the middle of the valley and onto the banks of the Nile where, to shelter themselves and their flocks from the periodic inundations, they were obliged to build their homes on artificial platforms." This would certainly have preceded by several centuries the foundation of the great civilization at Thebes, which therefore must have taken place at almost precisely the time (2500) that Fourier had calculated for the high point of Egyptian astronomy, marked by the carving of the zodiac at Esneh.[48]

The circle of evidence seemed to close. Everything seemed to point to the middle of the third millennium as the point at which Egyptian civilization had reached its magnificent heights. Though the dates for both Esneh and Dendera are distant from the tremendous antiquity that Fourier had unquestionably asserted shortly after returning from Egypt, there are hints even here of his continuing belief that a primitive form of Egyptian astronomy reached back very far indeed. He remarked at one point, almost as an aside, that at the latitude of ancient Memphis (effectively Cairo), the star Sirius would actually be invisible throughout the year as a result of precession during some prolonged epoch, but that this would never occur at all at the latitude of Dendera. "This coincidence is remarkable," Fourier continued, "but we do not have any reason to believe that the Egyptians knew it."[49]

Perhaps not, but the coincidence would have seemed remarkable indeed to Fourier. The dark bars in Figure 8.8, superimposed on a late

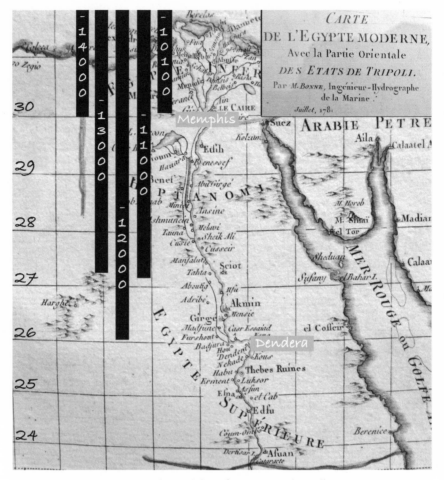

Figure 8.8. The visibility of Sirius over the millennia

eighteenth-century map of Egypt, cover the latitudes above which
Sirius remains completely hidden throughout the year at a particular
epoch. We see that in 14,000 BCE Sirius would barely have been visible at Memphis. Over the next two millennia the region of invisibility
moved southward as a result of precession, until it reached its southerly limit in about 12,000. And it does so almost precisely at Dendera's
latitude. Thereafter the region moved back northward, until by 10,100
Sirius would again have been visible at Memphis.

What might this have meant? Fourier wrote nothing more, but it
is not hard to see the implication. Dupuis had originally implied that

Egyptian astronomical knowledge dated back between thirteen and fifteen thousand years, which put it squarely at 12,000 BCE—the very time when Sirius would have been visible in Egypt only at Dendera and further south. Perhaps Fourier was subtly insinuating that the late-date temple at Dendera was constructed at an immensely old cult location. At this sacred place, so near the great center of Thebes itself, the brightest of all stars, though completely invisible throughout upper Egypt, appeared like a beacon. Here, it may be, in the most remote eras of antiquity, Egypt had first developed civilization, thousands of years before the biblical date of Creation itself. There is indeed evidence that Fourier had written a now lost memoir on the age of Earth that was intended to demonstrate the complete falsity of scriptural chronology.[50]

Of Fourier's three colleagues who also wrote about the zodiacs for this first volume of the *Description*, Jollois and Devilliers did not explicitly advocate such a great antiquity for the primitive origins of Egyptian astronomy, but the Orientalist Raige did. As we have seen, the former two, always collaborators, had briefly been in contact with Fourier about the zodiacs, or rather Devilliers had, soon after their return from Egypt. Several years later, just after the two men gave Jomard their contribution, but before it reached print, they (or one of them) spoke directly with Fourier, no doubt during his long stay in Paris from August 1807 through early February 1808 to work on the volume and to present his theory of heat propagation. Before that, they did not know what had guided his zodiacal dating, which makes it all the more interesting that they arrived at very much the same results—and using quite similar principles.[51] Raige took a different tack, and the disparity has much to do with the differences between the scientist or engineer and the philologist, despite (in this case) their common belief that Egypt had originated astronomy and all the arts. These differences would become more pronounced over time.

The Engineers and the Orientalist

Jollois and Devilliers had both trained at the École Polytechnique, though Jollois was four years older than his friend, who had taken the exit examination for the École in Cairo on October 6, 1798, under three formidable assessors: Berthollet, Monge, and Fourier himself. Born in Bourgogne, Jollois had studied at the same École Militaire in Auxerre that Fourier had attended. In 1792 he entered the first year of the Polytechnique, thereby becoming part of the famous "promotion" of Year 3 (1795), which included, among others, Biot and Malus. Jollois then joined the corps of Ponts et Chaussées (Bridges and Road Works) just when the expedition to Egypt was being organized, which he accompanied as an ordinary engineer under Jean-Baptiste Lepère and—much to his later regret—Girard, whom he accompanied to Upper Egypt. After Jollois returned to Cairo, General Menou assigned him to hydraulic works in the delta.

René Edouard Devilliers' life, unlike that of his good friend Jollois, had been much troubled by the upheavals of the Terror. He was born in Versailles, where his father, Marc-Étienne, was the commissioner of finances in the waning years of the monarchy. At the age of thirteen he saw his father condemned to death by the revolutionary tribunal in Paris. Marc-Étienne was, however, saved by an agent of the Committee of Public Safety, one Romainville, who sent him to the Hotel du Dresneux instead of to certain death at the Conciergerie, where prisoners awaited the guillotine. Alone now in Paris with his younger sister, René Edouard witnessed the September Massacres and "all the horrors" ordered by the revolutionary tribunals.[52] The two of them lived in the city as best they could, surviving from a small present that their uncle had given them before he had fled, from the remains of the family silver, and by selling precious books that their father had bought in happier times. Marc-Étienne was freed following Thermidor after many efforts by his now fifteen-year-old son. He found employ at the bureau charged with sorting title deeds that had been deposited at the Archives Nationales, assisted by René Edouard, whose studies were put on hold. Then, in 1795, René Edouard's uncle returned, occupied himself with his nephew's education, and after five months, in April 1796, succeeded in enrolling him in the École des Ponts et Chaussées, and slightly later in the new Polytechnique. René Edouard had studied there for fifteen months when he heard news of a secretive scientific and military expedition, which he was allowed to join. Once he had passed his exams (in Egypt), Devilliers was nominated by Napoleon himself to membership in the corps of Ponts et Chaussées. He became close friends with Jollois while both were with Girard in Upper Egypt. At Devilliers' funeral in 1855, Alfred Maury, speaking on behalf of the Society of Antiquaries of France, remarked of the friendship between the two that they had "the same tastes, the same talent for drafting, the same devotion to science, even admiration for ancient monuments." Indeed, that "once having set foot on the soil of Egypt," they seemed "thereafter to have formed but a single man."[53]

These two inseparable friends worked together to write an account of the zodiacs, in which they concentrated for the most part on linking the figures in them to ancient descriptions of the constellations. They had available the Aratus poem, Ptolemy's list in the *Almagest*, and—most significant in their view—the *Catasterisms* attributed to Eratosthenes (since the author actually remained doubtful, the work was usually assigned to "pseudo-Eratosthenes"). The original had long been lost, and what remained was a shortened version dating from first or second century CE Alexandria. It describes altogether forty-two constellations and specifies which ones rise and set together—data that could be used to find the place and epoch to which the descriptions refer. Jollois and Devilliers placed particular emphasis on this source because, as the librarian at Alexandria (a post to which he was appointed by Ptolemy III in 236 BCE), Eratosthenes "could consult the Egyptian books of which he was the guardian."[54]

Jollois and Devilliers were not interested in something that pertained to the time and place of Eratosthenes himself. They hoped that his descriptions (should they even be his) were derived from much older Egyptian sources far up the Nile. To investigate the possibility, they proceeded like the engineers they were by constructing a precession globe, very much the same kind of apparatus that Dupuis had had Fortin produce for him two decades before. Although neither of them had ever seen Dupuis' globe, they felt that the description of it in the *Origine* showed his device to be "inadequate" in that it would not have been sufficiently robust or easy to maneuver for the exacting correspondences that they intended to produce. The intrepid pair accordingly fabricated one themselves, after which they engaged Jean-Baptiste Poirson to make copies. Poirson was a well-known cartographer and globe-maker in Paris, who in 1803 had been ordered by Napoleon to produce a terrestrial sphere for the Tuileries. In 1811 Napoleon ordered another for the education of his son by Marie Louise of Austria. Poirson also fabricated astronomical globes, generally about thirty-three centimeters in diameter. He was later employed by Alexander von Humboldt to make the maps for his publications.

Unfortunately, neither Jollois and Devilliers' original nor Poirson's copies have apparently survived, and their brief description of the design is at best obscure.[55] In any event, with their device they could set a horizon and rotate the sphere in precession to see just which constellations would rise and set together at a particular latitude and epoch. They rapidly concluded that "between the epoch when the observations were made and that in which Eratosthenes lived, the sphere underwent a change which [Eratosthenes] did not perceive" because, placing the horizon at the latitude of Alexandria and precessing to the third century BCE, the set of risings and settings that occur together do not fit his descriptions, which must therefore derive from much earlier times.

To orient the sphere in the proper way, they used the pseudo-Eratosthenes remarks. Cancer, for example, was said to be rising when the extrazodiacal constellations Orion and Eridanus were completely above the horizon. Of course, just when Cancer is said to be rising required interpretation. Was it when the very first star in the constellation appeared, or some other point? Jollois and Devilliers decided that the Egyptians would not have accounted a constellation as rising until much of it could be seen, and so they assumed that the descriptions correspond to a moment when the middle of Cancer was just about on the horizon. They then adjusted the globe so that all of Orion's stars will have risen with mid-Cancer on the horizon. This, they found, will only occur if the globe is set to the latitude of Esneh and to a particular epoch, a period that they did not identify at this point. Instead they remarked in a note that the epoch in question had to be one in which Virgo was "the leader of the zodiacal constellations," the solstice having previously occurred.

The two polytechniciens now set out to compare the configuration with every pertinent remark in the *Catasterisms*. Working through the list, they claimed good correspondences throughout, whereas setting the device for Alexandria and Eratosthenes's own epoch produced major differences. "We therefore have the right," they asserted, "to conclude that the [table of pseudo-Eratosthenes] is not the result of observations done at the time of Eratosthenes, but that it was copied from Egyptian manuscripts that this astronomer could have consulted in the library of Alexandria." In fact, even the risings and settings mentioned in Aratus's poem fit. "Without doubt," they opined, "one ought to be astonished that the Greeks mechanically transmitted ancient astronomical tables without understanding them." Until about 255 BCE the Greeks, it seems, had remained completely ignorant of astronomy, just passing on what they had gleaned from Egyptian sages.

Jollois and Devilliers went on to divine the images of extrazodiacal constellations in the temple ceilings and to give reasons for their identifications. One example is enough to show their method. Consider the small constellation Corvus, the Crow, which appears in Ptolemy's list. It is located near and just to the south of Virgo (Figure 8.9). It can be

Figure 8.9. The constellation Corvus, highlighted below and to the left of Leo on a sky chart (*left*), and on Jollois and Devilliers' drawing of the Dendera circular

seen, they decided, in the circular zodiac of Dendera just below Leo's tail as a bird whose form, they averred, "differs little from that of a crow." Why is it present there? To answer that question, the polytechniciens brought in another textual source, using it as a sort of cartographic dictionary to translate between image and words. According to scholia on the Aratus poem attributed to Theon of Alexandria, and so dating to the fifth century CE, the dark color of the crow indicates the black soil of Egypt after the Nile retreats to its bed.[56] Since the crow lies just below Leo and in front of Virgo, for that correspondence to hold the Nile must already have retreated to its bed by the time that Virgo rises. The presence of the crow on the Dendera circular thereby places the inundation proper in Leo, and the rise of Sirius in Cancer. Why else would the temple zodiac's designers have included such a comparatively insignificant, extrazodiacal constellation?

By 1807 or 1808, if not before, Jollois and Devilliers had become certain that the original, primitive zodiac proper could not have been invented as long ago as Dupuis' fifteen thousand or so years. "How," they asked, "to admit such antiquity, when history, monuments, and myth itself are silent throughout so many centuries"? They believed instead that the zodiac must have been invented much later, and to press ahead with the claim they adopted a suggestion that Dupuis himself had offered in his *Origine* as a possible alternative to his own assumptions, though not one that he regarded as probable at the time. This "ingenious savant," they noted, had suggested that his correspondences between the zodiac and agriculture could be maintained for the "inventors of the zodiac"—those to which Esneh was presumably closest in time—if the ancient Egyptians had looked to the constellation that rises when the sun sets, instead of the one that rises with the sun.[57] For example, in Dupuis' original scheme, based on the simultaneous rise of asterism and sun, the summer solstice took place in Capricorn, which set the date of the zodiac's invention to about 13,000 BCE. Now suppose that Capricorn is instead the constellation that rises when the sun sets on the day of the summer solstice. In which case the solstitial sun must be somewhere in Leo. That in turn changes the epoch of the zodiac's invention to sometime during the third millennium BCE and would still maintain Dupuis' agricultural associations. Here is where Jollois and Devilliers diverged from Fourier, for he had not attempted to work with Dupuis' correspondences.

Jollois and Devilliers had work to do to make their understanding correspond to the zodiac at Esneh. There, they agreed with Fourier, it seems as though Virgo leads the parade, while Leo brings up the rear. Putting the summer solstice toward the head of Leo (with Capricorn rising as the sun sets with Leo) works perfectly well for Dupuis' associations, since then the end of Libra could naturally correspond to an equinox. However, if the solstice were to occur at the very front of Leo, Esneh could not date back much further than 1200 BCE, which was too late for Jollois and Devilliers' correspondences with pseudo-Eratosthenes to work. Consequently they had to make adjustments.

Suppose, they suggested, that the solstice did not have to occur at the front of Leo; suppose instead that it took place about halfway through the constellation. That was consistent with their claim concerning risings that the middle of a constellation was the point to which the Egyptians looked.[58] In that case, Esneh would date to about 2610 BCE—to near the very time calculated by Fourier on the basis of his quite different interpretation. Having conceived the idea, Jollois and Devilliers looked to the Esneh zodiac to see if they could find supporting evidence. Indeed they could: Virgo, they pointed out, is not actually depicted at the very front of the band, for it is preceded by a sphinx with the head of a woman and the body of a lion (Figure 8.6, second panel). In the panel above this one, two small lions bring up the far left. The paired little lions, they thought, signal that Leo should actually extend all the way to the left and, as it were, divide there, continuing below through the lion-headed sphinx. That might well have been done to signal that the solstice occurred midway through Leo.[59]

The young polytechniciens may nevertheless have continued to think that Egyptian astronomical knowledge dated back many more thousands of years than the invention of the zodiac proper. In their separate article on the monuments at Thebes in the same volume of the first edition of 1809, they divined evidence for the claim in the long avenue of stone rams there. Was it not built, they wrote, explicitly referring to Dupuis' original associations, "to recall the astronomical epoch when the celestial ram [Aries] occupied the autumnal equinox, when Capricorn was at the summer solstice, Libra at the spring equinox, and Cancer at the winter solstice, that famous epoch to which the primitive origination of the Egyptian zodiac had been referred?"[60] Or perhaps not. Perhaps Aries represented the vernal equinox, in which case the epoch would be much "closer to us" (in fact no earlier than 1000 BCE).

Jollois and Devilliers' interpretations were based directly on what men like themselves would have done in similar circumstances, men with attitudes forged in the drafting tradition of Monge. They reasoned backward from the Egyptian artifacts not merely to times and places in the distant past, but to the very methods, the very techniques, the very modes of work that the ancient Egyptian designers must have deployed. These could not have been men like Denon, who sought to capture the feel and sense of a place and a time, or men who figured their beliefs in stone. No, they had to be simulacra of polytechniciens, men who would have tried to sculpt the truths of the heavens by means of mathematical rules. Fourier had figured his Egyptians as the sage and knowledgeable inhabitants of a time that had scarcely ever been improved upon. Inhabitants of a world, as it were, in the mold of an early empire French ideal, a monarchy governed well and justly by the wise. In such a world, talented Egyptian savants would certainly have tried to represent the truths of nature, inadequate and undeveloped though their techniques might have been. The polytechniciens were not trying to see through an arcane and mystical symbolism, for they did not think there was anything at all occult about the zodiacs. To them, these temple ceilings seemed to be imprinted directly by nature, their carvers acting only as admittedly imperfect mediators between sky and rock.

Jollois and Devilliers took this to an extreme with the Dendera circular. That, in their view, was literally a projection of the heavens done using a simple sort of rule. Here they thought that the summer solstice was designated by the part of Cancer nearest to Leo, which, though they did not so write, would place it posterior to about 1500 BCE. This was certainly consistent with Fourier's claim that the Dendera rectangular had to postdate 2100 BCE. Mathematical exactitude was not to be expected at such an antique date, which, they averred, accounted for certain inconsistencies in the plot. The great circular zodiac was accordingly an object on which the relationships among the stars in the heavens were maintained according to fixed but imperfectly applied rules—a true planisphere. And the rectangulars were also planispheres of a kind, albeit "constructed according to a different method." What those rules were for the circular zodiac, and whether Jollois and Devilliers had thoroughly set them out, was to be a major point of contention when it reached Paris in 1822.

They did not arrive at their high estimation of ancient Egyptian astronomical knowledge after long and careful reflection; it was present from the very moment that they and others of like background had first set eyes on or heard about the astronomical ceilings. Another member of the expedition, the young chemist Hippolyte Collet-Descotils, was particularly excited when he learned about the ones at Dendera directly from Denon. On June 6, 1799, the *Courier de l'Égypte* printed Descotils' letter, in which he gave a brief but good description of the zodiacal figures. "These objects," he enthused, "give an elevated idea of the astronomical science of the Egyptians, and make one regret the loss of the hieroglyphic language." Unfortunately, many of the innumerable figures on the temple walls had been defaced. By whom? There was no doubt: "Drawings [on the walls] similar to those one sometimes sees in France on walls, and which represent men with crosses, would make one think that the mutilation of one of the most beautiful monuments on the earth was due to Christian fanaticism."[61] Here, in the deserts of Egypt, the savants spied the destructive workings of religion in the very place whose stones stood silent witness to a long-lost, immensely ancient science. But astronomically minded savants were not alone in pondering the mystery of the zodiacs; nor were they the only ones who had absorbed Dupuis' way of thinking about antiquity. Philology in the person of Remi Raige could compete with astronomy in deciphering the zodiacs of ancient Egypt.

Raige was brought on the expedition as a translator for his purported knowledge of Arabic, which al-Jabarti thought execrable. He was certainly no polytechnicien, having been schooled in the methods and ways of philology. In this respect, Dupuis, despite his expertise in languages and rhetoric, resembled more the outlook of Jollois, Devilliers, and Fourier than he did that of Raige. Recall that Dupuis' father had been a mathematics instructor, that Dupuis' own career was sparked when Rochefoucauld found the boy measuring with a surveying device, that Dupuis collaborated with the instrument maker Fortin both in producing a precession globe and in working a semaphore telegraph, and that he avidly followed Lalande's course in astronomy at the Collège Royale. These were hardly ordinary occupations for a philologist even in late eighteenth-century Paris.

Although Raige did not have Dupuis' training in astronomy, or his interest in devices, he was nevertheless an absolutely convinced adherent of Dupuis' system.[62] The ancient Egyptians had invented the zodiac fifteen thousand years ago, when the summer solstice lay in the constellation Capricorn. This had to be so—this was Raige's own contribution—because the very *names* of the constellations constituted a sort of linguistic zodiac in themselves. Take, for example, the Egyptian month Athyr.[63] It must designate "the month of the bull, because athyr in Egyptian signifies ox, bull, as Hesychius attests." Ancient Egyptian was of course long lost, but Hesychius was a fifth-century CE grammarian in Alexandria who produced a lexicon of obscure Greek words that, though extremely corrupt, had been printed in Italy early in the fifteenth century. It included occasional remarks about Egyptian words as well, though certainly by Hesychius's time ancient Egyptian had long since ceased to be spoken. The ox would naturally signify the labor of that animal in working the soil, which takes place after the Nile flood subsides, after, that is, November. The constellation Taurus must have been rising with the sun (according to Dupuis' way of thinking), which immediately produces (at least) his fifteen thousand years, and Raige made the dating explicit. Raige was no astronomer, and he thought that at that time also Sirius's heliacal rise would have corresponded with the summer solstice, which is very far from being the case.

If Athyr occurred in ancient times in November, then the Egyptian month in which the solstice took place must have been Epiphi. And so Epiphi had somehow to connect terminologically to Capricorn. However, at the time of Herodotus, the year began with Thoth, two months later. How could this be? Through Raige's conviction that "the Egyptian language must differ little from Phoenician and the dialects which have not ceased to be in use in Syria and Arabia." For Arabic has the verb *hebheb*, which in Edmund Castell's 1788 *Lexicon Syriacum*, used by Raige, has among others the Latin meaning "*qui evigilavit, qui experrectus fuit è somno*," that is "which is wakeful, which bestirs itself from sleep." And that, according to Raige, must certainly refer to the rise of the Nile. He went through the rest of the month names, building his case on similarly questionable grounds.

The zodiacs of Dendera and Esneh must have been constructed millennia after the naming of the constellations because in them, Raige thought, the solstice occurs, respectively, in Leo and in Virgo (this last point would push Esneh back to the fifth millennium BCE). He remarked that his own aim was not actually to establish zodiacal chronology per se, but rather to link the names of the Egyptian months to the constellations. The placing of the solstices at Dendera and Esneh was, he wrote, "due to M. Fourier." Either Raige had not seen Fourier's work when he wrote this, having perhaps just spoken a bit with him, or else he did not understand it (which is equally likely given the abbreviated character of Fourier's piece for the *Description*), because, as we have seen, Fourier located the inundations, not the solstices, in Leo and Virgo. Nevertheless there is little doubt that Raige would have deferred to Fourier on this point, and no doubt either that Fourier, Raige, Jollois, and Devilliers continued to share Dupuis' quarter-century-old claim (if not his specific chronology) that a more primitive astronomical knowledge had to date back many centuries before the construction of Esneh and then Dendera.

Though it would not have been wise explicitly to embrace Dupuis' assertions, Napoleon's discord with the church had eased the situation somewhat. Raige nevertheless took care, since at the very end of his piece he added the following note. "I must forewarn the reader," he wrote, who has "without doubt . . . sensed that reasons of a higher order make it necessary to regard these results as hypothetical; I have presented them in the affirmative in order to avoid repetitions and incidental questions." These "reasons of a higher order" likely refer to the dangers of offending the sensibilities of the religious Right.

The Naturalist

Although Napoleon's attitude toward the church had certainly changed by 1808, in that year an influential member of the Parisian scientific community had encountered religious disapproval. Thirty-nine at the time, the naturalist Georges Cuvier, head of the National Museum of Natural History and one of the two permanent secretaries of the Institut National, had written a report on the contemporary state of science in France. The report failed to make appropriate mention of the deity or even of religion at all and had as a result encountered opposition.[64] It finally appeared in 1810, the same year that the first volumes of the *Description* also reached print. Two years later Cuvier's great compendium of comparative anatomy was itself published. In it Cuvier attacked claims for the remote antiquity of Egypt, which he well knew to be associated with Dupuis. And yet Cuvier was neither profoundly religious himself nor interested in providing evidence to support the assertions of scriptural chronology. Here, then, we have for the first time a disagreement with the savants' claims that does not seem to have been motivated by religious conviction, or for that matter by phil-ological commitments that would be at odds with an immense age for Egyptian civilization.

Cuvier was born in Montbéliard, a Francophone area that formed part of the Lutheran Duchy of Württemburg and that would eventually be incorporated into France. Educated at Stuttgart, where he learned fluent German, Cuvier became tutor to a Protestant noble family in Normandy, pursuing natural history in his free time. Always ambitious, like the young Fourier he sought recognition in Paris, where he sent zoological papers for perusal by eminent naturalists. At first enthusias-tic about the revolution, he was disgusted by the violence in Caen. In 1795, following the end of the Terror, he made his way to Paris. There he obtained the support of Geoffroy Saint-Hilaire at the museum, who had replaced Lacépède when the latter left Paris, sickened by the Terror's massacres. Cuvier lived at the museum, at first lodging with Geoffroy and thereafter in the house of Jean-Claude Mertrud, the elderly profes-sor of animal anatomy, to whom Cuvier was appointed understudy. His own interest at the time was the anatomy of invertebrates, in particular the marine fauna that he had observed in Normandy. Working with Geoffroy, he expanded his concerns to include mammals. Toward the end of the year Cuvier achieved appointment to the section of anat-omy and zoology of the new Institut National's First Class, becoming thereby its youngest member.[65]

During the years between 1795 and 1803, Cuvier became increas-ingly powerful. He had great oratorical ability in a period when it remained necessary for those who aspired to power to impress the public. It was necessary as well to assemble a broad array of patrons among whom the aspirant to power and influence had to maneuver carefully. Cuvier successfully "confronted all the conflict which sur-rounded the gap between the apolitical ideals of science and the mak-ing of a scientific career in a rapidly changing and highly political metropolitan society."[66] This changed as the empire's structure stabi-lized, with the consequent clarification of the channels of authority. Cuvier's insight into the mechanisms of influence, and the fertility of working with the existing collections in Paris, are illustrated by his decision not to accompany the Egypt expedition, though his friend and supporter Geoffroy did. Remaining in Paris, Cuvier was able to consolidate his position, and to argue as well for the superiority of his sort of anatomical probings over the field observations undertaken by others, in particular by Geoffroy, who would write plaintively from Egypt asking that he not be forgotten—letters to which Cuvier seems never to have replied. In 1803 Cuvier even refused to support Geoffroy for the position at the reorganized institute that became vacant when Cuvier left it to become permanent secretary. Years later the enmity between the two men would emerge in public as they sparred over whether, as Geoffroy would eventually contend, species transformed over time or whether, as Cuvier had long insisted, species disappeared in cataclysms, somehow to be replaced by new ones.

Cuvier's approach to comparative anatomy had begun to gel in 1796 when he was asked to report on a large animal skeleton from Spanish South America. Arguing that the fossil was distinct from any known species, Cuvier showed that the skeleton exhibited structural characteristics similar to those of modern creatures like sloths and anteaters, though this fossil was immensely larger. He evolved a scheme according to which anatomical structure obeys strict laws of "combination" and "subordination" of characters and in which function connects directly to form. Confident that his method produced a rigorously invariant result, Cuvier argued for the subordination of geological speculation to the precise consequences of comparative anatomy. This also meant that the museum's geologist, Faujas de Saint-Fond, ought to be intellectually subordinate to Cuvier.[67]

During the eighteenth century and the last part of the seventeenth, speculations on Earth's history had often popped up. Many of these were in the vein of "geotheory"—vast theories of Earth that were only tangentially grounded in evidence. Most of them invoked cataclysms of one sort or another. During the last half of the eighteenth century a new form of Earth history appeared, one that was much more closely tied to evidence. This new form, argues Martin Rudwick, was based on the appropriation of techniques for investigation that had been developed for local histories by antiquarians, such as the interrogation of coins, of monuments, techniques of epigraphy, and early archaeological digs. Pompeii and Herculaneum were, for example, first excavated during these years. The new approach concentrated analogically on details such as rocks and fossils in order to construct, as it were, the "geohistory" of a locale. This could, and did, shade into grand geo-theory, but of a sort that was grounded in observations.

Jean-André de Luc worked a geohistorical vein based on fossil evidence that merged easily into a geotheory. We have already encountered his influence in the early career of Domenico Testa, the Vatican's forceful defender of scriptural chronology. De Luc was decidedly religious, but his geotheory did not envision a mere six millennia for Earth's history. Not at all: he was convinced that Earth had gone through many "revolutions" over immense periods of time, of which the most recent separated the prehuman from the human world and separated as well the present state of the continents from Earth's prior structure. This last revolution he identified with the Noachian Deluge. His views were developed in published letters and became widely known during the early 1790s, while De Luc lived in England for reasons of employment, having obtained a burdensome but lucrative situation at court. Crucial to De Luc's argument was the claim that the continents could not have existed as dry land before the last revolution, for if they had then humans might well have existed as well, thereby utterly violating scriptural chronology. Earth itself might be immensely older than six millennia (it was simple to reinterpret Genesis's days in appropriate ways), but once humans appeared biblical chronology stepped in and, for someone like De Luc, had to be accommodated.

De Luc's views were particularly congenial to British Tories, appalled as they were by the radical developments in France during the 1790s. One Tory journal, the *British Critic*, opined of De Luc that his work gave "demonstrative evidence against those who delight to calculate a false antiquity to the world, inconsistent with the sacred records."[68] In 1798 De Luc was appointed professor of geology at Göttingen, the renowned university town in George III's Hanoverian possessions, where he explained that his geological views supported religion and countered the radical horrors of the revolution.[69] When claims for the Egyptian zodiacs' great age appeared in the *Moniteur*, De Luc reacted vehemently. "This conclusion," he insisted, "given with a tone of assurance, may easily deceive, and make us suppose that it is well founded, though it can rely only on conjectures or mistakes in the application of astronomical calculations; but in such speculations, as they are restrained by no religious persuasions, they follow their own ideas, without ever inquiring whether they can be reconciled with nature." These impious reflections, he accused, derive in the end from the nefarious Voltaireans "throwing out their sophisms and their sarcasms against the account given by Moses." The *Edinburgh Magazine*, which approvingly cited and translated De Luc's remarks, was especially enraged by these sorts of claims and urged that "every attempt to discredit revelation, whether it derives its force from the relics of ancient times, of from the present appearance of nature, deserves to be particularly exposed." Particularly obnoxious was "the celebrated argument taken from the obliquity of the Ecliptic" for it served only "to shew the *credulity* of *infidel* philosophists"—a clear reference to Dupuis.[70]

Cuvier certainly did not share De Luc's deep-seated religious convictions, but he did agree with De Luc on the shape and development of Earth history—namely, that Earth had undergone a series of catastrophes over immeasurably long periods, with the present shape of the continents and humans having emerged at the close of the last one. Rudwick suggests that the match between Cuvier's views and those of the decidedly counterrevolutionary De Luc meant that Cuvier "could hardly avoid being suspected in Paris of having counter-Revolutionary sympathies." Cuvier nevertheless managed to find his way through these potentially dangerous obstacles.

The "preliminary discourse" to Cuvier's *Fossil Bones* was a public, comparatively accessible brief for his basic outlook, an exercise in refined albeit frequently overblown rhetoric. Every work of science is rhetorical in the sense that it is designed to persuade, but not every work is designed to do so through the artful use of language. Cuvier's "preliminary discourse" certainly was, just as the one by D'Alembert to the great *Encyclopédie* had been. Based on his public lectures, the "discourse" framed Cuvier as the central figure, presenting him as a "new species of antiquarian."[71] He would probe the fossil record to establish a natural analog of historical accounts. But history entered as more than a methodological analogy, for Cuvier intended to use antique textual records, corrupt though they may be—as corrupt as a fragmented fossil—to buttress his account of the most recent cataclysm. Peel away the layers of legend and fantasy from the historical record. "What then remains, is a body of convergent textual evidence that the earth's surface had indeed been ravaged by a 'catastrophe' of some kind, only a few thousand years ago, back in the infancy of human civilization if not of humanity itself."[72]

Not the least among the several historical records that Cuvier deployed was the Mosaic account of the Deluge. "The Pentateuch [the five books of Moses]," wrote Cuvier, "has existed in its present form at least since Jeroboam's schism, for the Samaritans accept it like the Jews;

that is, it is certainly more than 2,800 years old. There is no reason not to attribute the writing of Genesis to Moses himself, which would take it back another five hundred years."[73] Unlike the Egyptians at the time of Herodotus and afterward, whom Cuvier accused of greatly exaggerating their antiquity, Moses "had no motive for shortening the duration of the nations; and he himself would have been discredited among his own nation if he had taught a history quite contrary to what he must have learned in Egypt." In those remote days, the powerful Egyptians had no reason to overstate their antiquity, and the Hebrew slaves in Egypt would have learned an appropriate chronology from their masters. Moses's account of an "irruption of the waters" placed that catastrophe "fifteen or sixteen centuries before himself according to the texts that extend that interval furthest, and consequently to at least five thousand years before us." This Deluge occurred no later than 3,200 BCE, which is about eight hundred years earlier than most biblical chronologists had placed it. Though Cuvier was not concerned with scriptural apologetics, his widely read "preliminary discourse" happily deployed textual accounts of a massive Deluge (including one by the Babylonian priest, Berossus, that dates from the time of Alexander's conquest and which survived only in excerpts and excerpts of excerpts).

What then of the Egyptian zodiacs, or for that matter of Dupuis' fifteen thousand years? The zodiacs were artifacts that might have posed a particular sort of problem for Cuvier. They were, on the one hand, material evidence of a kind that would necessarily grip "antiquarian" interest. But, according to Fourier and others, they were much more than that because the zodiacs were thought to imprint the ancient sky onto stone, evidence rather like the bones that Cuvier had used to reconstruct the anatomies of vanished creatures. He could hardly have been unaware of the issue since it was prominently developed in the recently printed *Description*, and since he would certainly have heard about it from some of the expedition's members, and perhaps even from Fourier himself (though there is no direct evidence of their having discussed the matter at this time). The *Description* project was after all a major effort of the period, one with which Cuvier had early involvement, and he was well aware of the expedition's collections of animal mummies, having with Lamarck and Lacépède signed a report concerning them.

Cuvier did not cite anyone about the zodiacs, but his remarks show that he was responsive to the earliest available accounts, though perhaps not to the details concerning either the Dendera or Esneh depictions as discussed by Fourier and by Jollois and Devilliers. Though he ignored Esneh altogether, we will see in a moment that Cuvier may have had its zodiac at least obliquely in mind, for he had an argument about the rate of scientific discovery that looks to be pertinent. That zodiac was, as we have seen, dated by the *Description* authors to about 2500 BCE, which would seem to provide little enough time since Cuvier's last cataclysm for astronomy to have reached such a high level of development. As for Dendera, Cuvier might easily have dismissed its relevance to his considerations on the grounds that even Fourier had dated it comparatively late, but he did not, which very likely indicates that he had not taken the time to read through the *Description* accounts more than superficially, if at all.

To undercut the danger posed to chronology by Dendera, then, Cuvier relied on the 1802 critique of the original reports concerning it by Visconti. "Nothing proves," Cuvier wrote, "that [the Dendera rectangular's] division into two bands, each of six signs, indicates the position of colures resulting from the precession of the equinoxes, and doesn't correspond simply to the start of the civil year at the time it was designed."[74] And since the circular zodiac started with Virgo, it must be vastly later than the date assigned to it by the savants—provided that it had calendrical rather than astronomical significance. These were Visconti's, and only Visconti's, arguments. Cuvier would likely have met him at the famous salon run by Denon, or perhaps at his own, and he may well have consulted him on zodiacal issues. Visconti was after all well-known in Paris. Though Napoleon had chosen Denon over him as general director of the new Musée Napoléon, Visconti had become director there of Greek and Roman antiquities and had shortly thereafter been appointed to the Institut National. Throughout the Napoleonic era, Visconti's house was moreover a meeting place for Catholic eminences. It would have been entirely natural for Cuvier to approach him about arguments against the remote antiquity of the zodiacs, even though he did not actually need anything to preserve Dendera's comparative youth in view of the *Description*'s own accounts. Later versions of his "preliminary discourse," renamed the *Discours sur la théorie de la terre* (*Discourse on the Theory of the Earth*) and printed separately, had extensive remarks on the Dendera zodiac, as well as on Dupuis, that explicitly mentioned Visconti.[75]

Cuvier would certainly have known Dupuis' claims for Egypt's remote antiquity, and he could hardly have missed the enthusiasm for them on the part not only of Fourier, Raige, Jollois, and Devilliers but of Destutt de Tracy and others. Even a quick perusal of Raige's article in the *Description* would indicate Dupuis' continued influence. And that would, above all other claims, have been disastrous for Cuvier's dating of the last cataclysm. It is hardly surprising to find that, without mentioning the now-deceased Dupuis by name, Cuvier deliberately undercut his claims. "It remains to be known," he remarked, "whether our zodiac does not contain in itself some proofs of its antiquity, and whether the figures that have been given to the constellations had any connection with the position of the colures [in effect the equinoxes and solstices] at the time they were conceived. Now all that has been said in this respect is based on the allegories that have been claimed to be seen in these figures: that that Balance [Libra], for example, indicates the equality of days and nights" and so on through what are indubitably Dupuis' (and Raige's) associations.[76] All this should be rejected, Cuvier went on, because it is purely allegorical, meaning there is no material evidence of any sort for the connection, and because, were it correct, then different associations should have been produced "for each country, such that the zodiac would have to be given a different date, according to the climate to which its invention was assigned." This argument in itself indicates that Cuvier knew about Dupuis but had not read his immense work (who had, given the length of the turgid volumes?), or perhaps even de Tracy's or Dupuis' own shorter versions. For Dupuis' central argument had been more precise, namely, that the associations would work specifically and uniquely for Egypt fifteen thousand years ago if constellation risings with the sun were involved. No other set of links, he had claimed, would work in other climes.[77]

Cuvier simply rejected altogether any possible chronological significance of the zodiacs, whether materially inscribed on temple ceilings or iconographically associated with the seasons via allegory. "Who knows," he commented, "whether the names were not given very long ago, in an abstract manner, to the divisions of space or time, or to the sun in its different states, just as astronomers now give them to what they call signs." Words themselves had little historical value because they were in Cuvier's view entirely arbitrary. Here we may detect an echo of disdain for *idéologie*, the doctrine held by men like De Tracy, for whom words were hardly arbitrary but bore the necessary imprint of the sensations that gave rise to them. The stronghold of the idéologues at the institute had been abolished by Napoleon in 1803 at the very same time that Cuvier had become one of the institute's two permanent secretaries. Though he may not have had a direct hand in the reorganization, nevertheless the abolition certainly enhanced Cuvier's influence and power.

The year that Cuvier's work appeared, 1812, was a fateful one for the empire. In June the half-million soldiers of Napoleon's Grande Armée invaded Russia. The Battle of Borodino on September 7 produced huge losses and ended without a clear resolution. A week later the French entered a silent Moscow, nearly empty of inhabitants; the next day the city was in flames, set deliberately by the Russians under Count Rostopochine, who denied his role years later. The events of the next months are too well-known to repeat in detail: the calamitous retreat of the French in horrific conditions, the devastation of the army and, with it, of Napoleon's carefully tended image. At the end of December, Prussia tore up its treaty with France and joined Russia in an alliance against the empire. French success at the Battle of Lützen five months later was undercut by the inability of Napoleon's tired and young troops to carry through, followed during the next months by the assembly of massive allied forces, subsidized in part by British money. Another French success at the Battle of Dresden in late August was also vitiated by failure to press the advantage, and in early October the Austrians, funded by the British, added 150,000 men to the coalition. Two weeks later the French suffered a massive defeat at Leipzig. Events moved rapidly thereafter, and in March 1814 the Allies were at the gates of Paris. On April 11 Napoleon formally abdicated, and, after a failed attempt on his part to commit suicide, he was exiled to Elba, which was placed under his charge as "Emperor of Elba." On April 26 Louis XVIII left England, entering Paris on May 3. A week later the Treaty of Paris was signed, formally restoring the monarchy and returning France to its boundaries in 1792. Six months later the Allies met at the Congress of Vienna to settle the political structure of Europe. Napoleon's abortive return from Elba in March 1815 ended with his final defeat at Waterloo in Belgium on June 18 and permanent exile to the cold and rainy island of St. Helena in the South Atlantic.

Throughout the nearly ten years between the appearance of Cuvier's remarks on the zodiacs and the arrival in Paris of the Dendera circular, the questions about them that had generated so much heat between 1802 and 1809 attracted comparatively little attention. Which is hardly surprising given the upheavals that accompanied the end of the empire, followed by the arrival back in France of a monarch and aristocrats who had not set foot in the country for a quarter century. There was, moreover, little further to argue about, certainly before the printing in 1817 of the *Description* plates of the Dendera zodiacs; until then only Denon's depictions of them were broadly available. However, during the early years of the restoration attitudes and beliefs evolved that were to have a substantial impact on the reemergence of the zodiac controversies following the Dendera zodiac's arrival at Marseilles on September 9, 1821. To understand what happened when it appeared in Paris, we turn now to the restoration.

9

Egyptian Stars under Paris Skies

Who could count the numerous products of impiety that have multiplied to an unbelievable degree in our times?

—Abbé Duclot, 1816

Restoration Paris

Napoleon was hardly content with the reduction of his realm to the island of Elba in the spring of 1814. Neither were his clients reconciled to their reduced circumstances, despite the fact that Louis XVIII's government introduced a charter that was designed expressly to keep the peace in France. This *Charte* was in many respects quite liberal, providing for freedom of religion and of the press, and permitting the holders of property taken by the state during the revolution to remain in possession. Although the peerage was restored, appointments to it were now in the hands of the monarchy. An elected Chamber of Deputies was also established, albeit with extreme limitations on voters based on property. After a short period of relief that Napoleon's wars had ended, many among the French nevertheless grew bitter and resentful. Much to the consternation of Parisians, English tourists flooded Paris during the summer of 1814. By early November the police noted that "hatred for the English is growing daily. They are regarded as the destroyers of French industry."[1] The exiled emperor's return from Elba in the spring of 1815 generated renewed enthusiasm, and so the rout at Waterloo was even more devastating than the first one had been. Not so much, perhaps, for the final defeat itself, as for what eventually followed.

Paris was again occupied by allied troops, this time including the British. At first the boulevards, restaurants, theaters, and cafés were filled with Parisians as well as with their occupiers. When thousands of soldiers entered the city in May, an Englishman by the name of Fellowes dyspeptically recorded that the "Parisians appeared to view this military parade as a mere *spectacle*: a word which they apply to every species of shew. I heard them frequently exclaim, 'Que beau spectacle!'—'Que

bels hommes'—'Que magnifiques cheveaux'—"'Vraiment c'est char-
mant.'" Fellowes visited Denon at the Louvre, who, he wrote, "appeared
to be more hurt at the prospect of the Gallery of the Museum being
stripped, than interested in any of the political events going on." Fellowes had little sympathy for that, since "at the Musée Napoleon are to
be seen the rarest specimens of art, collected into one focus by robbery
and plunder!" He continued, "Let us hope that the allied sovereigns,
in arranging the peace, about to be established, will not lose sight of
that retribution and justice, which the late government owe to other
nations, in making them restore their plunder."[2]

Fellowes did not hope in vain, at least with respect to the restoration of the collections. The Prussians were particularly concerned
to recover their artifacts, and they moved rapidly to do so. Whether
royalist, imperialist, or republican, the French were indignant at the
ongoing removals. John Scott, another Englishman who witnessed the
immediate aftermath of Napoleon's downfall, told of the reaction as
the removals went forward amid crowds of foreign tourists who had
rushed to Paris to see the collections before they were dispersed. "The
Parisians," Scott reported, "stood in crowds around the door, looking wistfully within it, as it occasionally opened to admit Germans,
English, Russians, &c. into a palace of their capital from which they
were excluded . . . the agitation of the French publick was now evidently excessive. Every Frenchman looked a walking volcano, ready
to spit forth fire. Groups of the common people collected in the space
before the Louvre, and a spokesman was generally seen, exercising the
most violent gesticulations, sufficiently indicative of rage, and listened
to by the others with lively signs of sympathy with his passion."[3] Denon
himself had presided like a king over the Egyptian collections, and
they remained undisturbed, primarily because the English had already
expropriated what they wanted at the time of the French surrender in
Egypt in 1801.

While the purloined artifacts disappeared from the Louvre, a new
Chamber of Deputies was elected, filled with angry royalist Ultras out
for revenge and restitution. This Chambre Introuvable (Incomparable
Chamber of Deputies), as it was disapprovingly dubbed by the king,
intended nothing less than the complete restoration of what the Ultras
imagined to have been the prerevolutionary dispensation. It was so
extreme that Louis XVIII dissolved the chamber a year later, and the
newly elected deputies, the so-called *Doctrinaires*, aimed at a more
liberal monarchy. Newspapers flourished during these years, and the

press certainly remained much freer than it had been under Napoleon (though always under suspicion, as we shall see). The monarchy nevertheless attempted to infuse the nation—or at least Paris, where all important events were thought to begin and end—with a renewed religiosity. It revived the religious pageantry of the ancien régime, leading eventually to mawkish spectacles in which sanctimonious aristocrats paraded solemnly through the streets of Paris with lit candles— whereas police surveys of the day show that Parisians themselves had at nearly every level of society become more irreligious than ever before. Louis XVIII's director-general of police reported to him in 1814 on "the repugnance of a large part of the public for every sign of a return to the most respectable religious customs."[4] Another English visitor to Paris, John Hobhouse, remarked that the "religion of Louis and his family will hardly have many charms for the French. . . . It is the laughter of Paris."[5]

The monarchy's affinity for symbolic display, and its hatred for republicanism, was further exacerbated by the dramatic assassination of the only male heir, the duc de Berri (son of Louis XVIII's brother, the Comte d'Artois), at the Opéra in 1820 by an antireligious, anti-Bourbon saddler, whose only utterance, when brought before magistrates, was "God is only a word."[6] The Ultras returned thereafter to power, though even then they were unable to transform France into their imaginary ancestral Utopia. The press continued to remain comparatively unfettered, much to the Ultras' disgust. In fact, more copies of what reactionaries thought of as the great Satans of the Enlightenment—Voltaire, Rousseau, and others—were printed during these years than ever before. Between 1817 and 1824 over two and a half million copies of philosophe texts reached print despite the formal existence of censorship and Ultra claims that philosophie underpinned every evil that had struck France since the revolution.[7]

The twice-restored monarchy was nonetheless intent on obliterating all memories of the revolution, and especially any reminders of the royal execution, to pretend as it were that the period between 1789 and 1815 had never existed. The eleventh article of the Charter of 1814 had already specified that "investigations of opinions and votes expressed before the Restoration are forbidden. The same disregard is demanded of both the courts and the citizenry."[8] During the year of the first restoration, ceremonies of expiation for the deaths of the royal family contravened the article's spirit by reminding everyone of just what had taken place. Furthermore, in an effort to keep the peace the monarchy tended to overlook the public display of revolutionary and imperial symbols. This changed following Napoleon's final defeat. His Hundred Days of return made it imperative thoroughly to erase public memories of the forbidden past, and on November 11, 1815, a law was passed that banned all such displays. The prefectural police had already been ordered to remove every symbol of empire and revolution. Ownership of even small relics of the Napoleonic and revolutionary eras (cockades, printed matter, costumes) was prohibited; police seized these items and destroyed them in ceremonial bonfires that usually took place within steps of the local church, as civil servants looked on, obediently shouting "*Vive le Roi!*" The aim was to legitimize the Bourbon restoration by utterly effacing reminders of any alternative to it, by ideologically purifying the country.[9] Though two annual national days of mourning for the executions of the king and queen were mandated, the regime downplayed the events by having only Louis XVI's final words of pardon for his tormentors read in the country's churches on January 21, followed by the queen's denial of the accusations against her on October 16. There were to be no reminders of the events themselves.

Many of the aristocratic returnees were too young properly to have experienced the ancien régime, but their elders were angry and vengeful. The monarchy may have wished to underplay reminders of what had taken place, but the aristocracy and its supporters, as well as the church, continually disinterred the recent past. Since, however, there was no ideal lost world to regenerate, the religious and monarchist Right invented one. As extreme in its way as the creations of the revolution's early years, this imaginary world permeated the written words and actions of the right. It was during these years that the adjective *libérale* became a word of opprobrium, used to castigate those who were committed to the liberty of the press and to the tolerance of religions other than Catholicism. "The ultras," wrote the historian François Furet, "reinvented a cult of monarchy while they believed they were maintaining or restoring the old one; but the new cult, feeding on the Bourbons' misfortune, drew its substance only from the revolutionary years. It intensified the separation of king and people, instead of averting it."[10]

For its part, the church wallowed in mournful remembrances of the past. Missions were sent out to reinstill France with an appropriate level of belief. One among them built three immense crosses on a hill visible from Paris. The Abbé Rauzan (who became known as "the

Apostle of the Church of France, the glory of the Church of Bordeaux, and the model of priests") dedicated the monument in the early fall of 1816, praying to his deity that these symbols of power would "draw to him this guilty city, the center and source of so many iniquities."[11] In the provinces, the monarchy's attempt to forget the drama of regicide was trumped by the missions, which bathed in the gore of the iniquitous past as tens of thousands of spectators walked the paths trod during the Terror by the guillotine's victims, to end at huge crosses where they "were invited to sing canticles and pray for expiation for the sins that the nation as a whole had committed during the Revolution."[12]

These morbid and maudlin displays were echoed in the pages of the right-wing press, with its yearning for a past in which an obedient populace bowed gratefully before the indivisible unity of throne and altar. This was hardly an attempt at a return, since there never had been such a thing to return to. It was rather an effort to create an entirely new polity, thoroughly hierarchical, saturated with Catholic dogma, and utterly intolerant. Among the most vociferous propagandists for this ideal world was the rabidly anti-Semitic advocate of censorship, Louis de Bonald. Having returned to France with other émigrés, Bonald was at first considered dangerously suspect but in 1806 joined Chateaubriand and Fiévée at the *Mercure de France*. A member of the Chamber of Deputies for seven years from 1815, this visionary of the Right detested anything that struck him as *libérale*. More extreme than Chateaubriand, who never completely rejected the Enlightenment, Bonald nevertheless shared with his colleague (and with the Ultra Right in general) a deep suspicion of scientific reasoning. He admitted that materialistic science had its uses, but these were trivial in comparison to a proper science of the soul, which derived entirely from clear-sighted faith—Catholic faith. There was an additional cause for anger at science, one that others who did not necessarily share the Ultra predilection for a mystical union of throne and altar might appreciate. Bonald was a man of letters, as well as a bigot, and a persuasive one at that. He feared science as much for its displacement of men like himself from the center of French polity as for its nefarious effects on believers in the sacred mysteries of the church.

"All of these sciences," Bonald had already written in 1807, "of which matter, considered in its *quantity* or its *qualities*, is the subject, speak a *technical* language strange to literature properly speaking, and use stylistic forms unknown to it. The ones [physics, chemistry] proceed by *axioms*, by *theorems*, by *corollaries*; the others [medicine, the organic world], by nomenclatures and classifications of species and types." These things can be taught almost mechanically, and they do not touch the true depths of understanding. "But *theology, morality, politics, jurisprudence, history* can only be taught through words, in such a way that they are introduced into the soul." Indeed, "the moral sciences are the first and most beautiful of all the sciences, because they can only be produced by means of the most beautiful and elevated writing." They are necessarily joined to religion, Bonald argued, which is why "the philosophy that was in vogue during the previous century, hated religion and thought nothing of politics and morals."[13] This must be fought, this hateful philosophy that "confounds, in man, the soul with the organs; in society, the sovereign with the subjects; in the universe, God himself with nature, everywhere the *cause* with its *effects*, and which destroys all general and particular order in denying all real power to man over himself, to the heads of states over the people, and to God himself over the universe."[14] Most hateful of all would be any work that traduced history by marrying it to material science in order to demystify religion. That is precisely what Dupuis' *Origine* did, and what made it, and anything linked to it, an object of particular fear and hatred on the Right, who wanted it stopped.

The Sainted Bible Avenged

The *Origine* had cropped up in discussions on censorship during the empire. On April 11, 1809, a long meeting was held at the Tuileries palace on issues related to printing and bookselling, ever a fraught subject for governments fearful of what might be spread about. One of the participants, the Comte Treilhard, who had been instrumental in forging Napoleon's civil code, defined press liberty as the right to print what harmed no one. The Comte Boulay, who had aided Napoleon's rise to power and who had also worked on the civil code, defined it as "the right to write what is useful." A certain N*** demurred, saying that "this was not at all the idea that he had formed about the liberty of the press." He argued that press liberty "doesn't exist in France, because one cannot write on all matters." This caution was unfortunately necessary at the time (1810); the only question was the method of censorship, and who was to be in charge of it. However it was to be done, N*** argued, the process should not have the censors judge every work printed, giving them the power to pass only those they wished. "How many books would be stopped without real motive," he argued. Why, "it's not even certain, for example, that one would permit Dupuis' book on the origin of religions to be printed, even though it consists only of learned and systematic discussion."[15]

N*** brought up Dupuis' *Origine* again later that day. Portalis, director of the Department of Religion, asserted that where a state has a "dominant religion, censorship must without doubt repel all books that are contrary to its principles." However no religion, he asserted, was dominant in France, so that "among us writings contrary to religion would not become criminal." Still, need one permit direct attacks on religion"? "No," he continued, provided that "the attack is directed against morality." But "if it's [just] contrary to dogma, it's not possible to prevent it." Once again N*** objected to the logic because, he pointed out, "as a matter of fact, the Christian religion is the national religion, because it's professed by the greatest part of the French. The points that divide Protestants from Catholics don't forbid them from agreeing about the foundations of doctrine. Authority must therefore respect the Christian religion. Would it appear to do so if it approved a book that granted the world a much longer existence than Scripture grants it? . . . What would happen if, instead of a book that only wounded religion in a few points, there was one that, like the one by Dupuis, was directed altogether against it? Would the censor let it be printed? If it did, it would pronounce against religion. If the censor could reject it, then the censor would be dangerous. . . . These are the drawbacks of forced censorship."[16] N*** was not opposed to censorship, though he would have excepted a work like Dupuis'. He wanted instead to be sure that the procedure did not suppress a book *tout court*, but that it would point out where improvements had to be made.

Although these conversations took place under the empire, when formal procedures for censorship were being instituted, they were printed in 1819 during the restoration by the Baron Locré, who had been in charge of the secretariat of the Council of State from 1799 to 1815 and who reported directly to Napoleon. The issue of censorship was especially important to the Ultras, incensed as they were by the continuing presence of so many works of philosophie. Locré decided to print the earlier discussions because "when one undertakes to treat a difficult and important question, knowledge of the manner in which it had already been done is not to be disdained: it indicates at least a point of departure." There can be little doubt that the anonymous N*** was Napoleon himself. His name was redacted from the 1819 printing because it would have been unwise to refer directly to the deposed emperor. Nevertheless, anyone who read the text would have known perfectly well who N*** was and would have seen the exiled monarch's evident admiration for the *Origine*. That work accordingly not only was a reminder of the previous century's skepticism toward religious belief but carried as well the perilous memory of empire. Moreover, Dupuis' views had become especially widely known by 1812, when Francoeur, an early graduate of the École Polytechnique who had taught mathematics there in 1804, published the first of several textbooks, an elementary treatise on astronomy entitled *Uranographie*. The text proved popular and was widely used throughout France, with two editions in 1812, two in 1818, and one in 1821, the year the Dendera zodiac arrived in the harbor of Marseilles. The work continued to appear for years, and every edition of it contained a paean to the *Origine*, including one as late as 1838, nine years before Francoeur's death. "The sagacity of his interpretations," wrote Francoeur of Dupuis, "their striking accord, the happy reconciliations to which the results lead, are worth the attention of philosophers. This erudite professor was far from foreseeing, when he composed this beautiful work, that, 25 years later, the French would conquer Egypt, and would bring back documents that confirm his opinion on the origin of the zodiac."[17]

The *Origine* and works like it were vigorously attacked in a new organ of the religious right, the *Ami de la Réligion et du Roi*. It was founded in April 1814, as soon as the empire had ended (for the first time), under the direction of Michel Picot, who had studied theology, and who had refused to take the revolutionary oath for the civil constitution of the clergy. His 1806 *Mémoires* on the history of the church during the eighteenth century had been widely admired by the revived religious Right. Reprinted in 1815, Picot's *Mémoires* were more propaganda than history, for which Picot made no apologies, insisting that he had no intention of writing in a "cold," nonpartisan manner. Dupuis merited a particularly vituperative passage. His *Origine*, wrote Picot, took impiety to an altogether new level. "Two abridged editions of his work were made," he wrote, "in order the better to propagate the poison, and the better to mislead the inattentive and credulous young; and we saw, with shame and scandal, this dark compilation praised in the breast of the Institut by writers who later no doubt blushed with shame at this ignominious cowardice."[18] Here Picot certainly had in mind the idéologue De Tracy, author of one of the two abridgments of the *Origine*.

In pious contrast to the blasphemous history offered by Dupuis, Picot expounded the emerging Ultra vision of an idealized past that had never been. "What a beautiful spectacle, is it not," he wrote in 1816, "to see reappear in our midst that beautiful period of the golden age . . . a time of innocence and virtue, absolute mistress of the human heart, reigning sovereignly, where vice was known only through the horror it inspired, where probity and good faith were highly venerated." Picot's idealized falsification of the past projected, and to that extent shaped, a particular sort of future. He encouraged his readers to believe that a reactionary triumph was inevitable, and that opposition to it was not merely futile but indecent. At long last those obstinate opponents of this perfect dispensation, willfully blind to its true beauty, must surely "blush with shame at their perversity and envy a happiness that can only be found at the feet of throne and altar."[19]

That same year, on October 9, Picot's journal presented Abbé Joseph Duclot as the favored historical antidote to the poisonous Dupuis. Like Picot himself, Duclot was an avid antagonist of impiety. His six-volume defense of Christianity, combatively entitled *The Sainted Bible Avenged*, reached print for the first time in 1816. Translated into Italian in 1818, the entire treatise was reprinted in French in 1824. In it Duclot castigated the impious philosophes for their refusal to look more than superficially at religion, and he proposed to work through every possible argument against belief that he could think of. His object was "to avenge the Bible from the blasphemies of the impious" by bringing together "the victorious responses of that crowd of scholars who have so well defended the truth."[20] Among the particularly impious works that exercised Duclot was, of course, Dupuis' notorious *Origine*.

Duclot rejected the very foundation of Dupuis' theory, asserting that the zodiac does not in any era correspond to the climate of Egypt. Rather than argue his case, he simply insisted that Dupuis' several correspondences do not work, or fail to make sense, though he considered only three among them. Moreover, he mistook Dupuis' carefully constructed linkages. According to Duclot, Dupuis had set Cancer the Crab at the summer solstice, which seemed absurd, for how could such a creature, which "marches backwards," correspond to a time of the year when the sun "marches in completely the opposite way"—presumably, that is, as the sun has climbed to the highest point in the sky. Yet Dupuis had in the end set Cancer at the winter solstice, with Capricorn at the summer, precisely because the alternative could not link well to the seasons. Duclot seems to have known this, but he insisted that even Dupuis' reversal was impossible, for which he gave no reasons at all. He went on to say that an association between zodiac and climate could work, but only for Assyria, not Egypt. And, in any case, no linkage could possibly predate the Deluge because Genesis attests that before then the seasons did not exist at all.

If Duclot's engagement with Dupuis was superficial, it is likely because the learned prelate had his own theory to press: namely, that the zodiac must have been invented by rude peasants and shepherds who simply noted associations between the sun's position among the stars and natural events, such as when sheep give birth. Having dispensed with Dupuis, he advanced enthusiastically to a consideration of Cancer the Crab, which Duclot himself now connected to the summer solstice, because the Crab "marches backwards, and so designates the retrograde march of the sun." Just seven pages earlier, he had denigrated the very same association on Dupuis' part. As this contradiction shows, it is difficult, and probably impossible, to make Duclot's arguments coherent.[21] Possibly these difficulties involve misprints or mistranscriptions, but, whatever the case may be, Duclot's arguments against Dupuis are hardly cogent. They could only be persuasive to someone looking for any reasons at all to reject zodiacal antiquity.

Having "avenged the writings of Moses concerning the epoch of creation" from several attacks, especially Dupuis', the resolute Duclot went on to show that the records of the "Phoenicians, Chaldeans, Persians, Egyptians, Chinese, Indians, etc. . . . accord very well not only with the dates of the Creation and the Deluge according to Moses, but even with the chronology of the Hebrew text and the Vulgate." To do that Duclot agreed with a number of his predecessors that the Persian sage Zoroaster must have been a Jew, a disciple of Daniel in fact, and that Zoroastrianism was a form of "the religion of Abraham." The footsteps of Moses could be found everywhere, and there was every reason to take literally the Mosaic account of the Deluge.[22]

The Egyptian zodiacs did not escape Duclot's avenging pen. It is hard to see how they might have, given the hypervigilance with which Duclot and his sympathizers monitored popular and scholarly discourses about antiquity. Those stones which had elicited "victorious cries from the enemies of religion . . . reversed from top to bottom Mosaic chronology by their great antiquity, and, in natural consequence, the entire edifice of religion."[23] Duclot was particularly annoyed by Joseph Garat's funeral oration for Kléber, which had pronounced the great antiquity of Egypt. But, he went on, there was no need to worry, for Visconti and, especially, Testa had shown how utterly wrong were claims for the zodiacs' antiquity. And Testa, after all, was the "secretary for Latin letters of the sovereign Pontiff." For good measure, Duclot reiterated their arguments, but he had even more. For, he asserted, the zodiac did not originate in Egypt at all.

A stone had been found by the banks of the Tigris River in Mesopotamia by a Frenchman named Michaud. On it there appeared a serpent around which twelve figures circled that bore striking resemblances to the usual figures of the zodiac. There was something more. The figures on Michaud's stone were surmounted by others of an apparently antagonistic sort; on the Ram, for example, there seemed to be a wolf. This could only mean, Duclot opined, that the stone represented not just a zodiac but the contrasting forces of good and evil. Other figures indicated that it must have been made following Alexander's conquest. This would not seem to give Duclot much by way of new evidence, but he forged straight ahead.

Obviously, he went on, this zodiac is more elaborately contrived than the one that antiquity bequeathed to us. On the principle that the more complicated design should follow, not precede, the simpler one, Michaud's stone had to represent an elaboration of the original zodiac. Since the stone's designs reflect a system of belief founded on the opposition between good and evil, and since it was uncovered in Mesopotamia, the very site of Babylon, how could one doubt that it represents a religion formed "after the confusion of languages, the dispersion of peoples and their isolation on the earth"? In Genesis, Duclot reminded his readers, we can read the events that produced this monument to a debased faith. Babylon, the first city built after the Flood, had erected a tribute to false belief, a Tower that so angered the biblical deity that he dispersed mankind and forever set human tongues apart. But before that—this was Duclot's point—before the Flood itself, an original zodiac must have existed, one consonant with the simple, pure religion of the patriarchs. The original zodiac must have been "traced by the descendants of Noah, if it wasn't by Noah himself, after the Deluge." Just as Picot had created the politically useful fantasy of an erstwhile, and extremely pious, golden age of France, Duclot recurred to the primeval heavens as embodying true religion, a position supplemented by the reassuring suggestion that the cleansing power of truth had at last swept away the debris of false belief.

All of which meant, he concluded, that the first zodiac dated from the time of the Flood, making it only several hundred years younger than the oldest age attributed to the Esneh rectangular by the French savants. But Egypt, the home of zoomorphic idolatry, could no more than early Babylon have produced this simple image, for it had come into being through the pristine communion of the biblical deity's heavenly creation with the clear vision of the patriarchs.

Duclot's farrago mightily impressed the scions of the restoration. On October 9, 1816, Picot's journal printed the first part of an encomium to it, continued on October 26. The anonymous reviewer began with a florid homage to the saintly works bequeathed by the ages. Of all the horrors committed by those who attacked Christianity, the reviewer grumbled, none surpassed those directed against the holy texts, against those "divine oracles, where the very truth makes itself heard." These texts, "constantly revered by the Church," embody "the approbation of the ages." It is among them, gushed the reviewer, that the doctor finds "true science," the ignorant "the most solid instruction," the sick their consolations, and so on. But take care: the "heretics" (Protestants, that is) utilized the sacred texts to ill ends because they refused the priest's guiding hand. Worse still, "our century has crossed even these limits, and tried to take from Christianity the warrants of its foundation and the proofs of its divinity."[24] Insulted, ridiculed, bitterly criticized for forty years, the "sainted books" had suffered attacks that "omitted nothing in order to destroy the chronology of Moses." Particularly noteworthy among these enemies of the faith was "that Dupuys [*sic*] and his book on the *Origine des cultes*." But he was hardly alone. Traditions, fables, sciences, and systems—all were invoked against the holy books. The attackers availed themselves "of modern discoveries, and the progress of physics and geology to catch out Genesis, and to prove that the world was much more ancient than one seemed to imagine." These impious authors had even gone so far as to "establish injurious parallels between Christianity and monstrously false religions."

The *Ami de la réligion* was particularly taken with Duclot's calculation of the size of Noah's ark, of how "it was sufficient to contain all the animal species." These sorts of reckonings fit well with the latest results of "natural history," which had redeemed the science from the charge of impiety. He quoted from a decade-old article in the *Moniteur* concerning a report that Cuvier had written for the Institut National on De Luc's theory of Earth's surface. De Luc, recall, eschewed global claims to concentrate on local specifics but invoked a massive deluge in the not too distant past. The report castigated those other authors of unsupported geological speculations that often contravened Genesis. So much, effused the newspaper report, for "those systems which combat our holy books, and which lead so many astray. . . . Thanks to the Institut, the apologists of religion are avenged, and the science of natural history, which had been so very abused for the purpose of attacking religion, proves in the end its [belief's] truth and confounds its enemies."[25]

The bubbling rage of the religious Right, illustrated by Duclot's excoriation of the impious, hardly calmed even years after the empire's fall. It was if anything exacerbated because the monarchy failed to put in place the draconian censorship of bad books that conservatives deemed essential. The assassination of the duc de Berri in 1820 further stoked Ultra anger. The king was thereafter forced to accept the resignation of his favored but insufficiently conservative prime minister, Elie Decazes, while a bill went before the chamber that increased the power of the upper tier of the already limited ranks of voters and instituted press censorship. Royalists and Bonapartists went at one another's throats for several months. Louvel, the nationalist and atheist assassin who hated the Bourbons, was guillotined on June 7 at the Place de Grève. That day and the next the streets of Paris roared with crowds of demonstrating students. Fear of revolution kept the bourgeoisie in place; moreover, the economy was in good shape, unemployment was low, and the Garde Royale remained loyal and restrained.[26] The new minister, Richelieu (who had headed Louis' government after the Hundred Days through 1818), though a devoted royalist, lacked sympathy for the Ultras and opposed the institution of censorship. Defeated in that, he left office at the end of 1821 and died the following May. He was succeeded by Jean-Baptiste de Villèle, who shared Ultra sympathies and moved rapidly to stifle the press. The legislature, stacked with new, right-wing peers, supported Villèle's authoritarian program. Repression increased further following Louis' death in 1824 and the accession of his brother, scion of the Ultras, as Charles X.

The period between 1822 and the July Revolution of 1830, when Charles X and his family were forced out and Louis-Philippe, duc d'Orléans, took the throne, was marked by partially successful Ultra attempts to secure their imaginary version of the prerevolutionary past. Draconian laws passed the legislature. One, the Anti-Sacrilege Act of 1825, strongly supported by Bonald, stipulated the death penalty for blasphemy and other acts thought to cast religion into disrepute, though it was never applied. A stringent press law was proposed, as well as a law on inheritance that would have regenerated the massive holdings of the aristocracy. Despite its right-wing cast, the legislature rejected these two measures as a result of the extraordinary, antagonistic reactions to them in France at large. Villèle was ejected in 1827 and replaced on January 4, 1828, by a moderate royalist, Jean-Baptiste de Martignac, who abolished press censorship. Though Martignac favored various positions of the Ultras, which did not endear him to the liberals, the Ultras never trusted his comparative moderation, and he was himself forced out in April 1829, to be replaced by an extreme Ultra, the Prince de Polignac. He pushed severely repressive ordinances that suspended the press altogether, dissolved the Chamber, and excluded the bourgeoisie from future elections, measures that led directly to the July Revolution that ended the Bourbon monarchy.

The censorship provisions that Richelieu failed to prevent in 1821 had not been stringent enough in the eyes of the right. On April 20, 1822, Picot's journal reported that Denis-Antoine de Frayssinous, grand-master of the university and bishop of Hermopolis, had complained that "dangerous books are received everywhere." These works were often the more dangerous for their seductive style, since "whoever is charmed will soon be persuaded: one begins by admiring the writer, one ends by adopting his ideas." Frayssinous was, it seems, somewhat behind the times because on March 16 new printings of Dupuis' *Abrégé* and his *Dissertation* on the Dendera zodiacs had already been seized by the police at their publisher's, Adolphe Chasseriau.[27] A little over a year later, on June 26, 1823, Chasseriau was arrested by order of the Royal Court of Paris to "prevent outrages to the state's religion and to public morals, as editor of a work whose destruction had nevertheless been ordered."[28] He had in fact printed four editions of the *Abrégé*, the first in 1821, and three others in 1822—the last, or fourth edition, after the police had seized the third.[29] The trio of printings in 1822 enraged reactionaries and no doubt stimulated the police seizure. Two other publishers (Ledoux and Tenré) had also printed the *Abrégé* in 1821, and another, Bossange, had done so in 1820.

The seizure of the *Abrégé* likely had little to do with claims for Egypt's great antiquity, since it made none at all, at least not directly. But it claimed something even more disturbing to the religious Right, and it is worth quoting Dupuis for the flavor of his decidedly overt antagonism to Christianity.

We will not examine whether the Christian religion is a revealed religion: only fools today believe in revealed ideas and returns [from the dead]. Philosophy in our day has made too much progress for us still to argue about communications from divinity to humans, other than those that occur through the light of reason and the contemplation of nature. Neither will we begin by examining whether either a philosopher or an imposter named Christ has existed who established a religion known by the name of Christianity; because, even if we were to cede this latter point, the Christians wouldn't be satisfied unless we recognized in Christ an inspired man, a son of God, a god himself, crucified for our sins: yes, it's a god they require; a god who once ate upon the earth and whom we eat today.[30]

Later in the *Abrégé* Dupuis explicitly identified Christianity as an astro-agricultural fable.

The fable of Christ, born like the sun at the winter solstice, and triumphing at the spring equinox in the form of the equinoctial lamb, has all the traits of the ancient solar fables, to which we have compared it. The celebrations of the religion of Christ are, like all solar religions, essentially tied to the principal epochs of the annual movement of the star that governs the day: from which we conclude that if Christ was a man, it was a man who strongly resembled the sun personified; that his mysteries have all the characteristics of the sun worshippers or rather, to speak directly, that the Christian religion, in its legend as in its mysteries, has as its unique purpose the cult of the eternal light, rendered sensible to man by the sun.[31]

The *Abrégé's* third and fourth editions both also contained Dupuis' dissertation on the Dendera zodiac, and so the connection between it and the reduction of Christianity to a form of sun worship had been securely forged.

Although Dupuis' *Origine* had always enraged believers, few had the talent to reply in the only way that stood the slightest chance of attracting the skeptic, namely, through satire. In 1827, six years after Napoleon died in exile on St. Helena, one Jean-Baptiste Pérès did so.[32] Pérès was a professor of mathematics and physics among the Oratorians in Lyon, eventually becoming a magistrate in the town of his birth, Valence d'Agen in Tarn et Garonne, ending his career there as the municipal librarian. Neither religious conservatives nor mathematicians were then (or certainly now) generally noted for their humor, but Pérès, who was both, had a wicked wit. He produced a pamphlet directed at Dupuis' *Origine* (though he did not mention it) in which Napoleon's very existence was challenged. After all, Pérès argued, just consider the coincidences. "Napoleon" sounds rather like "Apollo," deity of the sun, and "Apollo" also signifies exterminator, an apt name for a conqueror. Apollo's mother was Leto; Napoleon's was Letitia—an obvious coincidence. Pérès went on to point out that Napoleon was said to have twelve marshals, as the zodiac has twelve signs; that Napoleon had four brothers, as the year has four seasons; and that three of these were kings, as the seasons of spring, summer, and fall reign over flowers, rain, and fruits, all gifts of the sun. But just as one brother had no kingdom, neither does winter reign over anything, for the wintry sun sojourns in the South. And so forth. From all this, Pérès amusingly concluded that Napoleon, this "alleged hero of our era," is "an allegorical person whose attributes are borrowed from the sun," one who therefore "never even existed."[33] Had Dupuis been alive, he might well have replied that the very terms of the allegory prove that Napoleon certainly did exist, albeit Euhemerized as Apollo.

The Dendera zodiac reached Marseilles from Egypt on September 9, 1821. Saulnier traveled there early in October, while Lelorrain, his adventurous collaborator, who journeyed separately from the zodiac, arrived several days later. Eager to have the cost of transporting the monument to Paris defrayed by the government, Saulnier quickly notified various members of the Institut National and others about it. On October 5 Dacier, the perpetual secretary of Inscriptions et Belles-Lettres (an enthusiast, recall, for Dupuis), congratulated Saulnier on the "success of this honorable enterprise for France, a success one would have thought impossible."[34]

One of the passengers on the ship had fallen ill (always thought to be an ominous sign for ships coming from the East, given the possibility of plague), and so the Egyptian prize did not enter the harbor until November 27. Lelorrain also remained in quarantine until the end of the month. Once off the boat, the zodiac caused a commotion in Marseilles. Greeted on its debarkation by the commander of the local division, by the prefect of the department, and by the town mayor, the stones drew a large crowd that had to be kept back, with only a few notables being allowed near. Among them was Saulnier, who approached the "ancient monument" in a state of "nearly religious" anticipation.

However overcome with "religious" excitement he may have been, Saulnier had already managed to avoid the usual import formalities. The director-general of customs had written his local counterpart that "a priceless monument, such as the one in question, must, in some ways, be considered a public property." Public for the purposes of avoiding duties and problems, no doubt, but not public at all when it came to ownership because Saulnier had every intention of selling the zodiac for a good price. According to the self-promoting little book that he had printed once the object reached Paris, many who wanted to see it "made considerable offers to the incorruptible watchman" who stood guard at the consignors in Marseilles, Gil et Frères. In fact, Saulnier went on, an unnamed "stranger" offered Saulnier a "considerable sum" for the zodiac, which was soon followed by an "even more advantageous" offer.[35] Styling himself a patriot to the core, Saulnier refused to pursue negotiations. His patriotic rebuff was no doubt helped along by his having succeeded in catching the attention of the interior minister, Joseph Jérome. Unfortunately for Saulnier's plans, Jérome left office with the fall of the Richelieu ministry on December 12. Shortly thereafter, this Egyptian monument to French glory trundled its way to Paris and into a cauldron of popular excitement, right-wing fear, and academic controversy.

Egypt in Paris

For Parisians living on the city's southern edge, along the road from Marseilles, the new year of 1822 began with a clamor. Over the crackling of children's *papillotes* and cries of *bonne année* rose the racket of a loaded cart rolling over cobblestones. In the cart was Saulnier's great prize, the circular zodiac of Dendera.

At every stop along the slow, five-hundred-mile journey to Paris, people gathered to catch a glimpse of the mysterious antiquity.[36] Saulnier naturally took advantage of the possibly fleeting enthusiasm and carefully balanced his vigorous publicity effort with a nearly pious reverence for the object—a reverence that seems ironic in retrospect, considering the zodiac's apparent threat to religious orthodoxy. As ardently as Saulnier hoped to recoup his expenses (and to realize a handsome profit), he could not appear too eager to make the sale. Transparent greed was unlikely to elevate the zodiac's market value. During the first months of 1822, while Saulnier gingerly toed this line, Paris fell under the spell of *zodiacomanie*. The Dendera stone became a Parisian obsession, "the idol of the salons," the only fashionable subject of debate in the cafes and on the sidewalks. "My dear," went the typical conversational gambit, "have you seen the zodiac? And what do you think, Monsieur, of the zodiac?"[37] In the more rarified atmosphere of the Chamber of Deputies, members advertised their religious piety by denouncing the zodiac as a "monument to atheism and irreligion." Others, less articulate but equally alarmed by the zodiac's religious implications, cursed the appearance of the "ugly black stone" on French soil.[38]

The cultural blaze ignited by the zodiac's arrival was nourished by France's long-standing passion for ancient Egypt, which had gripped the country ever since Napoleon's campaign. The popular reception of Denon's *Voyage* and the *Description de l'Égypte* (of which a cheaper edition by Pancoucke had begun reaching print just the year before) had only fanned the flames. Inevitably, however, attitudes had shifted. Egypt—as fantastic ideal, as scholarly object—was not what it once was. No longer simply a code word for imperial power, nor the exclusive preserve of antiquarians, Egypt was now *popular*. Its remote past had evolved into a fertile source of myths and stories that could be recruited to produce dramatic articulations of the present concerns of ordinary people, and not solely of the Parisian intelligentsia, whose members had been debating the zodiac's religious implications for years and would do so even more vigorously once they had the stone before them. In a period marked by official distrust of the productions of the public sphere, from newspapers to paintings and plays, the myths, stories, and monuments of ancient Egypt were discussed with exceptional, if hardly absolute, freedom. Although there were limits, Egypt provided a rich lode of social, political, and religious ideas that did not always trigger the press censorship exercised by the Ministry of the Interior.

Although the conversion of Egypt's ancient past into modern spectacle was widespread, it took a variety of forms that, as it were, constituted an ideological arabesque. From the obelisks and faux pyramids that were installed around Paris, to the turbans and slippers worn by fashionable women, to the proliferation of Egyptian themes, symbols, and stories in theatrical performances and books, remnants of Egypt's remote past relentlessly pricked Parisians with reminders of the hard-to-swallow events of recent history, especially the embarrassments that came with the loss of republic and empire, a past that could be neither accepted nor forgotten, but continued to return in ever more exotic, orientalized, and unmasterable guises.

In 1822 Paris was, in Balzac's vivid phrase, a city of a thousand novels—complex, seething, putrid, fascinating, frustrating, immense. Massive waves of migration from the countryside had swelled the urban population, which reached 800,000 according to a census taken at midcentury.[39] In this city of newcomers, men outnumbered women until the end of the 1830s, young men were especially well represented, and, not surprisingly, prostitution thrived. Meanwhile, the foundling infants who were the tragic, if inevitable, results of this demographic situation succumbed to death at high rates. (Deaths in general annually exceeded live births.) The city's inhabitants and public institutions strained to keep up, but with limited success. While the population doubled, the housing stock in Paris expanded by only 30 percent or so, with rents, traffic, congestion, and pollution rising accordingly.[40] At least 10 percent of the population was receiving some form of poor-relief by midcentury.[41] In the shabbier arrondissements, where the city's working poor squeezed into tiny rooms that recalled the apartments of the decrepit Maison Vauquer, immortalized by Balzac in *Père Goriot*, many homes in wealthier neighborhoods, especially in the northwest of the city, went uninhabited while their owners traveled or lived elsewhere for long stretches. At Montfauçon, in the city's fetid Northeast, abattoir and cesspool combined to form a reeking mess whose piquancy was only sharpened by the historic, melancholy presence of the old gallows, immortalized as a center of "stench" (*puanteur*) by Parent-Duchâtelet, a public health–minded physician with an extracurricular interest in living conditions of the Parisian poor, especially prostitutes.[42]

Paris was hardly the rigidly planned city it would become under Haussmann. It was instead a collection of distinct neighborhoods shaped by the particular histories and interests of their inhabitants over the centuries and linked by accidents of propinquity. When Eugène Sue, in his *Les Mystères de Paris*, tried to describe the streets of pre-Haussmann Paris, he characterized them variously as a labyrinth, a maze, a Chinese puzzle, and finally, in a phrase that has more than a whiff of imaginative, perhaps even national, defeat, a "bizarre alphabet d'orient."[43]

Sue's characterization was not merely figurative. Thanks to assiduous efforts by Napoleon and his ministers, the official discourse of the empire had literally materialized in newly constructed reminders of Egypt scattered throughout the city. Napoleon cultivated the association, preserving it in Egyptian-themed public monuments to his campaign on the Nile—pyramids, obelisks, and other architectural embellishments. By the time Louis XVIII came to power, these signs of Egypt as a French imperial possession were part of Parisians' daily visual world. No fewer than six of the fifteen new public fountains decreed by Napoleon were, for example, adorned with Egyptian iconography in commemoration of his campaign, which the emperor and his publicists had alchemically transmuted from military failure into political gold. The 1809 Fontaine du Fellah, by Pierre Beauvallet, in the rue de Sèvres was dominated by a statue of a man wearing a *nemes*, the striped headcloth of the pharaohs, while a nemes-adorned Egyptian head rested at the feet of the heroic, nude commemorative statue of Desaix, alongside an original Egyptian obelisk of rose granite, erected in the Place des Victoires in 1803.[44] Reminders of the Egyptian campaign also cropped up in the names of streets and squares—the rue Aboukir, the Place du Caire. On the very seal of Paris itself for 1811, the Egyptian goddess Isis sat on the prow of the *Nautes* guild boat, a symbol of the city.[45] By midcentury, the imaginary interpenetration of Egypt and Paris was so complete that one writer affectionately referred to the city as "that sphinx known as Paris."[46]

As symbols of Egypt proliferated in the city's labyrinthine streets, beneath them an ever-ramifying network of catacombs and sewers formed a kind of urban unconscious, a system for disposing of what could not be managed within the city's visible limits. Balzac, in *Père Goriot*, described Paris around 1820 as "*un bourbier*," a mire, a swamp. Despite its dazzling glass-enclosed galleries and shop windows, the city's shimmering surface never failed to hint at the presence of something sinister lurking just beneath. Peering out from behind her perfumed handkerchief, watchful for the contents of chamberpots that were still routinely emptied out of upper-story windows, the celebrated writer Delphine de Girardin imagined an entire "*ville souterraine*" inhabited by people who lived like "reptiles in the muck." Others, of a less fantastical turn of mind, like Pierre Proudhon and Louis Blanc, resorted to overtly derogatory, quasi-sociological comparisons, characterizing the seedy underworld of the *bas-fonds*, with its small-time crooks, gamblers, prostitutes, and vagabonds, as "bestial," "barbarous," and "savage"—words that readers of Saulnier's account would not fail to associate with the colorful details of Lelorrain's picaresque adventures in the desert among *fellahin*, pasha, and swindlers.[47]

In Paris, corruption pervaded everyday life. Foul "miasmas" floated on the wind from the northeast, especially in summer; this was the stink for which the city was notorious. Crime, especially of the ambulatory sort, was rife: prostitutes on walkabout, thieves deftly lightening the pockets of self-absorbed *flâneurs* and other arcade-strolling members of "le Tout-Paris," the phrase itself of contemporary coinage and used to refer to fashionable Parisians and their world.[48] In Napoleonic France the propaganda that raised Egypt to iconic status deflected attention from the sordidness of daily life by substituting a narrative of cultural supremacy and success for the expedition's military debacle.[49] Napoleon may have been sent to exile in far-away St. Helena, but memories of Egypt as a symbol of French power and influence persisted long after his departure.

To introduce the Dendera zodiac into Restoration France was, among other things, to participate in an ongoing political game, one in which official efforts to forget some elements of recent history vied with efforts to resurrect others. Befitting an era devoted to the restoration of an imagined sociopolitical world, under Louis XVIII the past became all-pervasive.[50] Symbols of the Old Regime, including the calendar, were revived according to the official policy of *oubli*, forgetting; after the Hundred Days, this policy extended to a purge of civil servants perceived to have been too sympathetic to Napoleon, and also to the slashing of the pensions of Napoleon's ex-soldiers. As ancient royal corpses, popularly known as "mummies," were exhumed and relocated, the exiled emperor became a kind of revenant, haunting Paris in a thousand symbolic forms, whose pervasive influence had to be ferreted out, neutralized, and, if possible, cast outside the city limits.[51]

Amid this wave of official forgetting, political conservatives evinced heightened sensitivity to, and concern about, the historical record. While Picot and his ilk had waxed nostalgic in the pages of the press organs of the religious Right, they were not the only representatives of this trend, which not only pervaded the highest levels of state institutions but also appeared repeatedly in the streets. In 1821 the government minister and historian François Guizot founded the École des Chartes, an institution dedicated to the preservation of official memory, where archivists were trained in the latest methods for preserving crumbling official documents. Some years later, in 1833, De Montalbert joined other intellectuals in an open letter to Victor Hugo in following the counterrevolutionary call of the scholar De Gaumont to halt the destruction of historically important French monuments and to urge their preservation.[52] Royal bronze equestrian statues that had been melted for cannon in the 1790s were recast and returned to their original settings. Prerevolutionary place names were restored; the Place des Vosges became again the Place Royale.[53] Arguments raged over the possibility of returning the Panthéon to its original function as a church, which would have made it one of the many new churches that appeared in Paris in the 1820s, mainly in the wealthier arrondissements.[54] Liberty trees disappeared as crosses sprang up in their place.[55] Scores of miracles were reported from all over the city and from other parts of France.[56] Six months after the assassination of the duc de Berri, Benjamin Constant reported with disappointment that "the sovereign people had handed in its resignation," a remark that was widely repeated in fashionable society, behind the closed doors of salons and drawing rooms.[57]

Secret Societies

The monarchy's curtailment of public life through restrictions on speech and the press resulted, paradoxically but not surprisingly, in the flowering of forms of social life that could be conducted out of reach of the eyes and ears of the regime. To express dissatisfaction with the policies of the regime, to sustain allegiance, however coded and ambivalent it might be, to the apparently lost ideals of the republic, and to convey republican solidarity without provoking the ire of the censors, writers, scholars, and artists found ways to meet and to share their views in salons, theaters, and secret societies. Participants in these activities developed special ways to present (and publish) their views, including cunningly deployed symbols and mythologies linked to Egyptian material. In the Egyptosophic lodges that sprang up in Paris during the ideologically repressive and politically conservative years following Napoleon's exile, republicans went underground, literally and figuratively, many as Freemasons who made creative use of this politically resonant but puzzlingly mysterious material.

While participation in a Masonic organization was never the sort of thing one bruited about, secrecy was especially crucial for survival during the Restoration. Rumors that Masonic orders had sponsored the revolution circulated along with the more conciliatory notion, espoused most notably by the liberal Catholic and author in 1820 of a book on Masonry, Guerrier de Dumast, that the purposes of church, state, and Masonic organizations could be aligned.[58] For members of these societies, the long-standing connection of Isis with the city of Paris—furthered most notably by Dupuis in helping to save many of the sculptures on Notre Dame's façade—remained persuasive until as late as 1820, the same year Dumast claimed that a "fever for secret societies" had gripped the nation.[59]

Egypt was an especially powerful ideological resource for Masonic organizations since they could not express their politics directly for fear of political retribution. Egypt had often been thought the source of lost, esoteric knowledge, and the expedition's savants had revived the antique notion that geometry and astronomy had been born there as well. Masonic organizations fastened on the reputed lore of ancient Egypt to forge a powerful language for Freemasonry, at once vivid and conveniently obscure. These organizations included the Isis Lodge, allegedly founded in Cairo by Kléber and Napoleon (who, despite his avowed mistrust of Freemasonry, did little to dispel these rumors); the Loge Impérial des Chevaliers de Saint Jean d'Acre, which conflated nostalgia for the Crusades with Napoleon's invasion of Syria; the Souveraine Pyramides des Amis du Désert, founded by an archaeologist in Toulouse in the early 1800s; and the Disciples of Memphis, established in 1814 in Montauban as the successor, on French soil, to the Isis Lodge.[60]

The most prestigious of these Egyptosophic secret societies, as well as the longest lived, was the Sacred Order of the Sophisians (Ordre Sacré des Sophisiens). Founded in Paris in 1801, the society was one of several Masonic organizations concerned with "the mysteries of Isis." This tradition of belief linked ancient Egypt to modern France through the writings of Dupuis and Court de Gébelin, who had postulated the existence of an ancient Iséum on the Île de la Cité, as well as those of Alexandre Lenoir, a Freemason, apologist for the Egyptian origin of all civilization, disciple of Dupuis, and creator of the popular, and controversial, Musée des Monuments Français. Lenoir would later have his own say about the Dendera circular, which he thought to be a sort of solar calendar.[61]

While preoccupation with Isis-worship was part and parcel of secret society ideology, the Sophisians could boast a direct connection to Egypt, for the original members of the society were recruited exclusively from the ranks of Napoleon's Egyptian campaign. According to an account of the order's origins written by Cuvelier de Trie (playwright, one-time soldier with Napoleon's forces in Europe, and historian of the Sophisians), the central members of the order had met first in Egypt, where their shared experiences formed the basis for an enduring bond. One nineteenth-century historian even went so far as to characterize the Sophisians as a "secret society of the army." The most significant organizing force behind the order was provided by the expedition's savants, with its locus the Institut d'Égypte. The mathematician Monge joined the Sophisians sometime during, or just after, his adventure with Napoleon in Egypt. The optical scientist Malus was also a member, as was the engineer Lancret, who also served as the second editor of the *Description* until his death in 1807.[62] Other Sophisians with links to the Egyptian campaign included Saint-Hilaire, the zoologist Lacépède, Charles Norry, and General Cafferelli. Fauchet, the republican lawyer and one-time prefect of the Arno in Florence who assisted Denon in purloining artworks and evacuating them from Italy to the Louvre, was an "Isiarque" in the order. Neither was Denon immune to the society's lure. Although his official relationship to the Sophisians is unclear, he at the very least counted many Sophisians among his friends, and his presence and activities were recorded in detail at the 1819 initiation of the travel writer Sydney Owenson, Lady Morgan, into the anti-Bourbon women's lodge, Belle et Bonne, which was a special favorite of the Sophisians. During the rite, Denon silently kissed the hands of the leader of Belle et Bonne as she offered member-ship to Lady Morgan. But, Lady Morgan recalled, while she "at once entered into the spirit of the proposition," Denon was suddenly much less composed. "[His] face would have made a cat laugh," Lady Morgan wrote. "He is so *grimacier* [grinning]!"[63]

That Napoleon's savants and soldiers should turn up in the world of the theater may seem odd to the modern reader, but in France the worlds of theater and military had been linked ever since the revolu-tion. Under Robespierre, the French Ministry of War "teemed" with "formerly unemployed actors and playwrights" who, as part of mili-tary outfits like the Compagnie Franche, put their talents to use in the service of political power, staging propagandistic shows on or close to battlefields, in order to boost troop morale.[64] Keeping in mind that a good number of these veterans of Napoleon's wars were also long-standing Freemasons of one sort or another, it should come as no surprise that the worlds of theater and Freemasonry also collided during the empire and throughout the restoration. After the 1807 theater decree (part of the Code Napoléon) forced most theaters to close, a number con-verted their now-empty stages to Masonic temples. Perhaps the best known of these was the transformation, in 1808, of the foyer of the shuttered Cité theater into an outpost of the Grand Orient de France, where none other than Napoleon and Josephine were rumored to have attended a Masonic meeting.[65] Members of these organizations were still coming to grips with what they had seen in Egypt, particularly the mysterious antiquities and the baffling hieroglyphic script. Particularly compelling to society members was the ancient civilization's apparently happy marriage of civil affairs with the secret and the sacred, a union that pointed up the many flaws in the current regime's efforts to combine politics and religion. Using ideas culled from classical sources and the latest Egyptological scholarship, organizations like the Sophisians made Paris a stage upon which elements of Egyptian mythology and history came to life. As we have seen, this development had historical roots in the imperial period when, despite Napoleon's public aversion to Freemasonry, secret societies enriched their rituals and writings with imagery taken from the accounts of the Egyptian campaign. During the Restoration, some Masonic organizations attempted to curry favor with the regime by exploring ways to reconcile church dogma with Masonic beliefs. The Sophisians were not among them. Significantly, their understanding of the age of human civilization posed a familiar problem for the regime. While Masonic dating conventions, used to date official documents like attendance registers, typically added four thousand years to the present era, signifying the beginning of time according to the Bible, by 1819 the Sophisians were adding *fourteen thousand* years, a practice that underlined their anticlerical stance and linked it to an alternate history of human civilization, one that began thousands of years earlier, in ancient Egypt—a date that tallied remarkably well with Dupuis'.[66]

The Parisian Theater

Paris during the nineteenth century hosted three kinds of theaters: the state-subsidized, official theaters; the secondary, petit or "boulevard" theaters; and the *forains*, the "spectacles de curiosité," which included circuses and pantomimes. All three were popular, even among elites. During the empire, no less than Josephine de Beauharnais became an aficionado of the boulevard theaters, and this despite Napoleon's dismissal of their productions as "tragedies for chambermaids."[67] The newly minted emperor's general antipathy to an uncontrolled theater took official form in 1804, when he instituted a decree that abruptly shuttered all but eight official theaters in the capital: the four national theaters (the Comédie Française, the Odéon, the Opéra-Comique, and, of course, the Opéra), which were subsidized by the state, and four secondary theaters, the Théâtre de l'Ambigu-Comique, the Théâtre de la Gaité, the Théâtre du Vaudeville, and the Théâtre des Variétés.

Smaller venues devoted to popular audiences—the petit theaters and the *forains*—nevertheless continued to thrive.

All performances in the official and boulevard theaters were subject to censorship. As a result, authors and directors sought creative ways to elude official restrictions on performances, such as a ban on staging plays with dialogue. One way around the censor's restrictions was to use mute characters. What could not be said in dialogue could be expressed in gestures to which stage directions might allude without describing them in the sort of detail that would leave them vulnerable, once again, to the censor's pen. The voiceless characters of pantomime shows, as well as otherwise silent circus performers like acrobats and tightrope walkers, were stock theatrical players whose muteness was routinely used to circumvent official rules. As we shall see, the development of creative ways to resist censorship would bear significant fruit in 1822, when the Dendera zodiac's arrival in Paris became the occasion for a popular vaudeville show with potentially objectionable political overtones.

The traditional impermanence of the performers associated with the circuses and secondary theaters, combined with creative feats of resistance to censorship, lent an illicit aura to a scene that was already excitingly shady. The revolutionary abolition in 1790 of the privileges enjoyed by the grands théâtres had helped to level the playing field for secondary theaters. Nevertheless, the perception remained that these venues and their denizens constituted a demi-monde of illicit pleasure and crime. It took the form, among others, of religious complaints about "un-Christian" actors and actresses, complaints furthered by a lingering conviction that the theaters were liable to corrupt impressionable members of the audience, especially young people.[68] The boulevard theaters were less scenes of freedom of speech and thought than sites of arguments over the forms that freedom might take.

Since the middle of the eighteenth century theatrical activity in Paris had been shifting from fairgrounds just outside the city to the Boulevard du Temple, then on the city's northern edge, where fairground troupes mingled with street performers, puppeteers, carnival barkers, and the denizens of gambling dens and brothels. The street was nicknamed the "Boulevard du Crime" at the turn of the nineteenth century, a title that reflected not just the sensational preoccupations of the new dramatic form, melodrama, then on the rise, but also the melodrama's physical milieu.[69] Shows opened and closed in quick succession, adding energy to an already frenetic and rapidly changing scene. Theatergoing was a wildly popular form rife with possibilities for incidental intrigue. Social classes mingled in the dark, on the streets and in the audience; those who went to the theater went as much for the people-watching as for the shows. Certain theaters, including the state-sponsored ones, offered free performances, extending the experience beyond the literate and leisured.[70]

The lively social scene in the theaters, not to mention in the streets and cafes that surrounded them, owed its vivacity partly to the freer atmosphere that obtained after dark and partly to the simple magic that the theater itself produced, as the site of fantasies that managed to move audiences to talk, and sometimes to fight. The popular theaters on the Boulevard du Temple routinely employed the most sophisticated stagecraft mechanisms—pyrotechnics, trapdoors of all kinds, dramatic lighting, ornate papier-mâché sets, chariots that slid across the proscenium on hidden tracks, and even live animals.[71] The playwright Eugene Scribe recalled how, as a child, he habitually sneaked out of his parents' house during holidays, just to see the shows. Topsy-turvy transformations of all kinds—magic wand to bouquet of flowers, high-born aristocrat to prostitute—constituted the irresistible imaginary of the stage during the empire.

With the advent of the restoration and its revocation of Napoleon's theater decree, theater regulation was relaxed, older venues were restored, and new theaters were built. Demand for new works rose as opportunities to stage them increased. Between 1815 and 1830, nearly two thousand original shows, ranging from comedy to vaudeville to melodrama, opened in theaters on the Boulevard du Temple.[72] As the popularity of the petit theaters grew, the state-subsidized theaters insisted on restrictions designed to blunt the effects of this new competition. But the popularity of theatergoing was not much dimmed by these restrictions. Even the repressive activities of the police and the restoration government's antagonism to any expression that did not strongly support the regime did not fully suppress the lively forms of public life that were developing, forms that centered on enthusiastic theatergoing. The numerous cafes, brothels, gambling dens, and restaurants that popped up in the theater district gave theatergoers additional opportunities to see and be seen in public.

Behind the scenes boiled a shadow world of writers, actors, scenery and prop experts, directors, and producers, all scheming to ensure their own success. Not that success was hard to find: it seemed that, no matter how many shows went up every night, Paris could not get enough of the theater. Looking back from near century's end, the theater historian Maurice Albert recalled that "the Variétés is always sold out; every night the Gymnase has to turn away the world; l'Ambigu rolls on floods of gold; the Gaîte is a variation on the same; the Cirque-Olympique has the air of playing every night for free." Albert went on to claim that even a dozen official venues like the Opéra would still not be enough to satisfy Parisians' demand. "There's such a taste, a general delirium [for the theater] that, before long, to satisfy this hunger for spectacle, Paris will have as many theaters as restaurants."[73] Performances ranged from staid state-sponsored productions, operas and classical drama, to riskier, wilder entertainments, such as the vaudevilles and melodramas that packed the secondary stages night after night, to tiny, lively café-concerts that by and large evaded regulation, to circus acts, carnival sideshows, pantomime and other wordless genres that were all the more engaging because they were dangerously spontaneous and improvisational, giving the impression that when the curtain went up, anything could happen—and, in fact, often did. Audiences were rowdy. Police restrictions on theatergoers' behavior give an idea of the atmosphere. In Rouen, for instance, intermissions were to last no longer than twenty minutes, speech was forbidden during the performance even in the aisles and hallways, actors were not allowed to read any notes thrown up onstage or respond to any entreaties from the audience, and the auditorium was to be emptied and locked immediately after the performance. Theater riots were not unusual.[74]

By far the most popular of the secondary theaters' dramatic forms was vaudeville. Literally thousands of vaudeville plays were written and performed, a veritable river of entertainment that washed over the stages in a long and increasingly elaborate wave, until by the end of the century the vaudeville had evolved into its recognizably modern form, the situation comedy or farce. In their early days, vaudevilles required less ambitious staging than other kinds of performance, such as melodrama, and they were generally shorter as well, having perhaps only a few acts. And they provided flexible elements in an evening's program, for they could be rotated into the venue's schedule at short notice or combined with a longer melodrama to fill out the bill. Vaudevilles were usually comic, even farcical, reflecting their fairground origins, as well as being sketchy, light on plot, and consisting of both songs and dialogue, with the songs being set to well-known airs; at times, the dialogue was indistinguishable from the singing.[75] Frequently, vaudevilles were inspired by current events, not least, we shall see, the arrival of the Dendera zodiac in Paris. These *piéces de circonstance* were not so much pieces of social critique as hilarious spoofs of the bourgeois foibles of parvenu ladies and gentlemen. "At the Académie, and in certain publications, one sounds the alarm against the Romantics; on the boulevards, one contents oneself with tossing firecrackers between their legs."[76]

Many dramatic works drew on themes and imagery from the Egyptian campaign. The Paris stage, especially the raucous boulevard theaters, provided a platform from which to disseminate stories and images about Egypt through a variety of channels, from the sets, props, and costumes to the themes and subjects addressed in the plays.[77] On a visit to Paris, an Egyptian student attending a vaudeville performance in the 1820s noted that the French were avid for theatrical spectacles that featured scenery and costumes evocative of the exotic East.[78] When Napoleon graced a performance of Grétry's 1783 *Caravan du Caire*, the emperor's guest for the evening was "a young Mameluke" whose appearance in the audience not only merited mention in the *Moniteur Universel*, but also, as the article noted, received nearly as much attention as the play itself, "updated" to include admiring allusions to the Egyptian campaign which were met with waves of applause.[79] (The reaction of Napoleon's guest was not recorded.)

One important source of the nostalgia that kept Egypt on the Paris stage through the 1820s was the Sophisians. In addition to being deeply rooted in Napoleon's desert campaign, this secret society also had close and enduring ties to the Parisian theatrical world. Through their efforts, the worlds of the theater and the military underwent a curious inversion: as Napoleon's activities in Egypt became the stuff of nightly entertainment in Paris, his messy, tragic theater of military operations was sanitized and transformed into the reassuring idioms of vaudeville and melodrama. The protagonist of this transformation, the Sophisian Grand Isiarque Cuvelier de Trie, had once been a commissioner of public instruction in the Ministry of Education and Censorship, a position that gave him unparalleled access to the dramatists of his day, as well as to directors and theater owners. As a member of the Compagnie Franche under Napoleon, he had organized performances of morale-boosting plays on the battlefield. And, since the early 1800s, he had followed a steep upward trajectory as a dramatist. He became known as the "Crébillon of the boulevards" for the propagandistic, pro-Napoleon pieces he wrote for the boulevard theaters. He favored historical subjects, particularly reenactments of scenes from military history, and Egypt provided an especially attractive setting.

In 1819 de Trie recruited several prominent members of the Parisian theatrical world to the Sophisian order, including Audinot-Aussy, co-owner of the Ambigu-Comique, and his employees Varez and Piccinnini.[80] The latter, a composer, wrote the music for de Trie's 1811 *L'Asile du Silence, ou Gloire et Sagesse*, a play with heavy Sophisian overtones, performed to honor the birth of Napoleon's son in Rome; the leading female role was played by one of de Trie's favorite actresses, herself a Sophisienne, and married to the painter, set designer, and props specialist Ponce-Camus, whom de Trie had commissioned to illustrate the central text of the order, the *Livre d'Or*, with images inspired by ancient Egyptian iconography.[81] De Trie styled himself the chief custodian of the esoteric knowledge brought back from Egypt by Napoleon's generals and savants, believing the Sophisians should use this knowledge to make redundant all other secret societies. Reputedly discovered beneath a pyramid, the ancient lore comprised "the original blueprint for all mysteries and religions."[82] Though de Trie's plays reflected his interest in ancient Egyptian symbols and secret society rituals, they were less propagandistic and more purely entertaining. Nonetheless, he was hardly averse to framing scenes that used Egypt-inspired props from Sophisian initiation rituals, such as the purifying sarcophagus-like "well of truth" in his 1812 *La Fille Sauvage*, not to mention obelisks and underground voids like those found beneath the pyramids.[83]

The sheer ubiquity and variety of publications about the theater attests to the magnetic fascination it exerted over audiences. All sorts of dramatic works, from elite tragedies and operas at the official theaters to the melodramas, pantomimes, *féeries*, and vaudevilles that were the bread and butter of the secondary theaters, not to mention reviews of these works, found their way into salons and drawing rooms. On a visit in the 1840s to the Countess Duchâtel's house at Sceaux, the poet Deschamps was treated to an impromptu dramatic reading, staged by the count and countess, of the "Corneille of the boulevards," Gilbert de Pixerecourt's wildly popular 1810 melodrama, *Les Ruines des Babylone*. That a thirty-year-old melodrama could still command such enthusiasm testifies to the persistent salience of the romantic, exotic East in French culture. This suburban revival was an unqualified success. In a letter, Deschamps recounted the story to Pixerecourt. "I doubt that the three hundred performances of *Les Ruines de Babylone* in Paris and the reports of its successes in the provinces and on all foreign stages could, for you, have matched this first triumph, this elite matinee," he wrote.[84]

When historical themes appeared in the theaters of Paris during the consulate and empire, critics typically evaluated the work, however much of a spectacle it might be, for its correspondence to what they took to be matters of fact. This was especially true when it came to themes from military history, whether in Egypt or elsewhere. "History as it was professed on the boulevards during the Restoration was above all military history," observed the historian Albert.[85] When it came to recent military history, the details of which were still fresh enough in the minds of theatergoers, the least departure from historical fact was especially risky, and audiences were highly sensitive to deviations from their image of recent military campaigns. The ill-received premier of *Kléber, ou les Moeurs orientales* at the Théâtre de la Cité in 1800 provides a pertinent example. One reviewer, writing for *Le Courrier de l'Europe*, asserted that "the event itself is still too recent for one to alter the truth of history" to the extent that the playwright had.[86]

Attitudes were considerably different during the Restoration. A case in point: de Trie's *La Mort du Kléber*, a melodramatic recounting of the general's assassination during the Egyptian campaign from the perspectives of both the occupiers and the occupied, which went up at the Cirque Olympique in 1819 and played through the fall of that year, in an unusually long run that brought the play to audiences beyond the theater's core subscribers. Despite the play's inflammatory potential—it did after all recall Napoleon, though (it might be said) Kléber's death could be, and was by many, laid at Napoleon's feet for his having abandoned the troops in Egypt—it was apparently only minimally censored. At sole issue was the appearance onstage of a tricolor hat with cockade. The censor's ire, if it was raised at all, must have been sufficiently appeased by the play itself, which ends on a politically innocuous note by blaming Kléber's assassination on a Muslim fundamentalist cabal, rather than seeing it for what it was, the outcome of a violent and unwanted military occupation.[87] Critics now faulted de Trie for not making the play more entertaining. "A theatrical play is not a history lesson," one critic observed, writing in the pages of the Parisian daily, *Le Courrier de l'Europe*, even though, of course, "everything is historical now."[88]

The Zodiac on Stage

The zodiac's arrival in Paris echoed at every level of society, from the most sophisticated drawings rooms through angry political, religious, and scholarly *fracas* to the popular stage itself, for vaudeville saw the production of a piece about the Egyptian stone that played on European cupidity and the debates over Egypt's age. *Le Zodiaque de Paris* opened on September 2, 1822, at the Théâtre du Gymnase, a location that could not have been more historically and socially resonant (Figure I.5).[89] Situated on the Boulevard Bonne-Nouvelle, along the city's "theater mile," which stretches from the Opéra to the Place de la République on the edge of the Marais just north of the Temple enclosure, the theater was a short walk north of the Place du Caire and its surrounding warren of streets named in memory of the Egyptian campaign, then renamed in the Restoration's efforts to erase the past. The theater, which specialized in vaudeville, had the advantage of a powerful patron, the duchesse de Berri, wife of the duc who had been spectacularly murdered on the steps of the Opéra in 1820. The Gymnase was a newcomer to the Parisian theatrical scene, having been open for just over a year following a few months of frenzied construction. Its architects were Auguste Rougevin, who was also in charge of Les Invalides and developer of the "Folie Beaujon" near the Champs-Elysées, and Louis Régnier de Guerchy, whose greatest claim to fame, his work on the Théâtre du Palais Royale, would come later.[90] Because the theater's director had secured an exclusive contract with the premier vaudeville writer of the age, Eugène Scribe, patrons were assured of being the first to see Scribe's latest productions as well as other plays, presented in a luxurious setting with plush velvet seats and five tiers of elegantly draped loges adorned with gilded moldings, cornices, and cherubs in relief, made even more magical by the flickering light of the theater's chandeliers.[91] The Gymnase soon became one of the most frequented popular theaters in the city, with space for 1,282 spectators. The most expensive seats cost the large sum of five francs, but the parterre cost two, and a place in the galleries could be had for one and three-fourths. The doors opened at six, and the show began a half hour later.

The allusive richness of *The Zodiac of Paris* reflects the broad interests and learning of its authors, all of whom were busy men of letters, deeply involved in the issues of their day. The first writer, Marie-Emmanuel Guillaume Marguerite Théaulon de Lambert, known simply as Léon, was perhaps the purest playwright. Over the course of his life, he wrote or cowrote about two hundred dramas, comic operas, and vaudevilles, often staged at the Théâtre des Variétés and the Théâtre du Vaudeville.[92] The play's second author, listed only as "Ferdinand" in the published version of the play, was mostly likely Joseph Adolphe Ferdinand Langlois, who went by various pen names including Langlé

and Eusèbe. The son of a musician and a librarian at the Conserva-
toire, Langlois trained initially as a doctor, following in the footsteps
of an uncle. But other familial influences, notably that of his famous
literary cousin Eugéne Sue, were at least as strong. As a *littérateur*,
Langlois' interests tended toward the nostalgic and romantic. In 1828
he edited *Les Contes du Gay-Sçavoir*, a collection of medieval bal-
lads and fables, printed in Gothic type with lithographed illustrations
by the British landscape painter Richard Parkes Bonington, a protégé
of Delacroix and Antoine-Jean Gros, who died that year at the age of
twenty-five from the Romantic disease, tuberculosis; and, with Emile
Morice in 1829, a volume devoted to medieval legends concerning the
figure of the *jongleur*, the itinerant storyteller who provided recita-
tions of *chansons des gestes* and other entertainments, *L'Historial du
Jongleur, Chroniques et légendes françaises*. In time, he became the
administrator of funeral rites for Paris—later, for all of France—and
in 1841 he composed an account of the repatriation of Napoleon's
remains from St. Helena, *Funérailles de l'Empereur Napoleon*. Little is
known about the third author, Mathurin-Joseph Brisset, except that he
wrote a comedy, *Le coureur de venues*, which debuted in March 1827
at the Théâtre des Nouveautés, and, in 1841 a broadside of sorts, *Les
enfants à Paris*, which began with a bold address to the city of Paris,
apostrophized as "the Eldorado of opulent women [and] the proving
ground of husbands," and which appeared as part of an eight-volume
portrait of contemporary France.[93]

Like the clamor that heralded the zodiac's arrival in Paris, *The
Zodiac of Paris* began with an extravagant noise. As the curtains rose,
the theater filled with the wails of a "choir of mummies" ranged along
the rear of the stage, a chorus of insinuating background voices that
may well have brought to mind both Sophisian rituals (which included
choirs of the sort) and the menacing background chorus specified by
Cuvelier de Trie in his 1818 adaptation of *Othello*.[94] The noise abated
with the appearance of Osiris, anxious to disclose some important
news. Jupiter himself had promised to deliver a gift, "a *jolie petite
curiosité*"—a pretty little curio—from Europe that would be "compensation for all that the Europeans, in their cupidity, have taken from our
ancient Egypt." As Osiris wonders about the form this compensation
might take, each mummy offers a different idea of what the gift might
be: perhaps Jupiter will arrange to relocate the Pont-Neuf so that it
crosses the Nile, or deliver to Egypt the Folie Beaujon, or the Fountaine

des Innocents, or the towers of Notre Dame. In a crowd-pleasing stab at the pretensions of the classier national theaters, where vaudevilles such as this could never have been staged, Osiris muses, "I would take the Panthéon, but I fear he'll send the Odéon."

At this, Mercury, played by one Mademoiselle Fluriet, arrives waving a magazine. She reports that the press is exceptionally excited about the arrival of the Dendera zodiac in Paris. But what, after all, can be truly *new* about an *antiquity*? Osiris, in disbelief, replies: "My zodiac, news! . . . But it was made before the deluge."

If this offhand remark were not enough to remind the audience of the zodiac's controversial religious significance—not to mention the specific connection of the zodiac with the Flood in the fight Duclos had picked with Dupuis—Mercury then makes the controversy over the zodiac's age the explicit subject of a ditty:

> Children, mothers, gentlemen and others:
>
> Ignorance and science are each holding fast
> They've taken the long view
> But more than one ingénue
> Has said: Bah, it's not
> So big a deal as that.
> Already we've heard advice
> From our most astute
> They're looking for the words
> To rouse the Institute
>
> One scholar says the zodiac's of recent vintage
>
> Another, just as smart,
> Affirms, for those who wonder,
> That the world is even younger.
> Children, mothers . . .

Mercury has more in mind than the provision of mere entertainment. Just arrived from Paris, he has come to Egypt with a souvenir, selected especially by Jupiter for Osiris. Is it jam? Osiris wonders. Paté? Perigord truffles? No, Mercury replies, it is none of these. "Just as Paris possesses the zodiac of Dendera," the winged-messenger explains, "so should Dendera possess a zodiac of Paris." Osiris asks, "But just what sort of thing is this zodiac?" To which Mercury simply replies: "I shall leave you the pleasure of surprise." The pleasure, of course, will be the audience's as well. The play's premise established, Osiris, the mummies, and the audience are left to wonder about Jupiter's Parisian gift.

At this point, the gift, whatever it is, has been wheeled onto the stage in a large case suggestive of a sarcophagus, perhaps the one in which Osiris was thought to have been interred, or an ark, which would have recalled, at least for the adepts in the audience, an important Masonic symbol from the Sophisian *Golden Book*. Osiris, who has been instructed by Mercury to open the case, and to call the names of the zodiac's signs as he does, wonders whom to call first. According to Mercury, a character representing each sign will appear with a special message for Osiris. The remainder of the play consists of interviews between Osiris and the dramatic embodiments of various signs of the zodiac who each, in turn, send up various aspects of bourgeois French culture, including intellectuals' efforts to make sense of the zodiac. As Osiris called for each sign in turn, the play catalogued the popular, official, and scholarly reactions to the Dendera stone while making sure that no one was safe from satire.

The Scorpion appears as a "pamphleteer" dressed "all in black"—an allusion to the inky fingers of the fourth estate's denizens. After the Scorpion touts his power to influence public opinion by means of anonymous screeds, Osiris warily compares the emblematic sign (and so their Parisian equivalents, journalists) to real scorpions in Egypt, which, he claimed, were just as black and dangerous. Then, when Osiris asks him to make Pisces, the sign of the fish, appear, the Scorpion declares that he has them on hand already, tucked up his sleeves. In this derogatory allusion to newspapers as fish-wraps, one imagines two actual fish appearing from the depths of the Scorpion's coat, though the stage instructions have been lost and would likely not have been written down in any case to avoid the censor's pen (another Scorpion if there ever was one). After comparing the fish to various establishment writers (he names Racine and Moliére) and government ministers, all of whom come to seem as interchangeable as fish, the Scorpion concludes: "But when the flame of genius / Throws a spark from our hands / It illuminates our country." At the height of this patriotic aria, Osiris interrupts to grouse: "So long as [the spark] doesn't consume it."

Aquarius, the water-bearer, arrives next in the guise of a wine merchant, a figure whose importance to French social life hardly needs explanation. Declaring the pursuit of amusement the sine qua non of his existence, Aquarius inquires after what Osiris does for fun at home in Egypt. Osiris replies, tongue lodged firmly in cheek, that "fun has certainly been the order of the day, ever since we've strangled the pasha," a sentiment that brought to mind the consequences of the French invasion and occupation. During this scene, the playwrights further tweaked both French and English imperial ambitions by finding a reason to stage, as a play-within-a-play, a scene from *Othello*—recast as farce. Although Shakespeare would become an emblem of appropriate emotional expressiveness to French Romantics, such as Hugo and Berlioz, at the time many thought his talent wildly overstated.[95] Shakespeare's passionately entangled characters and bawdy humor were largely lost on French audiences, whose expectations for polite entertainment excluded the buffoonery, intense emotionality, and bloody plots characteristic of the major plays. Political antagonism flavored French reactions to Shakespeare during the Restoration. Less than two months before the *Zodiac* premiered, a troupe of English actors had tried to stage *Othello* in Paris. They were met with cries of "Down with Shakespeare, Wellington's aide-de-camp" and pelted so violently with apples and oranges that police were called to quell the ruckus.[96]

The *Zodiac* performance rang its own changes on restoration attitudes to Shakespeare—and to Ottomans. Upon hearing about the play-within-a-play that is about to unfold, Osiris cries: "The English tragedy! . . . Now here is a very delicate subject for the Muslims [*Musulmans*]. Do you think they'll like it?" The wine merchant deadpans in reply: "I'm sure they'll enjoy themselves like Turks." A fragment of *Othello* followed. Although the playwrights did not specify which scene should be presented, Desdemona became a common shopgirl opposite the wine merchant's *garçon,* who played the lead.[97] As the interlude ended, the wine merchant wandered offstage, hawking refreshments: "orgeat, lemonade, ice cream, and roast beef." The last item was of course a crowd-pleasing poke at the English, colloquially known as *les rosbifs.*

After the fragment from *Othello*, Osiris summons Gemini, the sign of the twins, who appear on stage in the guises of "Panorama" and "Diorama." Their names alluded, of course, to two popular forms of public art in nineteenth-century France. Typically, panoramas were immense, static, enveloping paintings that depicted battles and other tableaux of national glory, including those related to the Egyptian campaign. The Diorama was an auditorium in Paris on the rue Sanson, right near the Boulevard du Temple, owned by Jacques Louis Daguerre (later known as the inventor of the daguerreotype), and Charles-Marie Bouton. Between 1822 and 1839 they displayed a sequence of immense paintings done on canvasses—twenty-two by fourteen meters—oiled to permit light to shine through and depicting famous places and historical events. The scenes were carefully lit, sometimes from the front, sometimes the back, for a maximally realistic effect that was enhanced by musical accompaniment and the inclusion of real objects, animals, or people in the displays. The lighting was cleverly designed to create the illusion of time passing. Daguerre's Diorama opened on July 11, 1822, just six weeks before the *Zodiac* was first performed.[98]

Next, Osiris called for Virgo, the virgin, appearing in the guise of an aging dancer who made her Paris debut in a different venue every night. In one way an edgy move given the inflamed religious sentiments of the day, in which the cult of the Virgin had become increasingly important, this was also a tongue-in-cheek allusion to the myth of Isis in Egyptosophic circles, which, we have seen, was especially persistent among the Sophisians. The playwrights had managed cunningly to insult both religious and Masonic orthodoxy at once.

Leo the lion appeared last. A French grenadier, the Lion had just returned from a mission in the Egyptian countryside. Not a single village, he reported nonchalantly, had escaped damage from bombing. The Lion poked fun at ancient Egyptian religion, devoting particular scorn to the large roster of deities, while also thumbing his nose at the close connections between church and state under the restoration.

Egyptosophic secret societies were easy targets, no doubt because they were so close to the world of the vaudeville and would have been familiar to the authors of *The Zodiac of Paris*. In its props, iconographic details, and stagecraft, the play took frequent aim at Egyptosophy's reverence for Egyptian gods. There was the gift from Jupiter to Egypt, a large, sarcophagus-like box, reminiscent of the ark central to Sophisian rituals. In the first moments of the play the box was wheeled to center stage, where it remained like a magician's hat, an apparently bottomless source of astrologically themed characters. Then there was the mummy choir which, as it smoothed the transitions from scene to scene with droll lyrical commentary emanating from the rear part of the stage, could not fail to evoke the initiation scenes described in the *Golden Book*, at least among those in a position to knows.

The caduceus—the "wand of Hermes"—that Mercury waves when Osiris makes his final call to the zodiac's characters provided another symbol with Masonic overtones. Then there was the scene from *Othello*, a version of which de Trie had put on in 1818 in the form of a three-act pantomime that included "the use of insinuating voices from the background"—much like the choir of mummies in the Dendera play and, not surprisingly given de Trie's familiarity with the Sophisians, like the use of disembodied voices in their initiation rituals.[99] The Sophisian emblems of the bull (according to the Golden Book, an embodiment of the chthonic Egyptian deity Ptah, the primal creator), and the egg (a symbol of the world, created by Ptah) are derided by the Lion in a song that celebrated his transformation of the "salad" of Egyptian deities into various French culinary specialties. The Apis bull is, for example, turned into "grilled meat" and served with a side of French onion soup. Egyptosophy was indeed a ripe subject for ridicule.

While the play parodied pro-Napoleonic secret societies, it also poked fun at the restoration regime and at the religious Right. But these barbs were better concealed and only become completely clear when we compare the printed play, which had passed through the regime's censorship bureau, with the original, which is still on file at the Archives Nationales in Paris.

Like all plays of the day, the Zodiac was submitted to the Ministry of Police for review prior to its first performance. Plays were censored for specific things, especially for insults to church and state and offenses to propriety; the censors were also sensitive, as we shall see, to attacks on the institution of censorship itself. Their remarks express by implication how they expected writers to imagine themselves and their work in relation to official power.

Beginning in 1800, both state-subsidized official and independent theaters were required to submit plays to a censorship bureau consisting of five men of letters, typically journalists or politicians, but occasionally members of the Académie. Their task was to suppress unflattering allusions to the regime and to defang overt criticisms of foreign or domestic policy. Throughout regime changes, these censors proved to be ideologically flexible, asserting their authority in defense of royalist or Bonapartist views with equal verve according to the political winds.[100] Nevertheless, of the hundreds of plays reviewed annually, few were rejected outright. Rather, the censors approved most of them after one or more rounds of editing and, sometimes, discussion among themselves and between the censors and the playwrights, during which offending passages were excised or rewritten. The forms of censorship varied, but, on the whole, if a play generated politically unacceptable representations of actual people, places, and events, the censor would insist on either removal of the problematic passages or a revision that would make it more difficult for audiences to connect what they saw onstage to historical events.[101]

The *Zodiac* reached the censor in August 1822, and the task of reviewing it fell to Alisan de Chazet, a sometime poet and ubiquitous presence at fashionable events, where his slithery obsequiousness earned him the scorn of the other guests.[102] Among them was the hack Lamothe-Langon who, remembering de Chazet as "the optimist of songsters," described him as "always pleased, always in a humor to praise . . . [he] had praised the Directory; at this moment he was praising Napoleon; and at a later period he praised Louis XVIII. He died, I believe, five or six years ago . . . I am not sure of that: but I am very certain that, if we were to have a new dynasty, or a new government, which Heaven forefend, M. Alisan de Chazet would rise from his grave to . . . praise it."[103] Reviewing the censors' notes on plays from this period, it is hard to avoid the feeling that the documents are the products of a constricted perspective, of lives made small and querulous by bureaucratic inertia, untroubled by the contradictions that service to successive regimes entailed. Chazet, who had a reputation for severity, was particularly sensitive to material that alluded to Napoleon and Egypt. He was, for example, instrumental in the 1824 censoring of a one-act play called *La Vivandiére*. Originally set in Kléber's Egypt, it was repositioned in Bohemia as a direct result of Chazet's objection that "Bonaparte's wars have been talked about enough for such a subject to be laid to rest."[104]

The copy of the manuscript on file at the Archives Nationales bears the censor's redactions, and from it we see that Chazet required two sorts of alterations: he demanded major changes involving the inclusion or exclusion of entire scenes and characters, and he insisted as well on minor alterations that aimed to mute criticism of the regime by the excision or addition of a few words. Changes of the latter sort blunted the original version's emphasis on France itself having despoiled Egypt. For example, in the third scene, after the audience learns that "the cupidity of the Europeans" in removing antiquities from Egypt will be compensated by a present to the Egyptians, Osiris, in an effort to guess what kind of gift this might be, imagines that it will be some characteristically French product. While the original version of the play specifies that he should only say "French comestibles," specific items were substituted for the original phrase: "jams, . . . patés, Périgord truffles." By replacing *French* with a group of products (French though they themselves were), the censor softened the emphasis on France proper as a despoiling agent. At the same time, the items suggested by the censor were small luxuries that one might conceivably bring home as

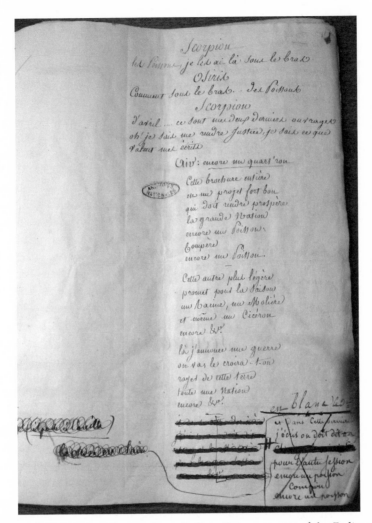

Figure 9.1. A page with censor's marks from the manuscript of the *Zodiaque de Paris*

souvenirs of a trip to Paris, thereby suggesting that the zodiac was itself not a spoil of war, but just one side of a fair and affectionate exchange of souvenirs.

The censor was the play's first, and in some ways its most important, audience. The picayune quality of the struggle between playwrights and censor, as it develops in the heavily redacted pages of the original manuscript, provides graphic evidence of the extent to which the censor's office viewed every element of a drama as a potential political landmine. This relentlessly obsessive attention was not misplaced. The play was, in fact, political down to the smallest details, and the censor had good reason to be attentive, for the playwrights seem to have written as much to twit him as to entertain the play's eventual audience.

Early in the play, bantering with the Scorpion, Osiris complains about how hard it is to convince talented hacks, like his interlocutor, to turn their talents to honorable pursuits, rather than rabble-rousing in print. In the version submitted to the censor, Osiris asks: "Why in this century must one take the role of censor? It is better to bring men back to what is sweet and good." Osiris's point not only poked fun at lowbrow scriveners who were not too different from the authors of *The Zodiac of Paris* themselves, but also, by explicitly questioning the institution of censorship itself, dared the censor to counter it. Which he did, by substituting *"Frondeur"*—a reference to the seventeenth-century Fronde, a civil war that unsuccessfully aimed to preserve the feudal privileges of the French *parlements* and nobility against monarchic absolutism—for the more inflammatory *censeur*.[105] This substitution had the nice effect of turning criticism of a royal institution into a reminder of royal power.

Despite the overtly political points in the play's original version, the authors apparently did not cavil at these minor, censorial changes. Their decision not to resist was sensible, for overt resistance would certainly have spelled disaster for the production. However, in addition to the smaller alterations, of which these are merely a representative few, the censor also required the elimination of an additional character linked to the astrological sign of Capricorn, whose presence in the original version was deemed too dangerous.

After Osiris tires of speaking with Virgo, he summons Capricorn, a move that causes Virgo to scurry offstage, oddly crying: "Ah, mon Dieu, that's my uncle!" Whereupon Capricorn appears on the back of a fat bay horse, sporting a pointed cap and a battering ram. In the original manuscript, Osiris then breaks into song, introducing Capricorn to the audience with this ditty:

He's a good bourgeois of Paris,

Born in *quartier* St. Denis,
But since to Egypt him we send,

The Nile will be for him the Seine.
There he may, free of fresh care,
Peacefully watch the water fare,
I see him already planning to retire,
All along, along, along my river.

The censor responded with ambivalence. On his first pass, he indicated that the song should be sung by both Mercury and Osiris; then he required that the passage be struck entirely.

In the next musical number, the original version required that Capricorn introduce himself to the audience as a cuckolded husband whose wife had left for France with "a great English Lord"—presumably one of the British expatriates, like Henry Salt, who controlled antiquities dealing in Egypt. Though rather difficult to spy through the censor's pen, the original seems to run something like this:

With a great *milord anglais*,

My wife's away to *les français*,
My wife is kind and pretty,
She moves one and all to pity,
But for the sparrow's tears aflow [*larmes de moineau*],
I'd rather watch the river flow.

The censor revised this passage by reassigning it from Capricorn to Mercury, by changing "*my* wife" to "*the* wife," and by eliminating the previous ditty about Capricorn, whose lines, though not his presence, disappear entirely. The changes subtly altered the meaning of the remaining passage.

In the "good bourgeois" Capricorn's mouth, the song is a skeptical comment on the French–English rivalry for possession of Egypt, its natural resources and antique treasures. The metaphor of the sparrow was, in a complicated way, both an allusion to England (sparrows were considered a typically English bird, as, later, in the "cockney sparrow") and an insulting pun (*moineau* also means "half-wit"). Although the censor's redactions make the original difficult to decipher, it is perhaps not overreaching to suggest that Capricorn's description of his traveling wife was intended to evoke the competition between France and England for possession of Egyptian antiquities. Only barely wrested from the machinations of Henry Salt, the zodiac's arrival in France could not fail to evoke the old but still incompletely healed wound of the Rosetta Stone. As sung by Capricorn—the "good bourgeois" of Paris, a cuckold gone native who could not care less—the song was a bitter reminder of the French obsession with England's earlier humiliation, and of England's continuing threat to French influence in Egypt.

It is not hard to understand why the censor might not have been happy with this. Capricorn's position reeked of defeat. The song hardly glorified a contemporary French presence in Egypt, recalling as it did an era that the Bourbons would have preferred to forget. Putting the song in Mercury's mouth altered the message. The change made Mercury responsible for the metaphor of the sparrow, and for the description of Capricorn as a defeated cuckold who would rather watch the Nile go by than chase after his English-loving wife. By changing Capricorn's pitiful first-person testimony to a third-person description of the sorry state of someone else's affairs, the playwrights invited the audience to identify with Mercury, to distance themselves from the Nile-obsessed Capricorn, the overwhelmed "bon bourgeois" from Paris, and to consider his plight from an outsider's perspective rather than as a sympathetic participant.

In the end not even this was enough. The printed version of the play does not include Capricorn at all, and the scene just described is missing altogether. From a dramatic point of view, the change makes little difference to the development of the play's (admittedly minimal) plot. Yet Capricorn may have had further significance that helps explain the excision, given the history of arguments over Egypt's antiquity. Dupuis, remember, had based his claim for the great age of Egypt on a novel agricultural interpretation of the zodiac, according to which the summer solstice at the zodiac's creation lay in the constellation Capricorn. What might the playwrights have known about this?

We may first of all be certain that the authors did pay attention to the controversy over the zodiac's age because they explicitly mention it. On the other hand, it is hardly likely that they would have seen much, if anything, of the discussions by Fourier, Raige, and Jollois and Devilliers in the expensive *Description de l'Égypte*, nor were they likely to know directly about the public remarks printed in the *Moniteur* and elsewhere two decades before. But much had taken place recently, and this they would likely have heard about. The play was authorized for production (with the censor's changes) in early August and first performed on September 2. Saulnier's dramatic account of the zodiac's acquisition had appeared not earlier than the previous mid-April,[106] after which, we shall see, negotiations had been opened with Louis XVIII to buy it for France. Discussions about the zodiac's age had heated up the preceding year, and it was widely talked about all over Paris by the time the play was being written during the summer.

A short pamphlet on the subject had reached print earlier in the year, likely much before the summer, and, together with Saulnier's book, would have been a useful source for the playwrights.[107] While the pamphlet was printed anonymously by Guiraudet at 315 rue St-Honoré, it was in fact authored by the otherwise obscure J. Chabert and L.-D. Ferlus.[108] That the publication was issued without naming its authors suggests that its appearance posed a problem for them, though the difficulty may have had more to do with a desire to hide their names from Saulnier than with any fear of the censor. Saulnier's "brochure," as he called his account of the zodiac's abduction, was available together with a new, and reputedly precise, engraving by Gau. The engraving could be bought for five francs at 52, rue de Rivoli. Chabert and Ferlus' pamphlet was also accompanied by an engraving, and it could be had at no less than three locations, including one (Le Gouvey Frères) in the heavily frequented Palais Royal and another (Neveu) in the "Passage du Panorama"—not too far from Daguerre's popular, panorama-competing Diorama. Saulnier's own brochure both touted Gau's design and denigrated another "small, lithographed design lately published by someone who had not seen the Dendera planisphere in either Egypt or in France." That design was "a very inaccurate copy of the one [by Jollois and Devilliers]."[109] Since no other engraving seems to have left a trace, Saulnier may well have had Chabert and Ferlus' in mind.

These two were apparently familiar with the early speculations by Coraboeuf and others concerning the age of the rectangular zodiac at Dendera, and no doubt with Dupuis' well-known claims for Egyptian antiquity. They also included details of the zodiac's dimensions that were not printed by either Fourier in his *Description* account (which had appeared in 1818 in the second text volume for antiquities) or for that matter by Saulnier.[110] The pamphlet advertised the circular zodiac's age at "4,400 years before Jesus Christ," when "the solstice took place in the sign of Leo," and mentioned that there was also one at Esneh, two thousand years older than that. The date 4400 BCE was merely the date of the temple zodiac that was embodied in stone at Dendera, not necessarily the beginning of Egyptian civilization, which must have been immensely older to have evolved such a sophisticated structure. And so the pamphlet concluded with the provocative declaration that "the 11,340 years [of Egyptian antiquity] spoken of by Herodotus . . . is incontestable."[111]

The playwrights would have noticed Chabert and Ferlus' 11,340 years for Egypt if they had seen it, which is likely since they were intent on composing a vaudeville with a catchy, provocative, and up-to-the-moment theme. Although the pamphlet itself did not explicitly connect Capricorn with an immensely remote Egyptian antiquity, the link had been made forcefully by Dupuis in his lengthy *Origine*, which, as it happens, was republished in full in 1822 in both a "popular" (inexpensive) edition and a pricier one. The popular edition contained Dupuis' dissertation on the zodiac and was printed by the "Librairie Anti-Cléricale," making the connection between anti-Christianity and the Egyptian stones as clear as could be.[112] It certainly does connect Capricorn to Egyptian antiquity, although the claim is buried well into the work. More accessible and equally notorious was Dupuis' own *Abrégé*, which, as we have seen, was reprinted in 1822, causing problems for its publisher, Chasseriau. The *Abrégé* itself mentions Capricorn three times, though it provides no clear discussion of Egyptian antiquity. Dupuis himself, then, was a possible source for the link between Capricorn and Egypt. Even if they had not read about it themselves, someone might have informed the playwrights once they made known that a zodiac vaudeville was in the offing.

The best indication that the significance of Capricorn was widely known comes from an extraordinarily popular book, namely, Cuvier's separately issued *Discours sur la théorie de la terre* of 1821.[113] The *Discours* included a much-expanded section on Egypt and on the age of the zodiacs that was stimulated by Cuvier's interactions with Delambre in judging a memoir by one Paravey, to whom we will turn below. In it Cuvier explicitly discussed (and rejected) Dupuis' claim that, at the zodiac's invention, the summer solstice was in the tail of Capricorn, and therefore that Egyptian civilization dated back fifteen thousand years. To anyone who had seen the critique, Capricorn would have stood for claims to Egypt's immense antiquity.

It is accordingly possible, though not certain, that the inclusion of Capricorn in the *Zodiac*'s original version indicates that the authors were aware of the symbol's potent status, and that they were deliberately using it to have their own say. The playwrights may thereby have intended to stake their allegiance to the anticlericalism that was so visible in Dupuis' agricultural cosmology of religion—and that only months before had led to the suppression of the reprinted *Abrégé*, to which they would certainly have been quite sensitive, subject as they themselves were to the censor. Moreover, by giving Capricorn a familial connection to Virgo (as her uncle), the playwrights were treading on even more dangerous ground because Virgo (the Virgin) might represent Christ's mother, in which case Capricorn would (according to Christian lore) have had to be Joseph of Arimathea. That would certainly have suited the play's anti-English tone because Joseph was reputed to have been among those who brought Christianity to England—portrayed here as a befuddled cuckold watching the Nile glide by, whose gift to an English *milord* is not Christianity but his unfaithful wife. Although these several links are certainly conjectural, still, given the inflamed climate of the day, the censor's elimination of Capricorn may well indicate his sense that the play subtly evoked dangerously unacceptable positions.

Despite (or perhaps because of) the excitement generated upon the zodiac's arrival in Paris, the vaudeville farce seems to have opened and closed with little fanfare. In an 1822 review, a critic complained that the playwrights had only managed to cobble together the various learned responses to the zodiac in the service of what the reviewer considered a tired critique of French geopolitical intrigues in Egypt. The review's author, Chalons d'Argé, was well positioned to comment on matters related to drama and imperialism, though unlikely to sympathize with a critical view of the latter. At the time, he was working on a two-volume study of the English envoy Hiram Cox's forced resettlement of Burmese immigrants to the Bazaar district in Bangladesh.[114] His two-page report, which appeared in a compendium that collected his impressions of a great number of plays that had opened in Paris the preceding years (a compendium that, he grandly claimed, would give "a just and true idea of the state of our dramatic arts"[115]), contained but a single positive comment about the play. Although he devoted the lion's share of his review to straight description of the plot, he did praise the representation of Gemini, the sign of the twins, as Panorama and Diorama, lauding the latter's performance as "very precise."[116] D'Argé wasn't much impressed with the pamphlets and arguments that had begun to flow like the Nile about the zodiac, calling them "forgettable and almost always ridiculous dissertations." Ridiculous they may well have been in some contemporaries' eyes, but these articles, pamphlets, and even books dramatically illustrate the Dendera zodiac's enduring power to summon controversy, which we will explore in the next chapter.

10

The Zodiac Debates

The Chevalier Paravey

Three years before the Dendera zodiac made its physical appearance in
Paris in January 1822, a one-third-size wax model of it that had been
produced in Egypt by the sculptor Castex had been exhibited in the
city.[1] Napoleon had originally contracted with him to produce a marble
version of the wax model. General Kléber then tried to obtain stone for
the project, but there was none available in Cairo. He heard that there
might be a supply in the storehouses of the Alexandrian customs, but,
again, nothing seems to have come of it.[2] Although in 1819 Castex did
produce a carving (Figure 10.2), the restoration government refused to
honor Napoleon's contract. After Castex's death the marble fell into the
hands of a British speculator who, working rather like Saulnier had in
1822, printed what appears to be an advertising brochure for its sale in
London in 1825.[3]

Many people came to see Castex's wax model in Paris, among them
one Charles Hippolyte de Paravey. A year later, in 1820, Paravey revived
the moribund zodiac debates by offering a new theory, which he presented in five consecutive sessions of the Académie des Sciences in July
and August, just two months before Lelorrain left for Egypt. Paravey was
later to claim that his work had been the original stimulus behind Saulnier's plan to bring the zodiac to France. True or not, there is no doubt
that his talks rapidly refocused attention on how the Dendera circular
might be interpreted as a representation of the celestial sphere on a flat
plane, one that had been produced in ancient Egypt according to some
sort of rule. Time formed an essential part of his argument, since Paravey
had every intention of undercutting any age for the zodiac much before
the era of the Ptolemies. The ensuing contretemps, we will see, involved
a galaxy of overlapping commitments, from conflicts between scientific
savants to arguments between some among them and historians and
philologists. Lurking behind these clashes was the brooding specter of
religious orthodoxy, quickened by political and social circumstance.

Born two years before the revolution in the town of Fumay in the Ardennes, Charles Hippolyte entered the École Polytechnique in 1803, a year before another young man by the name of Augustin Fresnel also enrolled. Fresnel became a central figure in French science a decade later, and he would come into conflict with Biot, whom we have met before, and who, we shall see, became the object of Paravey's burning hate when Biot injected himself into the controversy. Both Paravey and Fresnel graduated in 1806, and both joined thereafter the corps of Ponts et Chaussées as civil engineers.

We know little about Paravey's career for the next decade or so, including his activities during the first restoration and the Hundred Days.[4] To judge by his later views, he would not have been sympathetic to the emperor's reappearance. In any case, a year after Waterloo (1816) we find Paravey in the position of a subinspector for his old school, the Polytechnique. Here he seems to have played a role in protecting the school from an attempt to destroy it altogether at a time when the restoration government was convinced, not without reason, that the collèges were sites of antiroyalist sentiment and action. Suspicion of student groups preoccupied the regime, though this in itself was hardly new, since the empire had nurtured similar doubts and fears. Constant, even obsessive, surveillance exacerbated the growing antagonism of what became known as the "generation of 1820" to authority. "The discipline presumed appropriate to a community of healthy young males," writes Alan Spitzer, "was roughly analogous to what would be applied in a training depot of the Foreign legion. To these standard assumptions the Restoration added an insistent, meticulous religious regimen for the salvation of those corruptible small souls."[5]

Although few scientists and mathematicians were purged from the institutes and schools during the second restoration, with the exceptions of such major revolutionary figures as Monge and Carnot, nevertheless the atmosphere became corrosively antagonistic to overt expressions of hostility to religion and throne. In the spring of 1816, nine months after Napoleon's exile to St. Helena, the Polytechnique was riven with dissension when the students took particular umbrage at the overbearing attitude of one of the professors, whom they attempted to have removed.[6] The interior minister, Viénot, Comte de Vaublanc, took the opportunity of the ensuing fracas to send all the students home and to put the faculty on half-pay, pending the results of a commission headed by Laplace that was charged with the school's reorganization. A profoundly conservative royalist, and according to Decazes a tool of Artois

Figure 10.1. The shaded engraving of Jollois and Devilliers' drawing of the circular zodiac and surrounds

(the future Charles X), Viénot was later dismissed by Louis XVIII at the insistence of Richelieu, who threatened otherwise to resign.

In September an ordinance demilitarized the school and was used as well to dismiss politically suspect professors.[7] Particularly insistent on the deplorably liberal character of the school, Hughes de Lamenais, a priest soon to be famous for advocating the unity of the state with (Catholic) religion, printed a brochure in June to push Laplace's commission in a suitably conservative direction.[8] This influential, belligerent, and anti-Semitic apostle of belief had particular disdain for science, which he thought tantamount to atheism. "One worships human reason under the name of *science*," he bellowed in 1820, "for certain spirits it is the God of the universe; one has faith only in this

Figure 10.2. The marble sculpture of the Dendera circular, carved by Castex

God, one hopes only in him; his wisdom and his power must renew the earth, and, through rapid progress, elevate man to a degree of happiness and perfection that's impossible to imagine. This religion continues to develop, it has its dogmas, its mysteries, even its prophets and miracles; it has rituals, priests, missions, and its sectaries would substitute them for all others."[9]

Paravey's role in all of this remains obscure, but he was certainly involved in keeping the school alive throughout its difficulties. He had developed a string of enemies during the empire, and many of them remained in positions of influence during the restoration.[10] Among these was the head of Ponts et Chaussées under both Napoleon and

Louis XVIII, Louis Mathieu, the Comte de Molé, who apparently blocked his advancement in the corps, at least according to Paravey.[11] During his student days at the Polytechnique, the young Paravey may have expressed disapproval of the imperial regime, which would account for Molé's attitude since he was a strong supporter of monarchy, whether imperial or royal. Years later, Paravey claimed to have formed friend-ships at the time with at least two men who were decidedly antago-nistic to the empire. There was Mathieu de Montmorency, who had supported the revolution until the monarchy's fall, after which he left for Switzerland, returning following Thermidor. By the time of the res-toration, Montmorency had become a religious reactionary; at the end of 1821 he was appointed minister of foreign affairs. And there was Noailles, who had been forced to leave France during the empire and became active in furthering the Bourbon cause as Napoleon weakened in 1812. Both men had audited courses at the Polytechnique.

Paravey recalled years later having had "discussions" about the zodi-acs and deciding to "examine" them when he had free time. Whether this occurred during his student days at the Polytechnique, when the savants had only recently returned from Egypt and talk about the zodiacs was still fresh, or years later remains uncertain. In any event, his time apparently became sufficiently free following the reorganiza-tion of the Polytechnique that he was able to attend Arago's course on astronomy at the Paris Observatory and Remusat's on Chinese his-tory at the Collège Royale. Arago had entered the Polytechnique the same year as Paravey and, despite difficulties to which we will return, had managed to acquire his position at the Paris Observatory, while Remusat was the first holder of his chair at the Collège Royale. Paravey was by then interested in Chinese as well as astronomy because he had begun to forge universal connections between zodiacal symbols. By 1820 he was ready to present the theory that he had evolved dur-ing the previous four years. In July and August he presented it at no less than five consecutive sessions of the Académie des Sciences. One major consequence of Paravey's hypothesis concerned the period of the Egyptian temple zodiacs. He was certain that they had to be very late products, dating in fact to Roman times. That claim at once gener-ated quite a reaction, not least on the part of Fourier, reviving as it did the controversies of a decade and a half earlier.

Fourier had moved to Paris after the Hundred Days, having been relieved of his prefectural appointment. Although he had no funds to speak of, and he could hardly expect much sympathy from the restored monarchy, he had taught Chabrol de Volvic, now prefect of the Seine, at the Polytechnique. Moreover, Chabrol had been with Fourier in Egypt and had authored or coauthored three memoirs for the *Description* by 1813. He was a man "who devoted his career to the management, structuring, and organization of territory."[12] Fourier's connection to Chabrol provided him with a position as director of the Seine's Statistical Bureau, a post that he took seriously. However, his repeated applications over the years for a pension based on his prefectural responsibilities went nowhere, though he did eventually receive one, disconcertingly, for unspecified "services of information" to the Ministry of Police. Fourier also set about obtaining election to the Académie des Sciences, which was not a simple matter given the political circumstances. In a May 1816 election for two "free" positions (ones that had been specially created), Fourier did win in the second round of voting for the second position. Delambre conveyed the result to the interior minister, but the king refused to approve it, which is hardly surprising given Fourier's past associations and the ultra-royalist climate. The executives of the Académie nevertheless pressed ahead. A year later, when death opened up a regular position, Fourier was elected to it by forty-seven out of fifty votes, and Louis agreed, having dissolved the Ultra chamber the previous September in the face of increasing popular anger.[13]

The Académie committee chosen to examine Paravey's papers consisted, naturally, of Fourier as well as Delambre, Cuvier, Ampère, and Burckhardt. Cuvier had long before written about the Egyptian zodiacs, Delambre was an expert on ancient astronomy, Burckhardt had been the one to write Lalande from Egypt, while Ampère, who taught mathematics at the Polytechnique and would within a month become deeply involved in working through the consequences of Hans Christian Oersted's discovery that electric currents produced magnetic force, had encouraged Paravey. Delambre was elected to write the report, but Cuvier insisted on joining him, having (according to Paravey) sensed "the great philosophical and religious importance of our research."[14]

From the beginning, Paravey hoped that his contribution to ancient astronomy would give him the reputation of a true savant, which he desperately craved, and would confute those who set biblical chronology to the side, spreading thereby the noxious contagion of irreligion. Years later, in 1835, an embittered and ignored Paravey recalled how he had "established, *astronomically*, that the Egyptian zodiacs, those famous monuments which were to overturn biblical chronology, were no older than the epoch of the Romans." In this fifth year of the July monarchy that had replaced the Bourbon Charles X with the Orléaniste Louis Philippe, son of a regicide, Paravey reprinted his works in order to "combat those dangerous books which Philosophie prints once again."[15]

Heated arguments erupted among academicians from the moment that Paravey began to present his views, not least, he was convinced, because of his "political and religious" opinions. Soon after reading his memoirs, Paravey left for "the waters of the Pyrenees," no doubt to salve his nerves in the salts of local spas. While he was away, Arago, though not himself named to the committee charged with producing a report, "had demanded our manuscripts, took them to the Observatory, and, after a careful investigation, having discovered an error simply about the name of a star, announced that he would have the manuscripts rejected by the Académie." The following February 5, nearly six months after Paravey had made his presentations, the report prepared by Delambre and Cuvier, but written by Delambre, was presented to the Académie. There followed "a solemn discussion, in effect between M. Fourier, supported by M. Arago, our former schoolfellow, and M. le baron Cuvier, who didn't know us at all, but who, with that nobility of soul for which history will one day praise him, had the courage to defend the important thesis which we uphold." Heated arguments flew back and forth, and the report was nearly rejected, with Fourier and Arago being seconded in their opposition by the mathematician Sylvestre Lacroix and by a man whom Paravey referred to here as Arago's "intimate friend," Auguste de Marmont, the duc de Raguse.[16]

The report was in the end formally accepted, though without the lengthy appendage in which Delambre directly approved much of what Paravey had to say and offered some further thoughts of his own. Both Fourier and Burckhardt refused to sign the formal report.[17] Fourier had wanted to remove from the official statement every passage "which would tend to give a less than favorable idea of the knowledge of the Chaldeans and the Egyptians," for he had committed himself in the *Description* to the conviction that the Egyptians had possessed profound scientific knowledge by the middle of the third millennium.[18] To understand just how unsettling Paravey's work was, and how it set the stage for the arguments that followed once the physical zodiac arrived in Paris, we turn now to what he claimed, though even Paravey never printed his complete original manuscript.

Universal Symbols and the Zodiac as Planisphere

Paravey contended that primitive knowledge of writing and of numbers, as well as the development of certain simple symbols connected to the heavens and the cycles of the year, originated in one place millennia ago and spread from there throughout Asia. Before the Académie des Sciences, he spoke only about the much later evolution of that primeval knowledge, in particular about the origin of the constellations, reserving the wider claims for another academy, Inscriptions et Belles-Lettres. In his report, Delambre noted Paravey's assertions concerning the worldwide connections between stellar symbols, remarking briefly that Paravey's proofs were "so varied and numerous [that] despite the conviction of some savants, it seems difficult to deny that connections exist between the constellations of the Egyptians in particular and those of the Chinese and Japanese."[19] The basic scheme was this.

Shortly after the Noachian Deluge, new centers of civilization arose among those who had "escaped that Cataclysm"—a rather touchy point in itself, since arguments had gone on for centuries over whether the Deluge was universal or not, as well as over who might have survived it and where. As to location, Paravey settled on Assyria. Colonies spread out from there, he claimed, peopling the world while carrying far and wide the ancient arts and sciences that had survived the cataclysm. Paravey agreed with Cuvier about the date of the Flood, placing this last catastrophe near the beginning of the third millennium. The primitive constellations were themselves first delineated shortly after the waters had receded, which is why one finds (according to Paravey) such great similarities between the figures across Asia and the Near East and among the ancient Greeks.[20] For these are the descendants of those most antique of all signs. Although Paravey insisted, we will now see, on a comparatively recent date for the temple zodiacs, he averred that he had nonetheless "never denied that one could find positive traces of these same theories that now serve as the foundation of our sciences and all of our knowledge from the first epochs of the existence of man on earth."[21]

For a brief taste of Paravey's reasoning concerning the universal presence of stellar symbols, consider how he found traces of signs present in the "east Asian sphere" (by which he meant Chinese representations of the heavens) in the Dendera circular itself, or rather in Jollois and Devilliers' drawing of it. The Chinese "sphere," Paravey noted, places a sort of "fish-hook" near the north pole; it is represented by a

Figure 10.3. Paravey's "hook" in Chinese and on the Dendera circular

勾 *Kéou* 陳 *tchin*

symbol pronounced (in early nineteenth-century transliteration) *Kéou tchin*. Near the center of the Dendera circular, Paravey spied what seemed to him to be a hook with a foxlike creature perched upon it (Figure 10.3). He concluded from the resemblance that there must be a substantive relationship between the Chinese symbol and the Dendera image. There was more. Since the Chinese "hook" floats near the north pole, and the creature-topped one on the Dendera circular occupies nearly its center, then the Egyptian zodiac's center must also represent the celestial pole. And if that is so, he went on, then the circular must truly be a planisphere produced according to some rule, and Paravey thought he knew what that rule must have been.

Paravey was a graduate of the École Polytechnique, and he had also imbibed some astronomy from Arago's lectures. At the Polytechnique, he certainly learned a great deal about projective geometry from Monge, who would have introduced him to the several different methods for producing numerically useful representations of three-dimensional objects on a flat surface. In astronomy the oldest of these, dating to before the time of Ptolemy, worked in the following way. Array all the stars on the celestial sphere and imagine standing at its South Pole looking towards the north. Above you the equator will cut a plane through the sphere's center that is perpendicular to the polar axis. Above you as well is the plane of the ecliptic, on whose circumference are arrayed the constellations of the zodiac; that plane tilts at about a twenty-three-and-a-half-degree angle to the equator. Suppose we would like to represent the stars of the Northern Hemisphere on the plane of the equator. To do so in a way that is governed by a rule of projection, we draw a line from our position at the South Pole and through the equator's plane to end at the star we are interested in projecting. The place where that line crosses the equatorial plane will be the star's projection.

That, according to Paravey, was how the Dendera circular was produced, and his claim had consequences that could be examined. Consider the ecliptic, around which the sun moves during the course of the year. At the summer solstice the sun will be farthest away from your position at the South Pole, while at the winter solstice it will be as close to you as it can get. Since every diameter of the celestial sphere projects to a line that passes through the center of the equatorial plane, the solstices lie on the same line but on opposite sides of the equator's center, as we can see in Figure 10.4. The projection of the summer solstice S will clearly be located much closer to the center than that of the winter solstice W. Now consider the axis of the ecliptic itself, namely, the line that passes through its center and that is perpendicular to its plane. It will intersect the celestial sphere at some point, and we can project that point as well onto the equatorial plane, producing E in Figure 10.4.

Paravey applied this reasoning to the Dendera circular, having already associated the center C with the "hook." But where on the hook? It could not have been an arbitrary point, he thought, and therefore had to be one of its three ends, for which Paravey chose the long end closest to the head of the creature. Once given the center's projection, he had next to locate the line $SCEW$ that contains all four points: the solstices, the center, and the pole of the ecliptic. Here Paravey found a clue in the orientation of the Dendera temple itself, briefly described by Fourier in volume 2 of the *Description*, which had been printed in 1818.[22] There he mentioned that the "axis of the monument" lies at seventeen degrees to the north–south line. This was a bit vague, since Fourier did not also specify the orientation of the little room on the roof of the temple that contained the circular zodiac. Paravey reasonably assumed, correctly as it turned out, that its walls would parallel those of the temple proper, in which case a line drawn through the zodiac and perpendicular to its (more or less) east–west walls would also incline at seventeen degrees to the north–south line. He hypothesized that the line containing the colures would itself run in that direction, with the temple's orientation, as it were, having been designed to indicate their position.

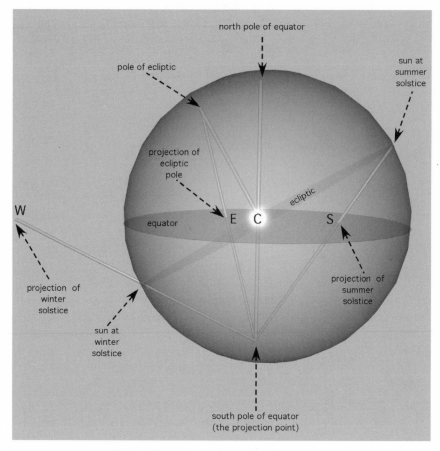

Figure 10.4. Paravey's suggested projection

Figure 10.5 is a graphic of the temple's orientation, illustrating the displacement of its longitudinal axis from true north–south.[23] Figure 10.6 draws the north–south line across the zodiac on the *Description* engraving by Jollois and Devilliers with points marked on the line according to Paravey's theory. The summer solstice must accordingly sit near the head of the twin Pollux in the constellation Gemini, while the winter solstice abuts the rump of Sagittarius. That immediately places limits on dating and in fact indicates that the monument represents a configuration of the heavens that could not have occurred before about the third century BCE and might even postdate the time of Augustus, which is essentially where Visconti had located it two decades before.

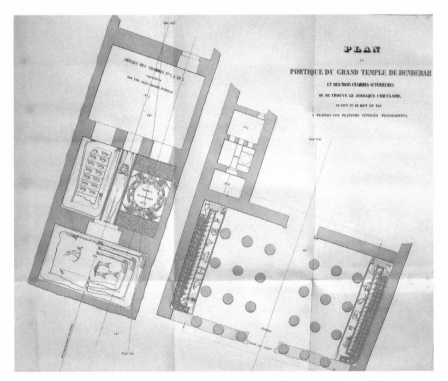

Figure 10.5. Plan of the Dendera temple published by Biot, as it would be seen from above if the ceiling were transparent

We have already seen how Paravey's talks produced quite a reaction among the auditors at the Académie, with Cuvier, Delambre, and Ampère (whom he thanked for having often "clarified and guided his research"), all of whom signed the report, defending him, while Fourier and Arago, with support from others, worked hard to remove any directly approving verbiage. They succeeded in that, and Paravey never forgot nor forgave the slight. Several months after Delambre's report was officially read on February 5, 1821, Paravey printed his own account, in which he first laid out previous claims about the zodiacs, followed by an aperçu of his own main points, and concluding with Delambre's official report and additional remarks.[24] In it, while remaining polite, he took particular aim at Fourier's claim that the "Egyptian sphere" (if not the Dendera circular itself) reached back to 2500 BCE, a time when, Fourier thought, Thebes had reached full splendor. Paravey recognized that Fourier had used the heliacal rising of Sirius in

Figure 10.6. Paravey's axis

his arguments, though no more than others did he see precisely what
Fourier had actually done with that. Nevertheless, he denied altogether
that the ancient Egyptians ever did link astronomical observations of
any exactitude to Sirius's rise. For, he challenged, just try to locate
a constellation containing the star among those found in the several
zodiacs. Furthermore, Paravey continued, Volney reported how hazy
the Egyptian horizon tended to be, and Fourier did not indicate how
that might affect the precision of rising observations. This last point
was quite acute because the state of the air does markedly affect such
things, and even today it is extremely difficult to take it into account.
Moreover, this and refraction are factors that Fourier seems to have

neglected in his own calculations. Paravey admitted he was no expert in the "high analysis" necessary to compute Sirius's rising, but that did not matter because he denied its relevance to the zodiacs altogether. Chiding Fourier for having once thought the zodiacs dated back at least 12,500 years, Paravey urged him to change his mind again and reduce the scheme by another two millennia or so.

The formal report by Delambre remained distant from Paravey's claim that the circular had been produced by the usual sort of stereographic projection. If it had been, then the figured symbols for the constellations should themselves show the effect, Delambre noted. Constellations further from the north pole should be larger in projection than those closer to it (see Figure 10.4), but they are not. "If it is a projection," Delambre wrote, "as it would seem permissible to think, it was done without any idea of geometry." Whoever did it had not "the least knowledge of Hipparchus' projection," but there was an alternative possibility.[25] Suppose, Delambre suggested, that the zodiac's designer placed the constellations at distances from the center of the representation that are in the same mutual proportions as their actual, angular distances from the north pole on the celestial sphere. This would not be a stereographical projection, but it would be a simple way to produce a workable result.

There was at least one obvious problem with that suggestion, Delambre continued, because Cancer is represented much closer to the center than Leo or Gemini, which should not be the case. But even that apparent defect had a possible remedy. Just below Cancer, and between Leo and Gemini, stands the figure of a man with the beak of a bird. Now the Ibis, or the head of a sparrow-hawk, "is the ancient sign for which the crab was substituted." For that tidbit Delambre provided no citations, but the connection had long ago been made by the seventeenth-century Jesuit and collecting polymath Athanasius Kircher. And so, obviously, the bird-beaked man takes its proper place as the representative of Cancer. In their contribution to the first volume of the *Description* in 1809, Jollois and Devilliers had, rather obscurely, suggested something similar to what Delambre had in mind, but, like him, they warned that it could not have been done with mathematical precision.

Paravey's revival of the zodiac debates soon generated considerable interest outside the Académie proper—not surprisingly given the unsettled temper of the times, in which anything that raised issues touching on matters of faith could generate controversy. Sometime during 1821 he met with the king himself, to whom he presented a

Figure 10.7. Delambre's bird-man highlighted at center

copy of his aperçu shortly after its publication. Louis, Paravey claimed years later, read the entire work (hardly likely), and "on his very table we unrolled all the drawings, and we received from him, eight days later, by way of one of the princes, the assurance of the interest that he took in our book."[26] Whether or not Louis had actually read the aperçu, it is striking that Paravey had gained access to him. The meeting probably took place shortly after the zodiac's arrival in France, since its presence had the potential to cause conflict over issues of faith, to which the royal administration was acutely sensitive.

Paravey's memoirs had already elicited favorable reactions from the religious press. On August 29, just before the zodiac reached Marseilles on September 9, that organ of altar and throne, the *Ami de la réligion*, published a glowing report. The admirable and learned Paravey, the journal averred, had demolished the fantasies of Dupuis and others, who sought to overthrow sacred chronology. "And so," the *Ami* concluded in triumph, "crumbles that system for the world's antiquity; daily it gathers fewer and fewer followers. The hypotheses of Bailly and of Volney are abandoned."[27] Some among the Académie members, the report continued, had been angered to see the world's great age so contravened. Indeed, so vociferous was the opposition that one suspected it "had its source in motives outside science"—obviously, in the opposing savants' penchant for anything that attacked the foundations of belief. But these savants could hardly be trusted in any case, the *Ami* implied, since they had so strikingly changed their minds about the zodiacs' ages.

The *Ami*'s review publicized Paravey's claims, and within a month new accounts of it reached the journals. On October 4 the *Quotidienne* headlined its report as "reflections on the scientific objections made against religion," and it continued by claiming with approval that Paravey had "destroyed" the "singular errors" of Dupuis and Volney and had confuted Fourier as well.[28] The day before, Aimé-Martin (by this time the controversial transcriber and biographer of the author, botanist, and friend of Rousseau, Bernardin de Saint-Pierre) had lauded Paravey in the pages of the *Journal des Débats*. Aimé-Martin was particularly thrilled to find the impieties of the likes of Dupuis and Volney so nicely repulsed. The recent arrival of the physical zodiac—"a conquest made in the interests of science which merits the more applause in that it was achieved without the shedding of human blood"—had given the impious false hope. Their "hatred for religion is so great that they consent to every error that would seem to destroy it." Indeed, "they find it easier to believe a lie than the truth," for, Aimé-Martin continued, "in their blindness they call light everything that demoralizes men." They have even "corrupted language to corrupt hearts, and affected false virtues to destroy true ones."[29] But Paravey had exposed these pernicious lies, for they were nothing but attacks on sacred chronology. Though still young (he was thirty-four at the time), Paravey had distinguished himself by his vast knowledge; young in age no doubt, but "old in study," he had no other ambition than to upend falsehood. Strong words indeed, but they help explain Paravey's ability to gain royal entrée, at least for a time.

Biot's Intrusion

Paravey's access to the court did not last long, for he soon found himself blocked at every turn. He was stymied by someone who was himself hardly on good terms with either Fourier or Arago, someone who had a well-established reputation for quickly and aggressively moving into a new area of research when he spied an opportune opening, whatever that might do to his relationships with those who had shown the way. That someone was Jean-Baptiste Biot. Unfortunately for Paravey, he found himself inadvertently in the midst of a continuing contretemps between savants at the Académie des Sciences that he clearly knew nothing about. As he saw it, Fourier and Arago were lined up against him, while Delambre, Cuvier, and Ampère were sympathetic. About that he was certainly correct, but, as an outsider to the savants' intrigues, he perceived only part of the picture. And the situation rapidly became much more complicated shortly after the zodiac reached Paris early in January 1822.

The zodiac was in Saulnier's possession at this point, and he had to decide where it could be displayed to best advantage in order to persuade the government to buy it. He quickly began gathering support, first by arranging for two (unnamed) members of the Institut National whom he seems to have known well to examine its physical condition. The stones could be profitably displayed at any of three locations: the Observatory, located not far from the Palais Luxembourg, the Louvre itself, or the King's Library. Lelorrain went to see the minister in charge of the King's Household, which included responsibility for the Louvre, while on January 25 Saulnier himself wrote directly to the new minister of the interior, Jacques Corbière, who would soon be made a count by Louis. Corbière, who had come in with the new government headed by Villèle, was sympathetic to the Ultras and committed to censorship of the press (recall that the printer Chasseriau's edition of Dupuis was seized by the police the following March).

The king already knew about the Egyptian zodiacs from Paravey's visit the previous fall, and Saulnier had arranged a great deal of publicity centered on the dramatic purloining of the Dendera circular. Corbière immediately appointed a commission to give him advice. It consisted of Fourier, Cuvier, and a member of Inscriptions et Belles-Lettres, Baron Walckenaer, who had studied at Oxford, Glasgow, and the École Polytechnique, and who in 1816 had become secretary-general of the Prefecture of the Seine. Quite wealthy and conservative (having recovered the family's prerevolutionary riches), Walckenaer had eclectic interests, ranging from insects to historical geography to literature (in 1820 he became known as a specialist on the works of the fabulist Lafontaine).[30] According to Saulnier, the commission urged the object's acquisition by the government, one among them arguing that should that fail then a public subscription ought to be brought forward because "the national honor" was at stake.[31]

Apparently Saulnier's demands were much too heavy, and the commissioners persuaded him to moderate them. Once the report was sent to Corbière early in April, Saulnier again wrote the minister asking for a decision. Why, Saulnier went on, he had already refused 200,000 for the zodiac at Marseilles; he did not mention in his public account that the commissioners had recommended a price of 150,000. To make matters easier, the generous and patriotic Saulnier was prepared to accept payment in two or three installments. If that was too hard, perhaps the cost could be split between the Ministry of the Interior and the King's Household. He heard nothing for at least two weeks, which prompted him to print his account of Lelorrain's dramatic voyage. In it he made sure to include the recent letter to Corbière, as well as sundry remarks about the zodiac's importance to France's honor.

At some point during the next month, Biot gained direct access to the king. He would certainly have known about Paravey's memoirs and would likely have been present when they had been read a year and a half before at the Académie. In any case, Paravey had sent him (and others) copies of his aperçu, and Biot decided to jump in. This was hardly the first time that he had moved onto terrain that someone else had recently explored, and we will see that his intervention was not altogether original. On Biot's own account, printed the next year, the zodiac had caused a sensation, with many "anxious to see with their own eyes this ancient monument that had escaped the ravages of the centuries, that mysterious work of a science which some people thought dated to the first ages of the world, and which must at least, without any doubt, reach to a very great antiquity."[32] Here we have a clue as to how Biot managed to see the king, for he could present himself as uniquely qualified to make a convincing mathematical argument for the zodiac's comparative youth, thereby stripping it of any danger to matters of faith. Which would have been Biot's own inclination, since he was becoming increasingly religious. He was also well-known as someone who had been antagonistic to Napoleon's regime. This had prevented his appointment to the Institut National during the empire, and that too would have ingratiated him with the restoration regime.

According to Biot, it was he who convinced Louis to buy the zodiac. "Without anything beyond a pure love for the sciences," he modestly recounted, "and all for the glory of France, I dared to do it," to persuade Louis to agree.[33] Whether or not Biot swung the affair, Louis did agree to pick up half the cost, the other half being borne by the Interior Ministry. Moreover, Biot managed to put something of a lock on access to the zodiac, at least where Paravey was concerned, because poor Paravey tried repeatedly to see it while the stones sat in the King's Library but was constantly refused permission.

Paravey asked others to intervene on his behalf, but to no avail. He was convinced, with good reason, that Biot, well aware of Paravey's work, tried to keep his own emerging efforts secret from Paravey. Biot's brother-in-law, one Tourneux, was apparently an acquaintance of Paravey's and had mentioned to him Biot's interest, whereupon Biot had (according to Paravey) complained "vividly" to Tourneux about his having done so. Although by 1835, when he wrote these remarks, Paravey had lapsed into near-absolute paranoia, seeing enemies and conspiracies against his work everywhere, nevertheless Biot's behavior here matched what he had previously done to Arago himself. Where Paravey likely saw a front of savants united against him, with Fourier, Arago, and Biot working together to block his every move, the latter two could not stand one another, while Fourier had long felt excluded from the circle around Laplace, who was Biot's major patron. In fact, there is some reason to suspect that he and Arago may have deliberately planned to let the untrustworthy Biot take the lead on this. To understand what was at stake, we need to have a brief look at the long-standing conflicts between Biot and Arago.

Arago's Revenge

The weapons that human imperfection always grants the critic to attack even the best of works.

—Jean-Baptiste Biot, 1823

Born in 1786, and so twelve years younger than Biot, Arago had studied at the École Polytechnique under, among others, Siméon Denis Poisson. Though only five years older than Arago, Poisson was an up-and-coming mathematician who had attracted the notice and support of Lagrange. We have already encountered Poisson as the one who, in 1807, would write a report so short as to be dismissive on Fourier's first work on heat. Early in 1805, at Poisson's urging, Arago was attached to the Longitude Bureau as secretary of the Paris Observatory, though he had not initially wished a career as an astronomer, having instead wanted to become an officer in the artillery. Late that year his first memoir, done jointly with Biot, was presented at the bureau and then at the Institut National the following March. This occasioned his first disagreement with Biot.[34]

They had together worked on the difficult and delicate task of measuring the refraction of light by gases, a question that had both theoretical significance and practical importance for astronomical and geographic measurements. All seems to have gone well, until Arago obtained proof-sheets of their joint memoir. "I was not a little surprised," he wrote years later, "in no longer finding my name on the title page." Instead, Biot put only his own name as author and merely thanked Arago for his "assistance." Angry, Arago went to Poisson and to the chemist Louis Thenard, who told him to protest but to do so with restraint. The protest worked, and Biot told Arago that he had only omitted his name "in conformity with academic usage" since Arago was not at the time a member of the institute.[35]

The contretemps apparently avoided, Biot and Arago together proposed setting out on a meridian-measuring expedition to complete the effort to establish the new system of measurement that had ended with the tragic death from malaria of the astronomer Pierre Méchain, whose previous work had been vitiated by an error that he had kept hidden and that he had hoped the new observations would obviate.[36] The two left for Spain in the summer of 1806. All went well for the next year and a half until, in the spring of 1808, Biot returned to Paris with the observations. Arago remained in Majorca to make other measurements, but he was trapped there two months later as a result of the rapidly developing war between France and Spain. A dramatic escape to Algeria was followed by the Spanish capture of the ship he was on when attempting to return to France, and then further escape and misadventure as Arago landed back in northern Africa, then in Algiers, finally to return to France only in July 1809. He had in the meanwhile been thought dead, and his nearly miraculous reappearance in Paris transformed him into a war hero, a change that soon had a decided payoff.[37]

On September 18, shortly after his return, the intrepid Arago, just twenty-three years old, was nominated to the Académie as a replacement for Lalande, who had died in 1807. Of the fifty-two votes cast, Arago received forty-seven. His competition consisted of Poisson, who received four, and the older astronomer Nouet, who accordingly had only one vote. And then something happened that had a decided effect on Arago's sense of his position among the Parisian savants. Laplace intervened directly and preemptively asked Arago to withdraw his name in favor of Laplace's protégé, Poisson, who was Arago's friend and mentor. "Laplace," Arago later claimed, "could not stand the idea that an astronomer, more than five years younger than Poisson, that a student, in the presence of his professor at the École Polytechnique, would become an academician before him."[38]

Figure 10.8. Laplace

Be that as it may—Laplace certainly was fond of Poisson, whom he saw, as he did Biot, as a fellow traveler through the wilds of contemporary scientific theory—Poisson was a fine mathematician and had the publications to prove it. The position was in principle for an astronomer, and by virtue of his appointment three years before, together with his experience in the field, Arago did nominally fit the bill. Yet what had Arago accomplished by this time? Not much in a strictly scientific vein. He had coauthored the paper on gas refraction with Biot, and he had made delicate but hardly novel or unusually complex meridian and related measurements. He was moreover extraordinarily young. Lagrange sprang to Arago's defense. "Even you, monsieur Laplace," he protested, "when you entered the Academy, you didn't have anything striking; you were only promising. Your great discoveries only came afterwards." Delambre, the mathematician Legendre, and Biot himself supported Arago, insisting especially on the "inextricable difficulties" he had overcome in Spain. In the event, Arago did obtain the position, but Laplace's intervention rankled even forty-two years later. What Arago had to do to justify his appointment was to produce truly novel scientific work. Two years later he did so, only to have the glory suddenly snatched away by no less than his old compatriot, sometime supporter, and original nemesis, Biot himself.

In the fall of 1808, just a few months before Arago's spectacular return to France, the first truly new discovery in optics since the seventeenth century had been made at the Paris Observatory by Malus, who had come to the institution years after his return from Napoleon's expedition, during which he had nearly died from plague. Laplace had been worried by reports of a successful experiment by the English chemist William Wollaston concerning a theory of light different from his own—a theory developed in the seventeenth century by Christiaan Huygens and based on waves, whereas Laplace held to Newton's view that light consisted of streams of particles. He arranged for the Académie to offer a prize in the area, to be awarded two years hence, and set Malus (who had already produced important work in optics) to investigate the issue. Malus undertook the task with the precision and mathematical acumen that was characteristic of leading Polytechnique graduates at the time. Greatly to Laplace's surprise, Malus confirmed Huygens's geometry, which led to some fancy mathematical footwork on Laplace's part to make the result consistent with his own approach—footwork that, however, Malus had himself first done, and that Laplace likely knew about, though in print he never mentioned the fact. Along the way something quite striking occurred, for Malus had serendipitously discovered that reflected light under appropriate circumstances exhibits a kind of asymmetry that had previously been observed only in light that had passed through certain crystals. Calling light affected in this way "polarized," Malus developed a device to produce and to measure the phenomenon, the original "polarimeter."

Malus' discovery opened an entirely new field of optical research. Like most fruitful scientific novelties, this one was rapidly developed in major part because it was accompanied by an instrument specifically designed to produce and to measure its effects. In fact, the most exciting and productive investigations for about the next decade or so concerned the properties of polarized light. Arago learned quickly about his fellow member of the Observatory's dramatic finding, and he knew as well that it was just the ticket that he could use to produce the kind of work that would justify his contentious appointment to the Académie. Malus himself was not as yet a member, though he shortly would be when, having won the Académie prize, he replaced the recently deceased ballooning pioneer, Joseph Mongolfier, in 1810. A further award followed, one that indicated the high esteem in which Malus' discovery was held, when the secretary of the Royal Society of London, Thomas Young, wrote him on March 22, 1811, that he had been awarded the society's prestigious Rumford Medal.

Several months before news of the medal reached Paris, Arago announced a discovery that he had made using Malus' polarized light. Quite familiar with lenses given his experience and presence at the Observatory, Arago decided to probe the polarization properties of light that is reflected and refracted between two lenses pressed hard together. In the seventeenth century Newton had explored the colored rings produced by such a configuration. Arago now found that the rings exhibit unusual polarization properties, unusual in the sense that they seemed to violate rules that Malus had recently produced. This caught the attention of both Malus himself and Biot, who had been a member of the commission that had awarded Malus the Académie prize the previous year.

Arago certainly had to know that he was onto precisely the kind of thing that would give him the scholarly cachet that he lacked, and whose absence had so embarrassed him in 1808. He continued to chase polarized light, this time placing a thin sheet of the crystal mica in the air gap between the pressed glasses. This generated extraordinary colors, which he rapidly pursued, yielding an entirely new set of phenomena that he named "chromatic depolarization." On August 11 he read a paper on his discovery before the Académie. And then disaster struck.

Biot was not in Paris when Arago presented his paper, but he did see a report about it that appeared in the *Moniteur* on August 31. That was enough to set him going, and over the course of the next seven months he pushed rapidly and effectively into the area that Arago had just opened. On March 12, 1812, Biot read a note on his work to the Académie, alerting an infuriated and surprised Arago to his intrusion into the one area that Arago had hoped to make his own. A nasty and bitter confrontation ensued, in which Arago accused Biot of claiming to have been the first to make observations that he, Arago, had already done.

Biot, ever prolific and mathematically adept, had produced over three hundred pages on the subject that contained elaborate experimental tables, formulas, and speculations. Arago had been stumped. Although Laplace effected a sort of peace between the two men, and though they did on occasion collaborate in future years, Arago certainly never forgot the slight, particularly as Biot rapidly became known as the major expert in the new field. Half a decade later, during Napoleon's Hundred Days, Arago found an opportunity to strike back.

Arago had left the Polytechnique in 1804 and had not likely encountered Augustin Fresnel, who was in the class below his. He had, however, become friends with Fresnel's maternal uncle, Léonor Merimée, a painter, chemist, and by this time perpetual secretary of the École des Beaux Arts. Fresnel had long dreamed of making a new discovery in science or industry, and to that end he had already initiated a correspondence with Ampère. In mid-December 1814 Merimée dined with both Ampère and Arago, and he asked Arago to have a look at a small piece that his nephew Fresnel had sent. Nothing much came directly of that, but matters changed about six months later when Fresnel was put under surveillance for having joined the duc d'Angoulême's resistance to Napoleon's return. Napoleon surrendered to the British at Rochefort in July, whereupon Fresnel returned home to his mother at Caen. On the way he passed through Paris and visited Arago, whom he asked about optics—Arago, though outfoxed by Biot, nevertheless being known as the original discoverer of chromatic polarization. Fresnel asked specifically about how light might be treated as a wave, and whether this might have something to do with the phenomenon known as diffraction. In subsequent correspondence, Arago advised Fresnel to read various papers, including comparatively recent ones by Thomas Young, who had done just what Fresnel was asking about. Fresnel, however, could not read English, though his brother, Léonor, could and apparently read to Fresnel parts of Young's *Lectures*, from which (or from remarks by Arago) he may have obtained a primitive notion of interference. In any event, back home and without access to the apparatus available in Paris, the young Fresnel rediscovered how to use wave theory to explain diffraction, and he produced a convincing series of experiments to back up his claims—experiments based on just the kinds of calculations and measurements that Arago himself had not been able to produce for his own discoveries. In September Fresnel wrote Arago about his work, and almost at once Arago recognized that Fresnel had opened a field of discovery that had even more promise than his own earlier detection of chromatic polarization. Fresnel then came to Paris at Arago's invitation, who arranged space for him in the Observatory. Over the next seven years, working alone and on occasion with Arago in the laboratory, Fresnel extended his wave theory of light right into the heart of the very territory that Biot had made his own.[39]

Arago certainly helped Fresnel a good deal, and there is little doubt that he was himself captivated by the new discoveries. But he was also quite clever in using the novel work to undercut Biot, for he urged Fresnel to develop the theory in ways that would directly attack Biot's own claims. His efforts in that regard became sufficiently obvious that Fresnel described an explosive situation in a June 1821 letter to his brother. "There was a great battle at the Institute," he wrote, "in the last two sessions, at the occasion of a report by Arago on the memoir I presented to the Academy nearly five years ago, in which I attacked by facts Biot's theory." Biot was upset, and the discussion "became very lively." At the next meeting Biot tried to regain the advantage, and, among other points, "he reproached Arago for having so long delayed my report [on chromatic polarization]." Arago remonstrated and condescendingly offered in reply that he understood why Biot might feel "chagrin" at seeing his theory attacked. Why, Arago asserted, he had been so concerned about distressing his dear colleague that he had held back reporting on Fresnel's work for five years. This was at best misleading. Five years before, Fresnel had not yet developed sufficient theoretical and experimental wherewithal to undercut Biot's own theory; that took time. Arago waited until he had what was needed.[40]

In 1818 Fresnel was awarded an Académie prize for an experimentally backed account of diffraction, though Laplace and Poisson, members of the judging committee, praised only his experiments and (to some extent) his mathematics, not the underlying theory. Fresnel's prize-winning memoir introduced and developed the conceptual and mathematical foundation for wave optics, a foundation that would over time evolve into the underpinnings of new instruments and elaborate theory. Just two years later, Fresnel's work had expanded, under Arago's clever aegis, to encompass all that Biot had done. Fresnel himself never went directly after Biot, but Arago, in reporting on what Fresnel had found, certainly did.

The resulting contretemps with Biot at the Académie took place four months after Delambre had read his report on Paravey's work on the zodiac, and just shortly after Paravey's own account, his aperçu, had reached print. Fourier and Arago, recall, had managed to remove

Figure 10.9. Biot (*left*) and Arago

approving words from the official report on Paravey. Now Arago turned his attention to Biot, whom he set up for the kill, having knowingly put him in an impossible position. For how could Biot possibly criticize Arago's claim that he had held back Fresnel's papers to avoid wounding Biot's pride without reminding the Académie members of his own questionable behavior years before in respect to both Arago and Malus? For not only had Biot twice set Arago to the side, he had also claimed to have anticipated Malus' discoveries.

Biot certainly had a reputation for playing in other people's sandboxes in ways that were not considered to be entirely fair, and Parisian savants knew it. Laplace, for one, had been quite angry over Biot's claim concerning Malus.[41] Arago's good friend, the German explorer and naturalist Alexander von Humboldt, who knew the Parisian scene intimately, and who had collaborated with Biot on a memoir concerning magnetism a decade and a half before, was also well aware of Biot's propensities. Now, six or so months after the dust-up over optics at the Académie, and with the zodiac having arrived in Paris, Biot decided to enter a field that someone else had just reopened after a decade and a half, again with hardly any acknowledgment of his predecessor's work. And he somehow made certain that his competitor, Paravey, could not have access to the object under investigation.

Although Biot would have had easier entrée to the royal administration than either Arago or certainly Fourier (given Fourier's close connections to Napoleon and his behavior during the Hundred Days), nevertheless it might seem puzzling that Fourier apparently never tried to examine the actual zodiac, nor did he ever publish anything more about it. He did frequently bore dinner parties with stories about the brilliance of ancient Egypt, but that was the extent of his future involvement. Why, one might wonder, did he let the area go when he and Arago had so stridently defended the terrain against Paravey, and when they both surely knew that Biot would make every effort, as he certainly did, to expropriate it?

It is possible that Fourier recognized at some point his own error in calculating Sirius's heliacal rise, after Biot had criticized him on the point. However, Biot never pinpointed, and likely did not know, Fourier's error (his having ignored refraction), and Fourier likely remained convinced that his own calculations worked well—as did his friends and supporters, Jollois and Devilliers. Another, though hardly certain, answer to Fourier's seeming indifference to the issue is not hard to find, given the intense competition for control of scientific institutions under the restoration. Arago and Fourier formed together a faction that aimed to prevent the seats of power from being acquired by savants closely associated with the now elderly Laplace, especially Biot. Years before, we have seen, Fourier had run into opposition when he presented his theory of heat to the Institut National, a theory that had taken a markedly different path from the one that the young Biot had sketched in 1804. Now, in Paris since 1815, he had been working hard at formulating a final account, which reached print the very summer of 1822 that the zodiac affair became white-hot, a summer during which Biot managed to keep him away from the zodiac while constructing a lengthy memoir about it: "After the arrival of the Dendera planisphere in Paris," Paravey complained, "M. Biot impeded us from seeing it."[42] Having months before allied with Arago to undercut Delambre's report on Paravey's memoirs, Fourier may have decided that it was hardly a good time to jump into a battle over zodiacs when he was anxious to push his newly printed heat theory. And keeping Biot, recently crushed by Arago in the contretemps over light, busy with something other than heat, something that would be comparatively marginal to Académie affairs, would do, with the added advantage that Biot was sure to attack the annoying Paravey. He would likely attack Fourier as well, but no matter, given the more important issue of keeping the wounded and therefore dangerous Biot occupied, for the moment at least. The Dendera zodiac perfectly fit the requirement.

Events moved rapidly following the zodiac's arrival in Paris. At some point Biot gained access to the artifact and began taking measurements. This likely occurred before the king agreed to the purchase in May because it was Biot who claimed to have clinched the deal.[43] His involvement in the details had a decided payoff. "When the monument was brought to France," Biot wrote two decades later, "a confluence of favorable circumstances gave me especially favorable opportunities to study its construction. Not only could I closely examine the details of the figures as much as I wanted to, but I was even accorded complete liberty to take all the precise measurements that I deemed necessary."[44] It is hardly likely that Fourier did not know what Biot was up to, because he was a member with Cuvier and Walckenaer of the commission that Corbière had empaneled to consider the purchase of the zodiac. If Fourier had himself wanted to become directly involved in studying the object, or if he had wanted to block Biot from doing so, he could have gone directly to Corbière. But, again, Fourier was hardly in good political odor even at this late date, so he would have had no direct access to the king, and Corbière would have been sensitive to the problem. We have seen how Louis had held back Fourier's appointment to the Académie in the spring of 1816, finally agreeing to it only the next year following the replacement of the royalist Ultra legislature and, especially, its extremist minister of the interior, Vaublanc.[45] Fourier had thereafter become extraordinarily active at the Académie, all the while pursuing his own research.

On August 19 Delambre, who with Cuvier had been Paravey's main support at the Académie, died. Jockeying for his position as permanent secretary for the mathematical sciences began almost immediately. On November 11 three names were brought forward: Fourier, Arago, and Biot. Four months previously, Biot had presented his two memoirs on the zodiac and was busy preparing a huge book on the subject that appeared in 1823. Fourier had few illusions about Biot but apparently felt that his opponent would be successfully deflected in the election, for he wrote at the time to his friend the mathematician Sophie Germain that he could not "doubt but that the wish of the greater number of my colleagues will be to choose me and that one of my opponents [certainly Biot] who flatters himself the most is very much mistaken. But he has resource to so many artifices that it would be imprudent not to fear him."[46] Arago and Fourier clearly had things arranged between them because Arago immediately pulled himself out of the competition, and a week later Fourier was elected, thirty-eight voting for him and only ten for Biot.

Fourier and Arago may not have directly encouraged Biot's engagement with the zodiac, but they did nothing to hinder it, and neither did they apparently make any remarks at the Académie meetings where he had presented his work in July. In fact on the very day that Biot made his presentation, which included the showing of Gau's lithograph that he had obtained from Saulnier, Fourier himself read a memoir on a completely different subject, namely, population statistics. The zodiac had by then attracted a great deal of attention in wider circles. The very next week the Abbé Halma, translator of Ptolemy's *Almagest* and now defender of Creation's biblical date, read a paper on the subject, which was eventually followed by a lengthy treatise, and the week after that Jollois and Devilliers chimed in. Paravey announced a new memoir on August 5, with Jomard having his own say on that same day. Two weeks later Paravey returned to the Académie to read his paper on Dendera, and not long thereafter he became increasingly incensed by Biot's intrusion. What had Biot actually done that so disturbed Paravey and that Fourier was prepared to ignore, for the moment at least, given other priorities?

Measuring the Zodiac

The week before Biot presented his memoir on the zodiac, a dinner was held at Laplace's country house in Arceuil. John Dalton, the English protagonist of the atom, was visiting France at the time in the company of some friends, one of whom, Benjamin Dockray, recalled the evening's events. After reaching Arceuil, the Englishmen were conducted to a terrace. Several moments later they found three of their French hosts sauntering through the garden, one elderly gentleman accompanied by two younger ones. "The two younger," Dockray recorded, "were Laplace's son, and the astronomer royal—Arago. Climbing some steps upon a long avenue, we saw, at a distance, Laplace walking uncovered with Madame Biot on his arm; and Biot, Fourier, and Courtois, father of the Marchioness Laplace." The group that gathered that evening included some of the most influential and creative scientists of the day, men who certainly had considerably different views of the workings of matter and force. Unlike Dalton, Berthollet was no fan of chemical atomism; Laplace and Biot disagreed with Fourier about heat, while Arago was the supporter of Fresnel, whose wave theory of light, as we just saw, countered Biot's extensive work on polarization and Laplace's own theories. And we have also seen that long-standing personal and institutional issues bedeviled the relationships among the French savants. Given the many exciting and contentious developments in science that had taken place in recent years, the conversation over dinner might have turned in those directions, though that might easily have bored the other guests or created uncomfortable disagreements. But it did not. Table-talk orbited instead about "the zodiac of Denderah and Egypt, Berthollet and Fourier having been in Egypt with Napoleon; the different eras of Egyptian sculpture; the fact that so little at Rome—of public buildings—is earlier than Augustus, &c."[47]

Dockray's recollections are telling because it was precisely over the issue of architectural style that so much ink had been spilled, Visconti having long before been the first to argue that Dendera must on that account be very late. No doubt Fourier had tales to tell about his trek up the Nile, while both he and Berthollet could regale the dinner party with romantic stories about Cairo and the vicissitudes of the expedition. Dockray did not record any remarks by Biot, but we can be quite sure that he had his own say as well. A week later Dalton would hear him on the subject at the Académie. And what Dalton heard was an elaborate attempt to turn the Dendera zodiac into a triumph of astronomical representation and measurement.

The idea behind Biot's calculations was ingenious and would have testified to his inventiveness if he had been the first to suggest it, which he wasn't, at least not in basic outline, since Delambre's report contained the essential idea.[48] Paravey, recall, had treated the Dendera circular as a stereographic projection, which Delambre, in his formal report, had found unpersuasive. But Delambre had suggested an alternative procedure, one that amounted to the same method that Biot adopted.

Suppose that the designer of the zodiac had no notion at all of stereography, but that he did have measures of the positions of the stars both along some great circle of the celestial sphere and along another perpendicular to it. It would then be a simple matter to draw a planisphere in the following way. Mark the center of the design as the celestial pole of the great circle in question—we will not assume a priori that this must be the celestial equator. Draw a line through the center and choose it as the one along which a particularly prominent star should be placed; mark the star on that line at a distance from the center that represents its actual distance from the equator. Do the same for all the other stars to be represented. Figure 10.10 illustrates the procedure. Suppose we have two stars, a and a', and that both are located, say, on the ecliptic, which is represented by the darkened disk in the figure. They will, respectively, have angular distances aP and $a'P$ from the celestial pole, P. Take these distances, stretch them out into straight lines, and place the resulting straight line, aPa', on the drawing in such a way that P coincides with the drawing's center, and such that the line is rotated around properly to represent the positions of the two stars in right ascension.[49]

This seemed to work well in Biot's opinion, though it is at once obvious that a great deal of leeway was available given the rather large sizes of the zodiacal images. But Biot had another thought. The lines that he had drawn turn out, he insisted, to be just about equal to the radius of the stone itself, which Biot had measured at 774 millimeters. Look again at Figure 10.10 and imagine that we rotate the dark plane that now represents the ecliptic until point a, say, coincides with the north pole, P. Then the arc aP vanishes, and we are left with arc $a'P$ to represent the entire line aa'. That arc now reaches from the north to the south poles and is equal in length to every other diameter of the sphere. Take this line and place its point a at the planisphere's center, P. Then its other end, a', will be at a distance from P that must equal in length the lines that were drawn in Biot's first test. The dark plane of Figure 10.10 now contains the polar axis, and we may rotate it around that axis so that the plane will intersect our chosen great circle—the one whose celestial pole corresponds to the center of the planisphere— at different points. If we now turn the line on our projection to the corresponding positions, then we will trace out a circle of radius aP that represents, in effect, the projection of the lower pole for all positions around the great circle. Biot's point was precisely that: if the projection were done according to the method that he described, then its outermost periphery should have a radius equal to the lengths of the lines through the center that connect diametrically opposite points of the zodiac. And, he asserted, "that's what we observe quite exactly."[50]

Perhaps so, but Biot's "quite exactly," here as elsewhere, overstates the case. This claim depends upon his identification of various points on the zodiacal images as the ones that should be used in plotting his lines. But why should this point and not another be chosen in a given figure? After all, the images are quite large. He was certainly well aware of this and other criticisms concerning his choices, and for every one of them he had an answer, usually a different one in each case. For Biot did

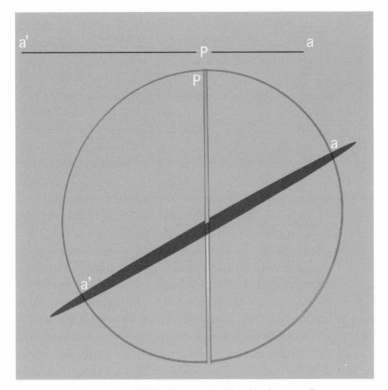

Figure 10.10. The "projection by development"

not merely draw a few lines here and there. He measured many points on the "projection" in an effort to identify the locations of specific stars, to justify which he gave a plethora of arguments. To cap it all off, he produced an elaborate statistical argument to back up the claims.

To grasp why Biot was convinced that his scheme could withstand almost any attack, it is critical to understand that he did not assume from the outset that the center of the "planisphere" must be the projection of the north pole of the celestial equator. If he had, it would have been a comparatively simple matter to calculate dates using precession since declinations and right ascensions could have been read directly off the monument, at least given a further assumption concerning the position of the solstitial or equinoctial colure. But he did not want his claims to depend on two debatable assumptions. He aimed instead to generate a convincing scheme by actually computing the celestial locus of the projected pole, and to do that he set about identifying specific

Figure 10.11. (*Left*) Biot's location for Antares on the actual Dendera circular; note Scorpio to the upper left. (*Right*) Biot's cow and hawk for Sirius

stars on the "planisphere." Four stars in particular were critical for Biot's claims, namely, Fomalhaut in the constellation of the Southern Fish, Arcturus in Bootes, Sheat in Aquarius, and Antares in Scorpio.

Consider Antares. The "star" in question is, according to Biot, the object that is barely visible immediately to the left of the "A" in Figure 10.11 left. It surely does not seem to be a star, but after "repeated examinations," Biot convinced himself that the object looked rather like a heart. This seemed persuasive to him because Ptolemy had placed Antares at the "heart" of Scorpio. Although the figure where he placed Antares was hardly Scorpio, the figure that represents it was nearby, and Biot insisted throughout his arguments that sometimes a star corresponding to a constellation figure would be depicted near it but not on it.[51]

This worked quite well when coordinated with Biot's identifications of symbols for his other three principal stars. Many of his exceedingly tiresome arguments—a single paragraph could go on for pages—were accordingly aimed at explaining why *this* symbol had to be interpreted as *that* star. There was, for example, Biot's insistence that the recumbent cow, which had long been taken to symbolize Sirius, did not represent the star's true position, even though the cow has a distinct star over its head. No—Sirius the star was represented by the symbol to the cow's right (Figure 10.11 right), a hawk perched on a lotus leaf. Why? Because a line drawn through that symbol and the center of the zodiac is parallel to the temple's long axis, and this Biot took to symbolize the importance of Sirius itself, incarnated as it were in the very architecture of the temple. The star-topped cow was placed nearby as a sort of pointer to the neighboring position of that important luminary.

Perhaps the most astonishing and, in the eyes of many contemporaries, outlandish of Biot's claims relied on probabilities. Having
located his four principal stars, he could directly measure their respective distances from its center. It was also a simple matter to measure
the angles between the lines that joined each star to the center. These
two measures were then used to calculate the six distances on the
celestial sphere among these four stars, and those distances could then
be compared with ones taken from contemporary measurements. In
doing so Biot found that his distances agreed with observation to about
the same extent that Delambre had calculated for the "observations of
Hipparchus, reported in his commentary on Aratus."[52]

What is the probability, Biot now asked, that such an agreement
could have occurred at random, that is, if the "planisphere" had not
been designed using the method of "projection by development"? To
answer that, draw a circle on the celestial sphere around, for instance,
Antares with a radius equal to the distance between it and Arcturus.
Next, allow for a two-degree spread in the observational precision of our
antique Egyptians. Then draw two other circles, concentric with the
first, with one closer by a degree to Antares and the other farther away
by a degree. Biot then calculated the chance that, with Antares fixed,
a star chosen thereafter at random will fall into the region embraced
between the inner and the outer circle. That chance, he asserted, is
equal to the ratio of the area embraced by the region to the area of the
sphere as a whole.[53] Now comes the pièce de résistance, Biot's table for
the resulting probabilities (Figure 10.12).

Take three of the principal stars, say, Antares, Fomalhaut, and Arcturus. Each of the corresponding probabilities is completely independent of the others, so the likelihood for all three of the distances among
them holding true is the product of their separate probabilities, or one
chance in 54,000. Now add in the fourth star, Sheat, and consider its
distances to Fomalhaut and Arcturus, for which the chance alone is
one in 14,060. Consequently the likelihood that these last distances
and the ones for the previous three stars should all occur is one in
759,240,000! Biot humbly concluded that his "geometrical proofs
thereby achieve a probability nearly equal to certainty."[54] To empha-
size the point, Biot included Gau's graphic of the zodiac, marked with
his loci for stars.

The primary issue around which the zodiac controversies had long
orbited, namely, the age of the Dendera stone, was nearly lost in all
of this. The distances among the stars did not alone determine that
because the distances remain fixed, or nearly so, over the millennia.

Arcturus Antarès....... $\frac{1}{60}$

Arcturus Fomalhaut..... $\frac{1}{30}$

Arcturus Sheat......... $\frac{1}{37}$

Antarès Fomalhaut...... $\frac{1}{30}$

Antarès Sheat......... $\frac{1}{1915}$

Fomalhaut Sheat....... $\frac{1}{380}$;

Figure 10.12. Biot's table of probabilities

Ever eager to calculate, Biot decided that the best way to find the date would be to compute the celestial coordinates of the pole of the great circle that the designer of the zodiac had used, for that pole moves around over the millennia as a result of precession. The procedure involved comparing measurements taken from the Dendera circular of what Biot took to be stellar positions with the coordinates of those same stars near Biot's time. Spherical trigonometry could then be combined with the rate of precession to find the date of the zodiac.

The procedure illustrates the degree to which Biot deployed seemingly convincing mathematics in chasing down the zodiac's date. Choose two of the zodiac's stars, say A and B in Figure 10.13. Now look up their coordinates in 1750, the epoch that Biot used. Next form a spherical triangle, ABE, on the celestial sphere consisting of these two stars and the pole of the ecliptic, E, whose coordinates Biot also had for 1750. He could then easily compute all the sides and angles of ABE, including the distance AB between the two stars. From the monument he could measure what had to be this same distance AB, since it hardly changes, and he could also measure the distances AC and BC between each of these stars and the center of the zodiac, C— to, that is, the pole of whatever great circle had been used to generate the projection. He now had all of the angles and sides of two spherical triangles: one, taken from 1750 coordinates, with pole E of the ecliptic as a vertex, while the other had the unknown pole C of the generating circle as one of its vertices. Both of these triangles shared a common side—namely, the known (and unchanging, or nearly so) distance AB between the two

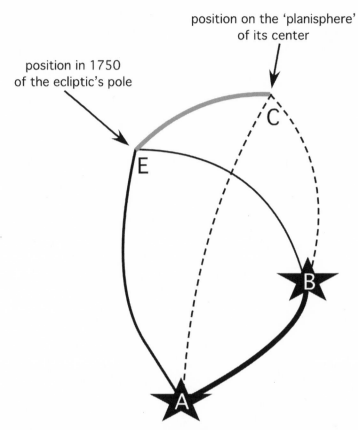

position on the 'planisphere'
of its center

position in 1750
of the ecliptic's pole

Figure 10.13. How Biot found the zodiac's pole, C. The thinnest solid lines AE and BE are determined from stellar coordinates for 1750. The dotted lines AC and BC are measured on the Dendera circular.

stars. Consequently Biot could now calculate distance EC, which gave him the position of the generating circle's pole, C, with respect to the ecliptic pole, E, in coordinates pertaining to 1750. Once he had that, it was a simple matter to choose any star and then translate its measured coordinates on the projection with respect to C into locations on the 1750 celestial sphere. And that, given the rate of precession, yields the zodiac's date.

Biot first used Fomalhaut and Arcturus for his two stars and then repeated the calculation with Antares and Sheat, averaging the results. A further step in which he searched for nearby stars yielded an astonishingly precise date: the Dendera circular represented the heavens as they stood in the year 716 BCE. The zodiac accordingly dated neither to remote Egyptian antiquity nor to the Greco-Roman period, and the ever-industrious Biot had proved it to be so by an irrefutable calculus of probabilities based on what looked to be a spectrum of mutually independent data points, namely, his identifications of four principal stars, and eventually of others as well. Or so Biot argued.

Jollois and Devilliers disagreed, not least because they felt that Biot had lifted the method from them.[55] They were sufficiently incensed that Devilliers wrote a letter of protest to the head of the Académie. That in turn angered Biot, who decided to print it the next year in a lengthy self-defense. The two savants, Biot retorted, asserted that, though they had themselves envisioned the projection method, it had proved to be unworkable. Moreover, many of the extrazodiacal constellations that Biot identified had, they unconvincingly claimed, also been noted first by them.[56] A decade later the two friends returned the attack in print. They had not done so years before for reasons that had nothing to do with the facts of the matter. In 1823, they recalled, it had not been "permissible to assign too remote an antiquity to the astronomical monuments of the Egyptians" given the political climate of the times, and so at the time they had not come out as forthrightly as they would have wished to. Others, more in tune with the political climate, took advantage of it, particularly since Fourier had by then become perpetual secretary of the Académie. Indeed, precisely because the now-influential Fourier had argued for an early zodiacal date, it had become (politically) "necessary" to counter his claims. That, they insinuated, had been Biot's main aim. He was after all the one who insisted that he alone had convinced the king to acquire the zodiac. They did not know what he had promised the king in exchange, but they suspected it must have been suitably to reduce the Zodiac's age. In any case, his tendentious and forced associations were unconvincing, and his probabilistic claims to certainty were absurd. Moreover, he had without warrant criticized what he thought to be Fourier's failure to recognize the millennia-long stability of Sirius's heliacal rise. About that, we have seen, they were wrong, since Fourier had overlooked an important factor in his calculations, though Biot had not seen what it was, preferring instead to call Fourier's vaunted skill as a geometer into question.

Jollois and Devilliers angrily rejected Biot's criticism of Fourier and castigated him for impolitely implying that Fourier, a fellow member of the Académie after all, was incompetent. "Was it because Fourier was

perpetual secretary of that learned society, and that dignitaries ever
have the privilege of being criticized and envied?" Biot certainly had
not hesitated to push in the knife in 1823, and it is unlikely to have
been a coincidence that Fourier had defeated him the previous year in
the election for perpetual secretary. Fourier had at one point thought
to reply, having written a letter to Devilliers on September 16, 1823,
hoping to collaborate with him in examining Biot's remarks, but either
that never took place or the reply never reached print.[57]

The two friends were not the only ones to criticize Biot, for Paravey
was also incensed, certain that Biot had kept him away from the zodiac
while working hard to steal Paravey's own theory. He was initially quite
mild in reaction, but that was before he had seen Biot's attempt in 1823
to expropriate the entire arena for himself, while scarcely mentioning
Paravey. Why, he was even flattered, at first, that Biot had in a way
"demonstrated the general principles which I had posed," though he
could not assent to the age Biot had found for the zodiac.[58] Paravey's
main critique at this early date (1822) concerned the position of the
colures. Biot had not used them, having located the date by means of
the distances among his chosen stars. However, once the date had been
determined, the colure positions followed, and they were not where
Paravey had placed them. In Biot's case, the solstitial colure had to
pass through the star-headed cow, right on the north–south axis. Paravey instead had the colure run parallel to the actual axis of the temple.
Where Biot put the colure near an emblem (but not the position) for
Sirius, Paravey ran it through the symbol that Biot (but not Paravey)
associated with Sirius's actual location. The difference in positions
translated into a difference of about five hundred years in dates.

After Biot read his second memoir on July 22, Jomard himself replied
at length, first at the Académie des Sciences on July 30, and three days
later at Inscriptions et Belles-Lettres. Ignoring Paravey, but defending
the priority of Jollois and Devilliers in recognizing the nature of the
projection (though it was really Delambre who had first suggested it
in a reasonably precise way, in his report on Paravey's memoirs), and
insisting on Fourier's ideas as the basis of "truly solid research" on the
zodiacs, Jomard drove to the heart of the matter. Although the projection had likely been produced according to a weak version of the method
Biot had used, nevertheless he had gone too far in trying to identify
specific stars. For if that were correct, if the "authors of the zodiac"
had intended to represent stars themselves in their correct positions,
why was "not a single one of the zodiacal constellations fixed by the

projection of even one star, not even Scorpio, even though M. Biot gave a position for Antares, which, it must be said, rests on no foundation whatsoever."[59] Only Arcturus, among Biot's four chosen stars, sits reasonably in place with respect to the symbol for its corresponding constellation. As for Biot's location for Sirius—well, the less said about that the better, Jomard insinuated. The artist who produced the zodiac did not intend to "make a rigorously faithful image, much less to compose randomly or capriciously; in a word, he put the *figures* in place, and didn't bother about *points* and *lines*."[60] Worst of all, Biot's romp through probabilities was itself utterly improbable because his calculations depended critically on Antares, whose identification Jomard rejected. With Antares suppressed, the probabilities "diminish enormously."

Biot took umbrage at Jomard's remarks, replying angrily the following year. Not only, he wrote, had Jomard not seen Biot's calculations, he had unsportingly made his critical remarks while Biot had himself been out of town. Years before Biot had also been out of town when Arago had read his papers on light, only in that case his being away had been used to excuse Biot's own failure to acknowledge Arago. Now it would be used to excoriate Jomard—a convenient reversal if ever there was one.[61] Biot now perceived an entire phalanx arrayed against him, consisting of the savants who had been with Fourier in Egypt, all of whom he accused of having joined together under the influence of Dupuis' theory to proclaim the Egyptian invention of the zodiac. Once returned to France, they had abandoned the extreme age suggested at first by Dupuis on the basis of constellations that rise with the sun, only to adopt his alternative, which used constellations that set when it rises, and which consequently placed the Esneh zodiac at about 2500 BCE. Fourier, fumed Biot, was responsible for all of this, since the other expedition members had just "rallied" to him. Of course, he continued, no one more than Biot himself respected the authority of the "great scholarly societies of France." But their independence, their diversity, "disappears when a small group of savants" knuckles under to a set point of view. Because when an outsider to the system—namely, Biot—muscles in with a contrary or even a slightly different opinion, they band together in united rejection. Biot went on in this vein at some length, defending his views in part by attacking the exclusivity of his opponents.[62] The anger and self-justifying tone of his remarks cannot be missed. It scarcely abated over the years, even long after his primary antagonist, Fourier, had died.

Paravey, Biot, and Jomard were not the only ones to engage with the zodiac shortly after its arrival in France. Other scholars took aim as well, some striving to save France from the dangers of impiety, others engaging with the issues raised by this latest invasion of astronomy into history. The Abbé Halma, addressing his work to Artois' wife, wrote that of "all the errors of which one has endeavored to convict the sacred books, none would be more grave than the epoch they [the impious] assigned to the creation. . . . One obtains warrant for so claiming by means of some ancient monuments that were furnished by paganism and interpreted by philosophie. But the same arms taken up to attack the truth of the Mosaic narrative will serve me in its defense."[63] Halma, translator of Ptolemy's *Almagest*, was thoroughly incensed by arguments that used "pagan" carvings to attack the indisputable truth given once and for all by Moses. For "superstition and immorality were the principal character of these monuments, much more than the state of the sky at the epochs of their construction." How could the Egyptians have produced anything but dangerous nonsense? What after all was their "spirit, this carnal, somber, and fanatical people" who "veiled the confusion of their morals with the cloak of their religion"? Why, Halma advised, just compare Egyptian works with that of the finest French artists—you will see in the former "the most shameful indecency united with all the horror of human sacrifice."[64] That Ptolemy had not cited a single observation from the Egyptians, the Abbé declared, in itself testified to their utter intellectual and spiritual bankruptcy. Halma's words about ancient Egyptians bear an uncomfortable resemblance to traditional Christian canards about post-crucifixion Jews, and he may well have seen little difference between the two. He nevertheless knew ancient astronomy well, and in the Dendera circular he spied only references to rituals and agricultural rites, not astronomical events.

Others with fewer qualifications jumped into the fray, each with his own views. One Leprince, a sublibrarian of the town of Versailles, published his somewhat obscure opinions in 1822, which apparently conceived the zodiac as a sort of agronomic calendar, though in the end Leprince came up with a date for it of 824 BCE, nearly the same as Biot's 716, and Leprince even included a chart with Biot's stars marked on it.[65] This coincidence is particularly ironic since Leprince's only other known publication was an effort to reject Newton's theories about light.[66] Biot was by this time widely known as Newton's apostle in optics.

Someone at the time took a sufficiently large interest in the debates that he accumulated and annotated a series of reprints on the subject by several of these authors. He had the pamphlets bound to form the book labeled *Zodiaque* on its spine with which our preface begins. One of the pamphlets bears the dedication "offered by the author [Dalmas] to M. le Comte de Chabrol, councilor of state, Prefect of the Seine, as a testimony of his old friendship." Dalmas had been in Egypt with the expedition, where he had acted as administrator of one of the provinces, and was by this time a civil engineer in Castelnaudary. He likely had known Chabrol (like Dalmas himself, an engineer in Ponts et Chaussées) in Egypt. All the articles were printed between 1822 and 1824 and so were put together during the high point of the affair's revival in France by someone who was clearly familiar with it. Whoever annotated the *Zodiaque* pamphlets was also able to calculate, knew some astronomy, was familiar with Egypt, and (we shall see in a moment) had access to the plates in the antiquities volumes of the *Description*. Though we cannot be certain, it accordingly seems probable that the collector and annotator was the prefect of the Seine and friend of Fourier, Chabrol de Volvic himself. Moreover, in addition to the three memoirs that Chabrol had worked on for the *Description* during the empire, a lengthy fourth on the customs of contemporary Egyptians was printed in 1822, the very year that the zodiac caused an uproar in Paris. He appended a note to it thanking "M. Fourier, perpetual secretary of the Institut d'Égypte, for being so kind as to communicate to us his notes on that country." Chabrol's previous efforts included a collaboration with Jomard in describing the area around Kom Ombos. In discussing the temple there, Chabrol and Jomard observed that its construction "confirms very well the tradition that attributes the invention of geometry to Egypt, and that gives Egypt the honor of having made the first geographic projections."[67] Consequently Chabrol certainly would have known quite a bit about the zodiac issues, and he also was convinced, like so many others on the expedition, that the Egyptians had millennia ago invented the foundations of geometry.

Chabrol, if it was indeed he, wrote extensively in the margins of three of these pamphlets, the ones by Dalmas, LePrince, and Saint-Martin. Most of the marginalia to Leprince's simply index the subject or claim made in the neighboring text and contain few comments, though at one point the annotator pointed out internal contradictions within the text.[68] The lengthy Dalmas pamphlet argues extensively against Dupuis and is deliberately set out as a brief for the adequacy of Mosaic chronology. Despite its dedication to his "old friend" Chabrol, the annotations to it are critical—"without force," "a gratuitous supposition," and so on. The annotations to Saint-Martin's article are less extensive but more respectful than those made either to LePrince's or Dalmas'. This memoir had been read before Inscriptions et Belles-Lettres on February 8, 1822, half a year before Biot spoke at the Académie des Sciences. Saint-Martin was a pupil of Silvestre de Sacy, under whom he had studied Arabic, Turkish, Persian, and Armenian, and had been a friend of the young Champollion during the latter's two years of study in Paris. With De Sacy's support, Saint-Martin was elected to the Académie in 1820. Nine years later, having mutated into a devotee of throne and altar, he joined with Rémusat (whose lessons on Chinese history Paravey had attended) to found a journal, *L'Universel*, devoted to the divine rights of monarchy. In 1825 he became secretary to the minister for foreign affairs but was removed, apparently for having published pieces attacking the Jesuits.

Saint-Martin's piece on the recently arrived Dendera zodiac set itself out as a model of even-handedness, and indeed it was, though he himself was apparently a rather ill-tempered sort by this time. Nevertheless, the memoir was restrained in comparison with the vituperative anger that was so characteristic of the period's academic discourse. This was evident in his assessment of Dupuis, the great bête-noir of the religious Right, for the now-devout Saint-Martin allowed that Dupuis' celestial-seasonal synchronies could work well, if reasonably applied. Unfortunately "Dupuis had not that sage reserve . . . and his work, as badly conceived as it was written, was a vast repertoire of fool-hardy, unreasonable, and false ideas" though still "much superior to the essays of his disciples." No one today (1822), he asserted, held any longer to the *Origine*'s details, which was something of an exaggeration. Saint-Martin went on to present the major claims concerning the zodiac that had appeared since 1800, including those by Testa, Visconti, and Fourier. He agreed with none of them, found points to criticize in all, and temporized that the Dendera zodiac was neither so old as the calculating savants had it nor so young as Testa and Visconti claimed. He doubted that much could be done by way of interpreting it as a "scientific" monument, but he also thought that, without a doubt, "the Dendera planisphere is a production of the art and science of the Egyptians," even if it had perchance been produced during Greek or Roman times (which he also doubted).[69]

Among the most interesting and significant of the marginalia to Saint-Martin's article were two little diagrams of Egyptian cartouches (Figure 10.14, top). Both were drawn directly from Jollois and Devilliers' plate in the *Description*, which in 1822 was still available only in the full-size, very expensive printing of 1817 (Figure 10.14, bottom).

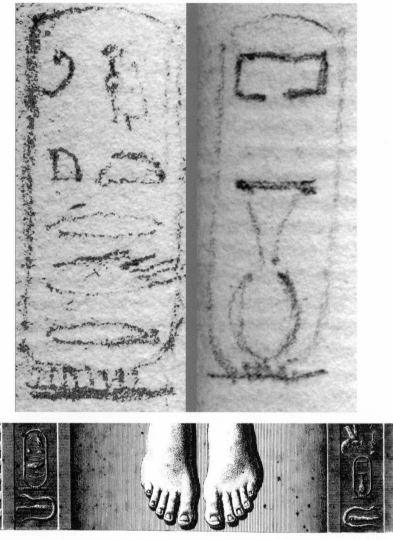

Figure 10.14. Chabrol's cartouches (*top*) and on Jollois and Devilliers' drawing of the circular zodiac

The part of the temple ceiling that contained the sky goddess and the side bars with the cartouches had been drawn by Jollois and Devilliers, outlined by Totard from the savants' illustration, and engraved with shading by Allais (Figure 10.1). The goddess and her surrounds had, however, not been removed by Lelorrain, and they remain in the temple to this day. None of the participants in the zodiac debates had ever seen the originals except for Fourier, Jomard, Jollois, and Devilliers, but even they would certainly not have remembered all the details.

The enclosed hieroglyphs were called cartouches by the French soldiers because they resembled cartridge casings, and it was early thought that they must contain the names of important personages. Saint-Martin speculated that the unreadable names in the two on the *Description* engraving held the clue to dating the zodiac itself. For, he suggested, they might refer to known Egyptian kings, indeed possibly to a king and his son. He pointed out that the cartouche containing a vase surmounted by a rectangle open at the bottom also appeared at Thebes, in the ruins of Medinet-Abou, and had been printed in the *Description* in 1812, five years before the volume containing Jollois and Devilliers' drawing.[70] In fact, a cartouche found in the Dendera temple itself and similar to the other of the two mentioned by Saint-Martin was printed in the same volume that contained the Jollois and Devilliers graphic of the circular zodiac.[71] Concerted attempts to read the hieroglyphs had been under way since the Egyptian expedition, and Saint-Martin had now added urgency to the effort since, it seemed, the extremely controversial question of the zodiac's age might be solved if only the names in the cartouches could be read. Chabrol, if he was the annotator of the pamphlet, thought the point sufficiently important to pencil drawings of them in the margins (Figure 10.14).

Saint-Martin's comparatively balanced account appeared months before Biot presented his own theory and had little effect on evaluations of it, particularly since Saint-Martin had asserted that attempts to read the monument as an astronomical relic were likely to go nowhere. Biot spent a great deal of effort trying to prove precisely the contrary. Most of those who took up arms against the astronomically minded calculators of zodiacal time failed to persuade because their counter-arguments depended for the most part on textual and stylistic interpretations rather than the seeming authority of images that could be transmuted into numbers.[72] Something more was needed to silence Biot, at least for a time, something like Saint-Martin's challenge to read names in the cartouches, names whose dates could be tied down without the ambiguities of interpretation. That something was already germinating while Biot was working on his memoirs, and he was warned by its author, the young Jean François Champollion, that astronomical calculation would collapse in the face of dramatically new evidence.

11

Champollion's Cartouche

Jean-François Champollion

One night in Grenoble, during the chilly final weeks of 1812, Jean-François Champollion, at twenty-two years old the university's youngest professor, placed an alabaster jar in a pot of boiling water. A little over a foot tall, the jar contained a thick, hard substance, like dried pitch. Its lid, carved in the shape of an ape's head, was on the table. The young professor of ancient history had good reasons to hope his experiment would not take long. Firewood was expensive, and his annual salary, never large, had recently been reduced by several thousand francs as a result of the death of Professor Jean Dubois-Fontanelle (translator of Ovid's *Metamorphoses*), for whom he had served as an assistant. Then there was the prospect of ruining the jar, a precious Egyptian antiquity that was most likely found in a tomb, an accoutrement for a mummified royal's adventures in the afterlife.

After half an hour, Champollion plucked from the jar's softened contents a linen-wrapped bundle. Sticky fluid spilled over the top. Unwrapped, the bundle was discovered to be an internal organ, perhaps a lung or a liver. This was a surprising result, for these so-called Canopic jars had previously been considered mere representations of ancient Egyptian deities, perhaps to be used by the dead as incitements to worship in the afterlife. Champollion claimed that his experiment had shown that the jars were something quite different. They were not "little gods, or symbols of the Nile's flooding, of water, of the origin of all things, as believed by our most celebrated scholars," such as "Visconti, Millin, Wolf, and so on, and so on."[1] Instead, the jars were "simply vessels, part of the embalming process." The fantastic shapes of the jar lids—heads of apes, people, jackals, and hawks—referred directly to the deceased's body, for each symbolized one of the "Four Sons of Horus," guardians of a specific body part.[2] Melting the jar's contents had revealed to Champollion an entire system of thought and practice, an ancient Egyptian funerary ritual. The jar was no longer an

CHAMPOLLION LE JEUNE
(Jean-François)
Né le 24 Décembre 1790 + le 4 Mars 1832.
Imp. A. Salmon.

Figure 11.1. The young Jean-François Champollion

abstract symbol of a religious belief or natural process but could only
be understood as an element of a structured whole, a system of beliefs
and related practices that were part of ordinary life in ancient Egypt.

Champollion is today remembered mostly for his decipherment of
hieroglyphics. But the decipherment was a late stop on a uniquely long
road, with more than one wrong turn and many setbacks along the way.
When Champollion melted the contents of the Canopic jar, he had been
at Grenoble for three years in a professorship that was hotly contested
by the other faculty. Their resentment of this younger, more gifted colleague was evident in the pittance that was agreed upon for his salary,
a disappointing 750 francs per annum. To have received a fair salary—
by extension, a vote of confidence—from Grenoble's academic power

structure would have come as a great relief to Champollion, who had long been dogged by money troubles. As an impoverished student in Paris in 1807 and 1808, he had reluctantly supplemented his education in ancient languages with bitter lessons of self-denial: in letters to his brother and chief moral and financial support, the young man reported that his ragged trousers made him a true sansculotte "without, however, having the principles or intentions," a complaint whose playfulness does not quite blunt its edge of real desperation. (For his part, Champollion-Figeac did the best he could, within his own, limited means, for he had a family and was ensconced in a business career to which he, a scholar by training and inclination, was not particularly well suited. And he complained, in turn, of his brother's spendthrift ways.)

At Grenoble Champollion was generally disliked, and although Fourier continued to protect him from the ever-present threat of conscription, Champollion found it increasingly difficult to attract his attention. Moreover, in 1812 Fourier came into open conflict with both brothers over the authorship of the preface to a volume of the *Description*. The following year, when Champollion needed Fourier to present his first book, on Coptic place-names as late versions of an ancient Egyptian language, for the ministerial approval required for publication, Fourier dragged his feet, and publication was delayed for several years. Champollion was neither passive nor merely unfortunate in all of this. For one thing, he did not go out of his way to make himself a politically unexceptionable presence at the university—an unwise move in a rapidly changing political environment in which safely nonprovocative aspirants to positions were the most attractive. His political views were, if not an opportune way to gainful employment, at least a useful source for wicked satire, not something to be expected from a scholar whose leisure activities included hours spent cataloguing items in Grenoble's municipal library. He went so far as to transform his disappointments in older colleagues into a satirical play, *Scholasticomanie*, that was performed to acclaim in Grenoble salons, much to the chagrin and resentment of those whom it pilloried. Although the political climate made it difficult to talk seriously about ancient chronology during his first years as a professor in Grenoble, Champollion insisted on teaching the subject, along with ancient history and geography. Even his pedagogy produced unease among his fellow professors. He gave hour-long lectures with the original texts, in their original languages, at the ready, followed by a classroom discussion in which he acted the role of the interested skeptic. This method (unusual at the time) with its lively classroom exchanges stood in stark contrast to the university's standard pedagogic model, according to which students were passive receptacles for learned monologues read by the professor from a prepared script. The popularity of Champollion's ambitious lectures probably did not help his cause, and he paid a price in appearing to be a sufficiently loose cannon that his colleagues were reluctant to support him.[3]

Despite these difficulties, as well as the blind alleys that Champollion stumbled down in seeking to understand ancient Egyptian (including forays into Etruscan, as well as the languages of India and Ethiopia), he became ever more deeply convinced that he, and only he, could unravel the mystery of the hieroglyphs. At the same time, he seemed unable to focus on the work. Neither could he keep himself out of politics. During Napoleon's exile to Elba, he fanned antiroyalist flames by penning scurrilous plays (performed in salons because the theater had been closed by the censors) and defiant ditties that became popular hits, sung all night in the streets of Grenoble to taunt the police. After the monarchy was restored and his brother had decamped for Paris, Champollion resisted wearing the royal lily, provoking his brother to complain, in the spring of 1814, that Champollion's stubbornness was threatening his efforts in Paris to obtain official approval to print Champollion's still-unpublished manuscript. As the year drew to a close, Champollion was still penning rude plays and songs to irritate the police.[4]

With the return of Napoleon from exile in the spring of 1815, Grenoble was once again plunged into chaos. The immediate problem for Champollion was how to toe a shifting line between showing too much sympathy for Napoleon, risking future retaliation should Napoleon fail to reinstall himself, and not enough sympathy, giving Napoleon a reason to suppress his work. A local solution availed: with the mayor of Grenoble and members of the university faculty, Champollion joined a group dedicated to the preservation and defense of the city regardless of what France's immediate future held.[5] Meanwhile, Champollion-Figeac was in Paris, making himself a great favorite of Napoleon—an investment that would not ultimately pay, despite his optimistic conviction that with Napoleon lay his brother's best chances for the scholarly acclaim he had long sought. Champollion himself met Napoleon and apparently made an exceptionally good impression, winning the returned emperor's trust by signing himself "Champoléon," causing Napoleon to murmur, "He must be all right, he has half my name."[6]

Supporters of Napoleon, Champollion and his brother suffered under the royalists when Austrian troops took Grenoble in July, following

Napoleon's defeat at Waterloo. Through the bureau of Inscriptions et Belles-Lettres, over which he exercised considerable control, Sylvestre de Sacy—his former mentor, now a competitor and always a convinced royalist—arranged to block publication of Champollion's Coptic grammar and dictionary, and he encouraged one of Champollion's rivals in the race to understand hieroglyphs, Thomas Young, the well-known secretary of the Royal Society of London, whom he kept informed about Champollion's progress. Nevertheless, De Sacy's perfidy was far exceeded in effect by the royalists' activities in Grenoble, for they purged the university faculty of all but their most ardent supporters. After the closure, at royalist instigation, of the literature faculty in January 1816, both Champollions found themselves unemployed. Although these developments were hard to bear, it seems unlikely that Champollion, at least, could have failed to anticipate them, for he was well-known as the author of antiroyalist ditties like this one:

Like another I have my loves,

And because myself I must explain,
I convey and ever will,
All good wishes to the Republic!
That alluring beauty,
Cherishes honor and courage,
And so from every Frenchman,
She covets homage.[7]

In March the brothers suffered another blow: both were exiled to sleepy Figeac, their hometown, where their two sisters eked out a living running the family business, a bookstore started by their father, Jacques, whose deepening alcoholism further encumbered their already burdened lives. By summertime, the tedium had grown so unbearable that, with the help of the local prefect, a passionate amateur archaeologist, the brothers contrived an intellectual adventure: to locate the ancient site of Uxellodonum, the mythical final stronghold of the Gauls, surrounded by cliffs and memorialized by Julius Caesar. Working together from clues in Caesar's *Commentaries*, the brothers conjectured that Uxellodonum must have been located at Capdenac-le-Haut, three miles outside Figeac. Champollion-Figeac supervised excavations at the site, which turned up ancient Roman artifacts that seemed to confirm the hypothesis. This happy project kept the brothers busy until November of 1816, when the travel prohibition was lifted for Champollion-Figeac, who left for Paris the following April. Waiting for a similar clearance, Champollion applied himself to the task of setting up a public school, modeled on the pedagogical principles of a British system called the "Lancaster method"—which, like his direct style in Grenoble's university classrooms, evoked Pestalozzian principles of learning by doing, as opposed to rote memorization, but with the additional twist that as students learned new skills, they were to teach them to the younger students in turn. This project was undertaken with the support of Louis XVIII's government, which wanted to promote literacy throughout the country, despite resistance from the Catholic Church and the Ultra faction, both of which were convinced that it would be difficult to control the populace in their usual ways, by indoctrination and control, if literacy were to become widespread in rural France. And so, a year after he had been digging in the fields outside Figeac, Champollion found himself before a classroom full of children, and lionized for it. About this triumph, he exulted to his brother: "I've been celebrated and embraced, even by people whom I knew neither from Adam, nor from Eve! The public and private opinion is for us, some masks have disappeared, we have won many new friends." Although his efforts to establish the school and to begin the literacy curriculum had borne fruit, Champollion's life was a shambles. Suffering from exhaustion, fevers, coughing fits, and bouts of anorexia, he had not touched his work on the decipherment in seventeen months.[8]

If efforts to materialize liberal ideals in new public institutions had banked the fire of Champollion's interest in hieroglyphics, they had not put it out. A favorable review of his *L'Égypte sous les pharaons*, written anonymously by Young, appeared in the *Monthly Review*, out of London. Tantalizing rumors of new decipherment efforts reached Champollion even in quiet Figeac. In the first half of 1817, he summoned the energy to comment on Fourier's memoir on Egyptian astronomy: "I certainly know now that what I'd earlier told him, which was the best of my knowledge at the time, about the names of the months and their meanings now appears to me to be less important than before. I considered the naming of months similar to the naming of deities, so that in Egypt, as in other areas in the Orient, the names of the months correlated with civil and religious purposes." Energized by these developments, Champollion arranged for the transfer of his notes on decipherment to Figeac from Grenoble, where he had left them. Champollion-Figeac, too, was growing restive. He left Figeac for Paris

eager to return to Parisian academic life now that he had secured the
support of a powerful patron, Dacier, at Inscriptions et Belles-Lettres.
Dacier, for his part, kept a fire burning under Champollion by providing him with news from Egypt, where Salt and his accomplices were
raiding tombs with their customary abandon, raising the exciting prospect of new archaeological knowledge and the ire of those in Europe
who disapproved of their methods.[9]

Nevertheless, Champollion's time in the wilderness was hardly over.
Just as his abandoned materials arrived in Figeac, he was tempted to
return to Grenoble by the installation of a new prefect who held out
the prospect of a university position. Though his chances were slim,
they were better than nothing—and certainly better than languishing,
isolated, in Figeac. Through the end of 1817 and the beginning of 1818,
while waiting in Grenoble for the prefect to make good on these prom-
ises, Champollion put his work experience at Figeac to good use. Years
of subjection to the narrow passions of intriguing adults had deepened
his appreciation for children, for their simpler enthusiasms and for
the promise they represented, if only they could be reliably spared
religious and political indoctrination in their schooling. By July 1818,
thanks to Champollion's work and talent, Grenoble boasted a new Lancaster school, as well as a Latin school for students interested in the
classics. Yet unfinished scholarly business pricked his conscience. "My
books sleep!" Champollion complained. "I work myself like a galley
slave . . . but this is how it has to be."[10]

Over the course of the following year, Champollion managed to return
to his study of hieroglyphics, although his work was sporadic and frequently interrupted by his pedagogical responsibilities, many of which
involved defending the new schools and their students from the depredations of the religious Right.[11] At the same time, Champollion-Figeac,
who had successfully lobbied for reinstatement in his former position
as librarian at Grenoble, discovered that he could not meet both his
local and his Parisian responsibilities and so foisted the former onto
his brother. In addition to his work in the new Grenoble schools and
his brother's responsibilities at the library (which included a complete
catalogue of its holdings, ordered by royal authorities), Champollion
had taken a new job, as a professor of history at the Royal College there.
Once again, Champollion found himself overworked, in poor health,
with no time or energy to devote to his Egyptian studies. Perhaps most
discouraging of all, after years of hard study and sacrifice he still had

not found an appreciative audience for his most important work. A description of Champollion by the Marquise de Maillé from about that time reports that, despite his ill health, "He is a worker of indefatigable zeal . . . kind and modest, who expresses himself with little facility and doesn't know how to withstand the brilliance of his work. He's strongly attacked, as are all those who are the first to explore a dazzling source of fame and utility."[12] And then, in February 1820, the moderate prefect of Grenoble, Choppin d'Arnouville, was replaced by the Baron Lemercier d'Haussez, an Ultra. Amid growing resistance to the new political order, Champollion-Figeac lost his library position, a development that at least had the merit of freeing the younger Champollion from the cataloguing responsibilities that he had undertaken for his brother during the latter's absence from Grenoble. Nevertheless, Champollion's health continued to deteriorate, to the point that his doctor put him under medical house arrest, and in March of the following year he was "temporarily" removed from his position at the Royal College (the removal would soon become permanent). This otherwise disastrous turn of events had a silver lining, for Champollion could devote his now-copious free time entirely to decipherment. With rising spirits, he wrote to his brother in Paris, "It is my Egyptian studies that shall win."[13]

The lull lasted three weeks. The Ultras could not hold their power in Grenoble, and by the end of March, Champollion had put his newly recovered health in the service of rebellion, leading a group of rebels to the top of a local fort where they raised the revolutionary tricolor flag. This mostly symbolic uprising had real consequences. The prefect, incensed, moved to call Champollion before a martial court for treason—with a possible sentence of death—and warned the government in Paris about his seditious behavior. A worried Champollion-Figeac hastened to Grenoble to rescue his brother from this latest predicament. In the end, the situation was resolved ambiguously. Both brothers lost their Grenoble positions, but Champollion retained his life and left for Paris, arriving in July 1821. The Dendera zodiac reached Marseilles eight weeks later.

Champollion vs. Biot

Champollion soon learned of the Dendera zodiac's appearance at Marseilles, and he, along with others, was not pleased. Ripped from its context, the monument had been denuded of its material connection to ancient Egypt. "Today, solely as a result of the zodiac's removal, the astronomical room is uncovered, and the rest of the ceiling is menaced with complete destruction; it's as though the allies had removed a part of the ceiling in the great gallery of Versailles to take some paintings; what will become of the rest of the roof and of the gallery itself," he protested.[14] The extracted zodiac stood alone, a singular and puzzling artifact, symbolizing an atomized conception of history, one whose past would now have to be reconstructed out of disconnected, unrelated objects. "Who knows," he went on (presciently, if ironically, as we shall see), "whether we won't argue about the place it occupied in the great monument from which it has been torn, about its orientation, about the sculptures that surrounded it, etc.?" Champollion was certainly not alone in objecting to the zodiac's removal. Two years later an American writing a review of one of the many pieces on the Dendera circular remarked that he had "strong doubts whether future travelers in Egypt will be particularly gratified with finding the roof of the temple of Dendera blown out by gunpowder and carved out with saws by M. Lelorrain."[15]

Champollion's protest was anonymous, which is hardly surprising given his recent troubles and the patriotic brouhaha that surrounded the monument's arrival. Now deeply engaged in the decipherment of hieroglyphs, he soon learned about Biot's interest in the zodiac. He had met Biot many years before in Grenoble through his brother and had developed friendly relations with him.[16] He may have warned Biot around this time that all attempts to construe the zodiac as an exact planisphere were wrongheaded. In any event, Champollion responded to Biot's claims in a signed letter to the *Revue encyclopédique* a week after Biot had read his second paper before the Académie on July 19, 1822, five months after his former friend, and now rival, Saint-Martin had read his own piece on the zodiacs. And the challenge that Saint-Martin had set out—to read names in the cartouches in order to date the zodiacs—was precisely what Champollion would shortly aim to answer.

"Here," Champollion wrote à propos Biot, we have "a new opinion on the presumed epoch of an astronomical tablet that, for twenty years, has been successively the occasion for and the subject of a crowd of systems, all contradictory, because those who wanted to explain it, and who claimed to draw from it rigorous consequences, were more or less well prepared, by the type and direction of their studies, to attempt this difficult enterprise." In other words the astronomers were rather like a twenty-first-century economist who, having only prices with which to work, perceives a world governed exclusively by money. Champollion insisted that it was not enough "to possess the learned theory of modern astronomy." One needed "also to have exact knowledge of that science as the Egyptians themselves conceived it, with all of its errors and in all its simplicity." This was necessary because "Egyptian astronomy was in its essence mixed with religion, and even with that false science [astrology], which claims to read the future of the world and of individuals in the present state of the heavens." Without that understanding "the courageous explorer of the Dendera monument will find himself on dangerous terrain; he chances mistaking a cult object for an astronomical sign, and considering a purely symbolic representation as the image of a real object."[17]

Champollion, who had always disliked mathematics, sidestepped Biot's calculations, and drove right to the heart of the problem as he saw it. His argument rested on Biot's identification of four principal stars on the monument. The Gau engraving produced for Saulnier and Lelorrain was available by this time, and Champollion had a copy of it. Biot published a version of the engraving the next year, marked with the positions of his putative stars (Figure 11.2).

Look at the zodiac as a whole, Champollion urged the reader. On it there are many images of men or animals, nearly all of which are accompanied by a small group of hieroglyphs either above or next to the image. Every such group includes a starlike symbol (except for the one that Biot identified as containing Antares). There are thirty-eight groups altogether, thirty-three of which sit on the circle held up by the kneeling figures. According to Biot, four among these represent the locations of his principal stars. But if that is so, Champollion argued, then it ought to be the case that the remaining twenty-nine should represent stars as well. Why? Well, why not? Why would the Egyptian designers have drawn so many similarly placed figures only to have just four among them representing physical objects? The "analogy of position," Champollion averred, argues for "an analogy of expression." But this cannot possibly work because all of the images are symmetrically arrayed on the circle, and nothing astronomical could correspond to the remaining twenty-nine in the array. Champollion did not need to know much about astronomy to reach that conclusion, because in Biot's interpretation the circle in question is centered on the monument's center, calculated by Biot to be the north pole. Consequently the circle itself must represent a parallel of celestial latitude, and even

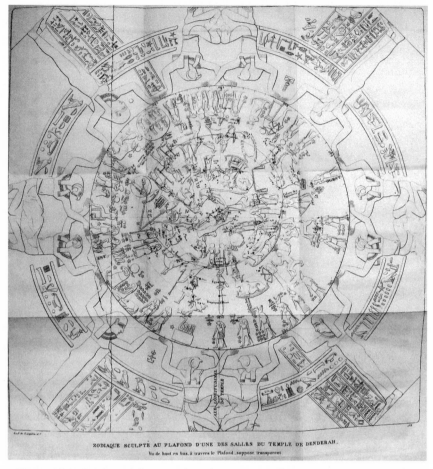

ZODIAQUE SCULPTÉ AU PLAFOND D'UNE DES SALLES DU TEMPLE DE DENDERAH.
Vu de haut en bas, à travers le Plafond ,suppose transparent

Figure 11.2. Biot's modification of Gau's engraving. This was a mirror-image of Gau's since Biot insisted on presenting the zodiac as though seen from above. The black dots mark his putative stars.

a quick glance at a celestial globe would show that nothing of the kind occurs: there simply is no latitude with visibly noteworthy and symmetrically disposed stars arrayed along it. Moreover, Champollion continued, the attempt to identify Antares with what was obviously a vase and not a star could not be accepted (Figure 10.11, left)—he would have none of Biot's forced argument concerning the "heart" of Scorpio as a way out of this conundrum.

This was all well and good, but if the objects that looked like stars in all but one of Biot's groups did not represent stellar locations, then what were they for? Champollion had an answer to hand, for a key element in his now rapidly progressing decipherment concerned precisely such a situation. "We recognized," he explained, "that every hieroglyphic group, placed on the head or by the side of the image of a god, a man, an animal, etc., expresses its proper name, or at least a particular qualification devoted to it." Moreover, each such group must be read from the direction toward which the heads of the figures face. The star-figured symbols were therefore the *last* in each group naming the figure. Their regularly occurring terminal positions implied that they were particular kinds of elements in the hieroglyphic script, in which case they should be considered "as a sort of letter, and not as the imitation of an object." Now, Champollion continued, his study of the three writing systems on the Rosetta Stone had led him to conclude that most of the proper names of individuals that belong to the same type are either preceded or followed by a sign that designates their kind. So, for example, the names of deities are always followed by a sign indicating the idea of a god. If this principle held here as well, then the starlike sign would have to indicate that what it follows is the name of a celestial object, presumably a constellation or the parts of a constellation. The tablet *names*; it does not *depict*. It is a text, not an image.[18]

Biot had considerable respect for Champollion, given his later positive remarks about Champollion's decipherment, but he never accepted the claim that the zodiac was not an image produced according to his "method of projection," that it had not been designed to represent the state of the heavens in about 747 BCE.[19] Nevertheless, Biot was soon faced with a result that seemed to make it certain that the zodiac could not actually have been carved at such a distant date. For in September the young Champollion produced a startling discovery that did not depend on stylistic interpretations or philological arguments in the vein of Visconti and Testa, as well as others. Instead, Champollion provided evidence that was even more powerfully convincing than the savants' astronomical calculations.

The Letter to Dacier

On Friday, September 27, an audience gathered at Inscriptions et Belles-Lettres for the usual meeting. Thomas Young was there, as were Alexander von Humboldt and Edmé Jomard. What they heard was far from the ordinary report of some minor philological discovery. Champollion had been invited several times before to read papers on his work with the Egyptian scripts, but this time was different.

Just thirteen days before, copies of highly accurate drawings by Jean-Nicolas Huyot of hieroglyphs at the temple of Abu Simbel had provided him with new material that cemented his conviction that he had the decipherment in hand. Rushing to tell his brother, Champollion had collapsed in exhausted excitement at his feet. It took five days for him to recover, after which his brother helped Champollion write a report on the system, which was dated September 22 as a formal letter to Dacier at the Académie. Copies of the "letter" were prepared according to the usual rules for distribution to the members at the next session. Rumors about it circulated for days among the city's scholars, and the audience that Friday anticipated a major event.[20]

The eight-page report that Champollion read before the Académie that Friday was published in October in the *Journal des Savants*—this, the oldest journal devoted to natural philosophy, had been suppressed during the revolution and then revived in 1816, albeit with a literary rather than a scientific purview.[21] Shortly thereafter, a forty-four-page version of the letter was published by, significantly, Firmin Didot, printer to the monarchy. From the outset, Champollion framed his work as a major contribution, not solely to the decipherment of the scripts per se, but to ancient chronology. He presented it, in part, so that savants might avoid "grave errors concerning the diverse epochs of the history of arts and of the general administration of Egypt." This was possible because those hieroglyphs that represented names, and that, exceptionally, "were endowed with the faculty of *expressing the sounds* of words," were the "*titles, names,* and *surnames of the Greek and Roman sovereigns*" that successively governed Egypt.[22]

Among the titles that Champollion read was one taken from hieroglyphs that were printed in volume 4 of the *Description*. There he read several variants of the imperial title Αντοχρατορ, *autocrator*, and he included three of these in the second plate that accompanied his letter (Figure 11.3).[23] Since the hieroglyph is read from the direction the symbols face, the bird symbol, or sparrow hawk, must be the first in the sequence for the first two, though not for the third. Champollion read this symbol as having the sound of Greek *A*. The curly symbol, rather like a backwards Greek *rho*, Champollion read as sounding a Greek *O*. The hand, he decided, sounds a *T*, the wedgelike symbol a *K*, and the lion a *rho*, for the sound *R*. Cartouche 48 in Champollion's figure accordingly sounds, in English transliteration, AOTOKRTR, reading the order of symbols in a multisymbol line in whatever way is necessary to make phonic sense of the word in Champollion's system. In

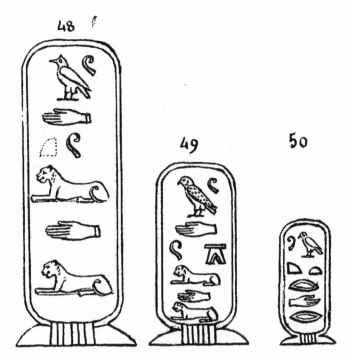

Figure 11.3. Champollion's variant hieroglyphs for *autocrator*

cartouche 49, the stool-like symbol replaces the wedge, so that this cartouche again sounds *AOTOKRTR*.

The third cartouche reads somewhat differently. The representation is, first of all, mirror-imaged from cartouche 48 (excepting the bird figure, which faces the same direction), a semicircle has replaced the second *rho*, and ovals have replaced the lions. Champollion read the semicircle as a homophone for the hand, and the oval as a homophone for the lion, so that this cartouche transliterated as *AOTKRTR*. His interpretations of homophones were central at the time to his decipherments, which were based on comparing different cartouches with one another. (The hand would later be read as sounding like a *d*, whereas the semicircle sounds like a *t*). Most significant of all, this third cartouche, Champollion's number 50, was taken directly from the one at the bottom of the leftmost of the two columns depicted on the *Description*'s plate of the Dendera circular (see Figure 10.14, bottom left, and Figure 10.1 for the entire engraving).

Champollion placed special emphasis on his reading of cartouche 50. Unfortunately, he complained in exasperation, "that important part of the monument is not in Paris." For "the stone was cut near that point because the only aim was to remove solely the circular zodiac, and so it was isolated from a bas-relief that in all probability referred to it." And the zodiacal cartouche was especially significant because it "establishes, in an incontestable way, that the bas-relief and the circular zodiac were sculpted by the hands of Egyptians under the domination of the Romans." The "infernal" stone, bane of the religious Right, was after all no older than the Roman Empire, just as Visconti and Testa had long before argued. "Our alphabet," Champollion concluded, "acquires by this sole fact high importance, because it greatly simplifies a question that has been so long debated, and on which most of those who examined it have presented only uncertain and often diametrically opposed opinions."[24]

Jomard, for one, contested Champollion's dating. That a skeptical response should arise from this quarter was no surprise, for Champollion, in his typical way, had already angered Jomard by disparaging the reproductions of hieroglyphs in the *Description*, which had been Jomard's primary responsibility since 1807. Jomard was particularly proud of representational precision, and he responded harshly to criticism of the *Description*'s accuracy. When it came to the Dendera circular, he would certainly not have sanctioned any critique of Jollois and Devilliers' drawing, printed under his auspices. And, to be fair, theirs was evidently more accurate than any drawing made previously, in particular Denon's. Like Jollois and Devilliers, he had sketched the sky goddess with the parallel columns that contained the important cartouches at their bottoms, though in his drawing both cartouches were curiously empty (Figure 11.4). But Denon's depictions were known to be inaccurate, and so their vacant cartouches would not have raised questions, and in any case Jomard certainly did not question the existence of the hieroglyphs that appeared in Jollois and Devilliers' drawing, and that Champollion had relied upon. He continued to insist that the Dendera zodiac had nevertheless to be quite old, older indeed than Biot had recently calculated, to say nothing of Champollion's late date. Reacting with defensive skepticism, Jomard challenged Champollion: "Which emperor does this isolated *autocrator* designate?"[25]

Apart from Jomard's, there do not seem to have been any persuasive challenges at the time to Champollion's reading of *autocrator* in the Dendera cartouches. Reactions to the decipherment were quick and

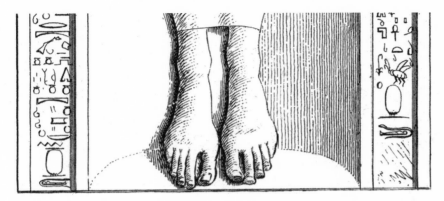

Figure 11.4. Denon's drawing of the Dendera cartouches

on the whole favorable, though Champollion had yet to work out many significant issues and had not elaborated the details. But it was also true that competitors abounded, and Champollion had to defend himself. His decipherment had gone substantially beyond what Young had accomplished, though the Englishman would eventually grow angry at what he thought to be the young Frenchman's refusal to acknowledge that he, Young, had built the foundation on which Champollion relied since he had previously identified several of the hieroglyphs within the cartouche-embedded names. Moreover, Saint-Martin, scholar and cofounder of the Société Asiatique, had, in effect, prepared the way for Champollion's reading of the Dendera cartouches to have the substantial effect that it rapidly did, because he had suggested the possibility of identifying the words within them as the names of kings.[26]

Among the first to discuss the point was Jean-Antoine Letronne, who had for some time been preparing a lengthy work on Egypt under the Greeks and Romans. Letronne had begun his career as an artist in the studio of David, but after 1801 he turned to the area of historical geography.[27] His interests were extraordinarily wide, and he sought always to assemble a broad array of evidence to reconstruct not merely the physical geography but the life of a period. Where, for example, Jomard saw antiquity through a prism of numbers that he thought were encoded in the remnant monument, Letronne, in this instance like Champollion himself, envisaged a vibrant and colorful civilization. He was particularly attuned to the limitations of ancient evidence and was highly skeptical of any overarching hypothesis about the past unless it could synthesize a very wide range of data. Probabilistic arguments struck him as especially doubtful in historical matters. He remarked late in life of Condorcet's and others' attempts to calculate the likelihood of historical evidence that "they have flattered themselves in thinking that they can calculate the odds that this event or that event occurred or did not occur. It is a misfortune that they do not see that this probability, resting as it does on an entirely arbitrary numerical basis, can only provide chimerical or illusory results. In no case could it replace that intimate and absolute conviction, which admits neither less nor more, produced by the examination of all of the diverse circumstances accompanying a real event."[28]

Although Letronne wrote these remarks in 1842, he certainly would have had Biot's probabilistic dating of the zodiac in mind. He had long been aware, and suspicious, of efforts to put astronomy in the service of history, efforts that threatened the autonomous authority of history. In 1823 and 1824 Letronne rejected all such attempts, supporting his claims in part with Champollion's reading of *autocrator* on the Dendera zodiac.[29] The primary thrust of his argument relied, however, on stylistic and philological considerations. These were based on the fashion of the temple and on the Greek texts that were carved here and there upon it. That the temples bore such inscriptions had long been known and had not persuaded the advocates of astronomical dating, who argued that the inscriptions were added later. They were after all not present on the zodiacs proper. And neither Fourier nor Biot, much less Jomard or Jollois and Devilliers, found stylistic arguments compelling—arguments that in their original form dated back to Visconti's claims two decades before, to which Letronne referred, regarding his own contribution as proving what had before been purely conjectural. Visconti's "simple and ingenious" system had not carried the day in its time, which saw "the appearance of many explanations of zodiacs, all contradictory, all mutually destructive."[30] Letronne was well aware of the counterarguments to Visconti, and he aimed his own work directly against them. He proceeded by comparing inscriptions among one another, by arguing against previous readings of them (including one by Champollion's brother), in short by marshalling the weaponry of a contextually aware philology against the claims of calculating astronomy.

Letronne paid particular attention to cartouches, filled or empty, and developed a series of assertions concerning them. First, he argued, empty cartouches signify that the monument in question must not have been finished. Conversely, finished monuments always have filled cartouches. Where one finds cartouches bearing the names of a Ptolemy or a Roman emperor, the monument is always in the "final style." Finally, one and the same monument often has the names of different kings.[31] On this basis he developed arguments for using the cartouches together with Greek inscriptions to reconstruct Egyptian history under Greco-Roman domination. Champollion's reading of *autocrator*, for example, "confirms the opinion founded on the character of the sculptural style" concerning its late date.[32]

Letronne's main concern in 1823 had not been with the zodiacs per se, but, when he turned directly to them the next year, he did not mince words. Some, he began, had used astronomy to date these "emblems." But "philologists and antiquarians approached the question from another side and arrived easily at its resolution."[33] Which they did through the careful examination of inscriptions in Greek characters and in "phonetic hieroglyphs." Biot, he claimed, had previously destroyed Fourier's theory by noting the stability of Sirius's heliacal rise. Of Biot's own work, Letronne said nothing, except that it involved "long calculations" and was "deep." But his own opinions were independent of the results of "that mathematical discussion." Which was a good thing, because Letronne averred that not a single zodiacal representation from Egypt was "purely astronomical"; all of them were tied to "some astrological, religious or mystical combination, and must be considered as the result of the singular development of astrology, and of the influence that it has exercised since the Christian era, or of the mélange of religious ideas from Greece and the Orient. Whence were born the most absurd superstitions and the most extravagant symbols."[34]

With Letronne we have, nearly for the first time in a quarter-century of controversy, a defender of zodiacal youth who was not also a defender of religious chronology. Letronne certainly did not admire Dupuis' work, vast though he admitted the latter's erudition to have been. He thought the *Origine* fatally contaminated by an addiction to a preconceived system that Letronne attributed entirely to Macrobius.[35] He lectured extensively about that at the Collège in 1838 and 1840. Portions of these lectures were printed in 1843 by one of his auditors in the *Annales de philosophie chrétienne*, a believer who lauded Letronne's demolition of the impious Dupuis—though there is not the slightest indication that Letronne himself was motivated by religious conviction rather than by an aversion to the creation of ill-conceived systems.[36]

Letronne instead defended the virtue of an archaeologically alert philology against the pretensions of calculation. In 1846, following Biot's re-presentation of his now somewhat modified theory, he continued his attack on all astronomical theories, this time without any specific concern with dating, which had been the hub around which the original controversies had swirled, given the implications for biblical chronology. Now, however, it was a matter of deciding once and for all, as Letronne put it (in italics), the following question: *"Are these representations, or are they not, an exact picture of the heavens?"* Biot was now willing to admit that the temples into which the zodiacs had been built dated from Roman times, indeed that the zodiacs themselves had been constructed at these same late dates. But, he insisted, they were still planispheres, albeit ones calculated to represent an era centuries past. That, Letronne insisted, verged on the ridiculous. Placing the zodiacs at the middle of the eighth century BCE located them in the same century as the founding of Rome itself, in which case they would have been constructed centuries later while Egypt was under Roman domination to represent the period of that empire's birth. "As if one were to say," he continued, "that the builders of a fifteenth-century church would have sculpted in it a planisphere for the time of Clovius, of Charlemagne, or of Hugh Capet."[37] And since Letronne was convinced that Biot had demolished all other astronomical interpretations besides his own, it followed that the zodiacs simply could not be planispheres. Their only link to the heavens had at best to be religious or mystical in character, not astronomical, no doubt connected to Egyptian funerary practices. Just so, he insisted, do mathematicians' dreams fade, when one "applies calculation too soon to facts that are inimical to it."[38]

A great deal clearly depended on Champollion's reading of the cartouche in the early 1820s, because it was the only strongly persuasive evidence from the vicinity of the circular zodiac itself. Yet the only source for his reading were the two plates in the *Description* that were drawn from Jollois and Devilliers' original (Figure 10.1). The fact that Lelorrain had not brought back the bordering panels that contained the cartouches had certainly been noticed. In England the *European Magazine and London Review* even hoped that the "remainder of the ceiling might surely be obtained," to be paid for by Louis, who would surely not "fail to possess himself of everything appertaining to this extraordinary specimen of the prosperity of ancient Egypt."[39]

The consequences for Champollion himself of his decipherment, and especially of his *autocrator* reading, were eventually positive, though

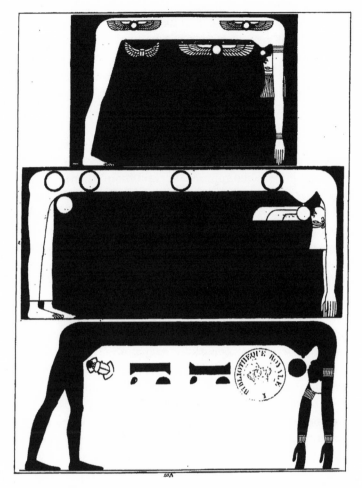

Figure 11.5. Champollion's renditions of the sky goddess

he faced several challenges and difficulties and had not as yet fully developed the final scheme. To put his stamp on the area, he began the printing in July 1823 of a collection of annotated plates concerning the "mythological characters of ancient Egypt," among which were his renditions of the sky goddesses from Esneh and Dendera (Figure 11.5).[40] This publication was succeeded in 1824 by a *Précis* of his system.[41] After a difficult first year following the publication of the letter to Dacier, Champollion found a supporter in the sympathetic and influential duc de Blacas, who superintended royal property and who was himself a collector of antiquities. On March 29, 1824, Champollion presented a copy of the *Précis* to Louis XVIII, who had been persuaded to meet him by Blacas, despite Champollion's reputation as an anti-royalist. At Blacas' urging, Champollion had dedicated the book to the king in an effort to obtain his patronage, which was successful, enabling Champollion to set off for Italy the following June to examine Egyptian collections in Turin and Rome. He reached Rome on March 11, 1825, having spent months with the Turin materials.

Recall that Biot himself was in Rome that very March, in the company of his son, and that Testa, one of the earliest defenders of zodiacal youth, had introduced him to the pope, Leo XII (né Annibale della Genga). This pope would, among other reactionary endeavors, require Roman Jews to sell their property; on November 26, 1826, he forced them to return to the ghetto. Uprisings and revolts throughout the papal states were the eventual consequences of della Genga's repressive policies. Biot was, however, enthralled. The pope seemed to him "endowed with an indulgent kindness, suave, charming, as though descended from heaven to earth." Why, he had the "serene tranquility of an old man's soul, allied to the dignity of a pontiff and a prince, ornamented and made higher by a superior spirit, which the princes of this world rarely acquire."[42] This serene and kindly pontiff was especially bothered by tolerance for "Deism and Naturalism," which he had attacked in an encyclical entitled *Ubi Primum* that had been issued on May 5, 1824. As intolerable as Rome's Jews were the "flood of evil books that are intrinsically hostile to religion." Among these infernal publications would certainly have been those that attacked sacred chronology, such as ones that claimed great antiquity for the Egyptian zodiacs. For della Genga undoubtedly knew about the dangers posed by the zodiac controversies from Testa, his personal secretary.

Champollion hardly shared Biot's awe of the pope or his beliefs, political or religious, though he had shared a coach to Rome with Biot's son. He had nevertheless learned by this time the ways of patronage. As he had for Biot the previous month, Testa arranged for Champollion an audience with the pope, whom Testa had, it seems, carefully if incompletely prepared beforehand. On June 22 della Genga greeted Champollion enthusiastically, to say the least, telling him three times that he had rendered "a beautiful, great and good service to the Church," for his identification of *autocrator* had proved that the zodiacs could not traduce biblical chronology. The pope told the French ambassador that Champollion "had abased and confounded the arrogance of that philosophy which pretended to have discovered in the zodiac of Dendera a chronology anterior to that of the sacred Scriptures."[43] So enthusiastic was he that on the spot Leo XII offered the married Champollion, whose daughter had just been born, a cardinalship. He demurred, remarking that "two ladies wouldn't agree."[44]

Champollion's reputation as a defender of the faith had gone quite far by this time. A rumor soon spread in Ultra journals that he had undertaken his Egyptian studies for the express purpose of saving sacred chronology. Champollion was immensely bothered by these assertions. He detested those "vile reptiles" for treating him, he wrote a friend, as "a true father of the faith, defender of Religion and of good chronological doctrines." He was comforted, however, by his conviction that the Ultras would surely think differently just as soon as his Egyptian "alphabet" was applied to "monuments whose antiquity will frighten them." More than anything, it paid to take the long view. After all, he counseled his friend, "he laughs best who laughs last."[45]

12

Epilogue

On May 15, 1826, a royal order created the Musée Charles X at the Louvre, comprised of a section devoted to Greco-Roman antiquities and one devoted to Egypt. Champollion was named director of the Egyptian section. Assisted by the sympathetic and persuasive duc de Blacas, who superintended royal property and was himself a collector of antiquities, Champollion persuaded the conservative king to help fund an expedition in collaboration with Tuscan authorities, whose interests were represented by Ippolito Rosselini. The group left on their fifteen-month adventure in 1828 and, having obtained the appropriate *firmans* from Mehmet-Ali, sailed up the Nile on ships Champollion named *Isis* and *Athyr*. They arrived at Dendera on November 16 at four in the afternoon. Alone, without guides, but "armed to the teeth," the party set out immediately for the temple (Figure I.6).[1]

Singing marching songs from the latest operas, the troupe encountered a man who, terrified by their appearance, tried to flee; they chased him down and persuaded him to lead them to the temple. This "walking mummy," as Champollion called him, not unsympathetically, guided them until the monument at last appeared. "I would not even try to describe the impression" the sight made, Champollion wrote his brother, except to say that it was "grace and majesty united in the highest degree." They sat for two hours in "ecstatic" contemplation of the temple glowing by the light of a nearly full moon against the dark blue desert sky. The group returned to base, only to come back at seven the next morning. Champollion canvassed the temple's columned interior, enthralled by the architecture but unimpressed by the bas-reliefs, which reflected the "decadent" period of their carving, as he saw it. Although the temple had itself been constructed during this same era, nevertheless its architecture, Champollion averred, being based on a "numerical art," had better resisted the vagaries of fashion.

Ascending to the temple's roof, he entered the small room that had housed the circular zodiac until it was ripped it from its millennial-old home. There, surrounding the void created by Lelorrain, Champollion

Figure 12.1. "Champoléon" inscribed on a column at Karnak

could examine the carvings that had been left behind, most especially the cartouche in which he had so influentially read the Greek word *auto-crator*. Turning his eyes to the ruined ceiling, Champollion must have frozen in shock. *"Risum teneatis, amici,"* he wrote to Champollion-Figeac—"hold your laughter, friends!"[2] For "the piece of the famous circular zodiac that carried the cartouche is still in place, and that same cartouche is *empty*." As indeed are all the others in the temple's interior (Figure I.7). Champollion decided that it must have been "the members of the *Commission* who added the word *autocrator* to their drawing, thinking to have forgotten to draw an inscription that doesn't exist:—that's called carrying whips to beat yourself." Still, Champollion wrote, Jomard, upholder of the zodiac's extreme antiquity, should not think to triumph, because the sculptures are "in the worst style, and cannot be older than the times of *Trajan* or *Antoninus*."[3] Remarkably, after all the ink that had been spilled over the decipherment of hieroglyphics, even Champollion, whose understanding of the ancient script had convinced so many, had in the end to invoke style.

Jomard did not have the chance to triumph, or to renew arguments about style, because Champollion told only his brother about the empty cartouche. In 1832, having at last gained admission to Inscriptions et Belles-Lettres and by then heavily overweight, overworked, and overwrought, Champollion died of a stroke at the age of forty-two.

Champollion-Figeac became the custodian of his memory, and he tended it carefully. Nineteen of Champollion's letters to his brother were printed as they arrived in the widely read *Moniteur*. The year after his death, they were collected and reprinted in Paris. The seventh letter, dated November 24 and written from Thebes, contained a version of the visit to Dendera, but the entire passage concerning the zodiac had been excised, replaced with an anodyne remark that the cartouches throughout the temple were all empty.[4] There is no doubt that Champollion-Figeac edited the volume, despite the absence of his name from it, and he, of course, had submitted the redacted originals to the *Moniteur*. An edition reprinted more than forty years later and containing a few remarks by Champollion's daughter, Zoë Chéronnet-Champollion, still omitted the passage, which became available only in 1887 when his nephew published the omitted portion.[5] Nevertheless, by 1845 enough time had apparently passed that Champollion-Figeac thought it reasonable to be more forthright. In that year, the fourth volume of drawings done during Champollion's expedition was printed, the first having appeared in 1835.[6] It contained an illustration done under Champollion's auspices during the expedition that shows the empty cartouches (Figure 12.2). The caption, written by Champollion-Figeac, reads "Planisphere or zodiac of Dendera. This plate was revised from the original monument, whether at Paris, where the zodiac, properly speaking, is, or at Dendera, where the part of the monument on which the figure of a woman, placed between two vertical columns of hieroglyphs, still exists. The two cartouches, at the base of these two inscriptions, are empty, and do not contain the Roman name that was inserted by error. This plate is the most faithful representation of all those that have been published of this celebrated monument."

Denon's original drawing, with its empty cartouches, was correct after all, whereas a hieroglyph for *autocrator* had somehow crept into the purportedly more accurate rendering by the Polytechniciens, Jollois and Devilliers, who never wrote a word about it, though they surely knew of the mistake by 1845 at the latest. How had it happened, this error that had been so instrumental in establishing Champollion's reputation as a decipherer and, to his regret, as a defender of sacred chronology? The errant hieroglyph appears in the two renditions of the Dendera circular that were printed in the *Description*. The first of the two is an unshaded, outline drawing that should closely have followed Jollois and Devilliers' original, which apparently no longer

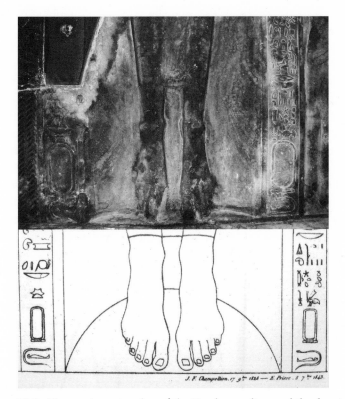

Figure 12.2. The empty cartouches of the Dendera zodiac, and the drawing of them done during Champollion's expedition

exists. Both the outline and the shaded engraving done from it (Figure 10.1) have the nonexistent hieroglyph. It must have been taken from another place, but it appears nowhere else in the *Description* in this precise form (Champollion's number 50 in Figure 11-3 is the closest). Since the symbols are in the unshaded engraving (which, being simpler to make, was surely produced before the shaded version), they were probably inserted either by the line drawing's engraver, Totard, or else by Jollois and Devilliers themselves. One or another of them must have taken the hieroglyph from another place that was not separately printed in the *Description*.

In 1825 there was an exhibition of Castex's marble sculpture of the Dendera circular, which he had completed at his own expense shortly before he died in 1822. (The restoration government, now in possession of the original, had refused to honor Napoleon's commitment to pay for the marble copy.) A pamphlet printed to advertise the exhibition remarked that an unnamed but "spirited English speculator" purchased it "in order that it might become the property of the British nation."[7] The pamphlet could be bought for sixpence, and there was certainly a charge for admission. This English "speculator," like his French counterpart, Saulnier, may have had his country in mind, but a nice price for the honor done the nation would no doubt have been appreciated. His pamphlet, like Saulnier's before it, was written as a sales brochure that also contained interesting information about the monument's history, in this case going into considerable detail about the arguments that swirled about it. This information would only pique a buyer's interest, raising the value of the marble, but there was too much of it, and it was much too carefully presented to attribute its inclusion entirely to the imperatives of the marketplace. Money and scholarship are ever uneasy bedfellows, particularly in the worlds of art and antiquities, but their marriage certainly did help to generate public interest in the zodiacs. As did Champollion's decipherment. Though it might equally be said that the decipherment generated the great attention that it rapidly did in no small part because of the brouhaha over the purloined temple ceiling from Egypt.

The warriors for sacred chronology had been thrilled to transform Champollion into St. George, the slayer of those Egyptian dragons, the zodiacs. Horrified at finding himself the object of such adulation, Champollion predicted that Egyptian chronology would eventually betray the "vile reptiles" of the religious Right; and indeed he was right, for the reign of the pharaohs dated to the beginning of the third millennium, more than five hundred years before Noah was said to have built his miraculously capacious ark. Still, at the time the calculating savants had fared little better than their pious antagonists. Their conviction that astronomy alone, unencumbered by the historical and cultural detritus that attach to language, symbol, and myth, could penetrate the remote past had collapsed before Champollion's decipherment of words that were never there.

The siren song of calculation, with its tempting Royal Road to the unknown past, continues to the present day, and the Dendera zodiac remains a favored attraction, as do the various monuments of Stone Age Europe and of South America. There is certainly warrant for pursuing astronomical connections since, after all, the stars and planets have always been there, mysterious and portentous presences throughout humanity's existence. The danger lies not in the judicious use of every tool available to the investigator of the remote past, but in too ardent an embrace of one above all others. Letronne put it particularly well. In 1837, reflecting back on the zodiac controversies, he wrote

> I think it superfluous to recall here the learned and consci-
> entious works to which the discussions of these monuments
> gave birth, those researches of the erudite, those extended and
> subtle calculations of the mathematicians, finally the lively
> controversy that agitated all of Europe to determine the epoch
> and the purpose of the zodiacs by means of astronomical char-
> acters that everyone strove to divine in them. It will suffice to
> say that all of the savants who took part in that memorable dis-
> pute, the defenders of the monuments' great antiquity as much
> as the partisans of a more restricted one, were about equally
> successful in proving their diverse opinions by combining the
> signs which they perceived to be there. The complete absence
> of fixed and determinate points on which everyone could agree
> excluded the possibility of a methodical and regular discus-
> sion. Each went ahead on his own, composing an hypothesis,
> or combating that of others, without worrying too much about
> the objections that his own was subject to in its turn. The spec-
> tators of this opinionated battle, utterly worn out by useless
> debates, developed in the end an unfavorable prejudice against
> all these attempts, and were disposed to apply Voltaire's words
> to the Egyptian zodiacs: "What can be explained in twenty dif-
> ferent ways merits no explanation at all."[8]

Letronne set astronomy altogether to the side where the zodiacs were concerned. If he erred too much on the side of history and phi-lology, his caution was understandable, as he pointed out, given the overwhelming number of conflicting calculations. Still, it may well be that the Dendera circular does encode some vestige of ancient skies as they were actually observed, since the circular's most recent inter-preter links planetary symbols on it to celestial configurations near the time of the temple's construction.[9] Or, perhaps, the monument is just what Letronne and Champollion thought it to be, a collection of reli-gious and astrological symbols from the heavens.

Though the evidence that Champollion had so successfully, and influentially, used to date Dendera simply did not exist, he was in the end right about the temple's date. Years later the reason became clear. The monument had been constructed during the interregnum between the death of Cleopatra's father, Ptolemy Auletes, in 51 BCE, and the coregency officially established in 42 between Cleopatra and her five-year-old son Caesarion by Julius Caesar (Figure I.8).[10] Built during the interregnum, the Dendera zodiac's empty cartouches, like those of every monument constructed during that period, forever awaited royal names. And so, in a final irony to this story that has so many, it is the very absence of words that dates this protean monument.

ACKNOWLEDGMENTS

This book began with the discovery of the *Zodiaque* collection mentioned in the introduction. A brief article about the subject by Jed Buchwald appeared shortly afterward. Later he invited Diane Greco Josefowicz to join him in further exploring the fate of the Dendera zodiac. She had obtained her Ph.D. degree at MIT under his supervision with a dissertation entitled "Gauss and the Study of Terrestrial Magnetism at Göttingen, 1818–1839" and has since become a short-story writer and novelist. The *Zodiac of Paris* is a thoroughly collaborative effort, since each of us commented on and edited the other's work in every chapter, in most cases several times.[1]

Our story involves several subjects that are not usually brought together, which made collaboration not merely a pleasure but a near necessity. As we pursued the zodiacs across time and place, we encountered antireligious ferment during the French Revolution, cultural clashes during Napoleon's campaign in Egypt, attempts by technically minded engineers and astronomers to expropriate ancient history, reactions by the religious Right in the early empire and during the restoration to attacks on Mosaic chronology, attempts to control the press, archaeological vandalism for profit and national glory, salon and literary cultures of the times, personal antagonisms and surreptitious plots, intricate calculations of the heavens, shifting institutional structures and alliances, the politics of vaudeville during the restoration, newspaper and print culture, philological judgment in conflict with astronomy and statistical argument, and other things besides. We have tried to present as much of these several aspects as we thought essential for understanding the power of the zodiacs in early nineteenth-century France.

Many individuals have given us advice and assistance, while two readers of the entire manuscript provided insightful remarks and suggestions. Darius Spieth's comments were exceptionally thorough, helpful, and astute. Over the years we have profited from discussions with

Charles Gillispie, Tony Grafton, Paul Legutko, the late John North, and Martin Rudwick. Nicholas Christopher, Linda Josefowicz, Matthew Josefowicz, Michael Josefowicz, and John Glusman all read portions of the manuscript in its earliest stages and responded with wise counsel and encouragement. In Paris, Chantal Grell helped locate materials at the Archives Nationales; her work on Nicolas Fréret has illuminated the world of eighteenth-century French philology. At Caltech, Moti Feingold and Noel Swerdlow helped enormously, while Mac Pigman offered his superb knowledge of Latin, Greek, and ancient texts. Diana Kormos-Buchwald provided the invaluable benefit of her critical eye as editor-in-chief of the *Einstein Papers*.

Our work would not have been possible without the assistance of Emily de Aranjo at Caltech, and of the Caltech Library staff, who reached far and wide to obtain many of the materials that we needed to complete our research. At the John Hay Library, Brown University, William S. Monroe and Ann Morgan Dodge helped locate images of the zodiac in obscure volumes. Anne Watanabe-Rocco and Robert Montoya of the Charles E. Young Research Library, UCLA, obtained high-quality images from the *Description* for us. Thanks are additionally due to Mary Warnement and Jayne Guidici at the Boston Athenaeum.

In France, Michael and Linda Josefowicz provided childcare and warm hospitality, as did Jean-Marc Martin and Nadine Frey. Jane Josefowicz politely endured numerous events in which the conversation strayed to zodiacal subjects unexciting to a five-year-old, even if she is a Euhemerist, and enlivened those occasions with her wit. Matthew Josefowicz knows better than anyone what this book's completion means to one of its authors; she hopes he likes the book enough to stick around for a few more. The Buchwalds visited Egypt twice, during which Jed was able to photograph the Dendera temple, and, thanks to its discovery there by Diana, the home of the Institut d'Égypte in Cairo.

We also thank Peter Dougherty and Rob Tempio at Princeton University Press for their interest in this unusual project.

NOTES

Introduction

1. Recent popular works concerning the Egyptian expedition and the savants include Burleigh (2007), Cole (2007), Meyerson (2004), Russell (2005), and Strathern (2008). The best modern treatments are Bret (1998), Laurens (1997), and especially Laissus (1998). Earlier sources include Charles-Roux (1910; 1925; 1934; 1935; 1937) and Jonquière (1899–1907).

Chapter 1: All This for Two Stones?

1. The *Minerve* counted among its founding editors Antoine Jay, who from 1803 through 1809 had served as tutor to the three sons of Joseph Fouché (whom we will encounter again, for he developed and twice ran the Ministry of Police until his disgrace and replacement by Savary). For more on both, see Hatin (1866).

2. Manley and Rée (2001, 194); Fagan (1975, 262 n. 6); for the final detail about Boghos, see Minutoli (1827, 29).

3. Demotic was a known but also undeciphered Egyptian script.

4. One of the Alexandrian obelisks, or "Cleopatra's needles," that was conveyed to London in 1802 qualified its gentlemanly tribute to the valor of Napoleon's defeated troops with the remark that "under Divine Providence it was reserved for the British Nation to annihilate their ambitious designs." The inscription was widely repeated by nineteenth-century British writers on Egyptology, including the physician Erasmus Wilson, who reprinted it in Wilson (1877, 185).

5. Saulnier (1822, 12).

6. Ibid., 11.

7. One much later, and rather suspect, account attributed the discovery to Desaix himself. See Roy (1855, 134).

8. Any book written today that follows the vicissitudes of an Egyptian artifact's meaning in the imaginations of modern Europeans must

necessarily, if not always comfortably, recognize Edward Said's *Orientalism* (1978). Despite its many flaws, *Orientalism* transformed a previously inchoate body of source material (Europeans' literary and popular accounts of their encounters with peoples, cultures, and landscapes beyond Europe) into a locus of legitimate scholarly inquiry. Since 1978 scholars have offered vital correctives to Said from a spectrum of intellectual and political positions, including Bhabha (1986); Clifford (1988); Irwin (2006); Lowe (1991); MacKenzie (1995); and Varisco (2007). A fair appreciation of these arguments would require a book of its own; too many have perhaps already been written. Here we note only that after three decades the debate still raises issues of authority: who is authorized to say what about encounters between East and West, what secures that authority, and what forms should such utterances take? Ironically, these very questions are also central preoccupations of our story's characters, for whom authority and dogma intersected in manifold ways. Recent scholarship has emphasized the variety, subtlety, and ubiquity of preconceived European attitudes toward the peoples and cultures they encountered on their travels, and we have striven to avoid uncritically repeating those opinions in our account.

9. Marsot (1984, 36–59).

10. Ibid., 37, 70–71.

11. On Ali's early life, see Fahmy (2009, 1–9).

12. On Ali's rise to power, see ibid., 27–37.

13. His dynasty ran through King Farouk, his great-great-grandson, who was removed in Nasser's coup of 1952.

14. Lelorrain's trip to Egypt coincided with a specific transformation of European travel to the East. With improved traveling conditions enabled by new rail and steamship routes, mass tourism began to replace the aristocratic grand tour. Travel narratives changed in tandem with this development. (Indeed, the word *touriste* appears for the first time in French in 1816.) Romantic accounts marked by the narrator's heroic exploits in a faraway land fell out of favor, to be replaced by travel guides written for readers who might actually make the trip. Volney's *Voyage en Syrie et en Égypte* (1787), about which we will have more to say in subsequent chapters, is typical of the earlier sort of book. In contrast, later travel guide writers oriented their descriptions toward the middle-class traveler's practical interests, including (and perhaps especially) how to obtain antiquities

to take home as souvenirs. In our use of travel narratives by early nineteenth-century Europeans in Egypt, we have tried to be attentive to this shift in tone and audience, following the work of Behdad (1994, esp. 35–52). (Behdad makes the point about *touriste* on p. 35.)

15. Mountnorris (1809, 3:368–69).

16. Madden (1840, 11).

17. Montulé (1821, 2:348).

18. Manley and Rée (2001,78–79).

19. Saulnier (1822, 9).

20. Manley and Rée, (2001, 78–79).

21. Montulé (1821, 348).

22. The "Vras! Vras!" episode is recounted in Manley and Rée (2001, 77–78). Their account is based on Madden (1840) and on an eyewitness report by Gally Knight sent to Stratford Canning, a diplomat in Constantinople who would later become ambassador to the Ottoman Porte.

23. Marsot (1984, 1–23).

24. On Ali's transformation from small-time governor to statesman, see Fahmy (2009, 39–60). On his reliance on foreigners to undertake and manage important projects, see Marsot (1984, 162–95).

25. Ali's mistreatment of ordinary Egyptians is detailed in Marsot (1984, 100–136); and Fahmy (2009, 104–8).

26. For these and other details about Drovetti's life, see Ridley (1998).

27. Jean Du Boisaymé to Drovetti, April 15, 1820, ep. 112, reprinted in Drovetti (1985, 143–44).

28. Quoted in Ridley (1998, 251).

29. Montulé (1821, 249).

30. Ibid., 194–95.

31. Forster (1961 [1922], 156). Forster is also quoted by Salt's biographers, Manley and Rée (2001, xiv).

32. Manley and Rée (2001, 5).

33. Quoted in Greener (1966, 120).

34. Quoted in Manley and Rée (2001, xiv).

35. Ridley (1998, 83–85, 124).

36. Henniker (1823, 5).

37. Minutoli (1827, 149).

38. Rifaud (1830, 116–17).

39. Montulé (1821, 285).

40. Saulnier (1822, 8).

41. Ali is quoted in Rhind (1862, 263).

42. Whatever its merits, the complaint was undercut by its Euro-centric bias, since the scientific knowledge in question was conceived as produced by universities and museums in Europe, rather than in the countries where the artifacts originated.

43. Ibid., 265–66.

44. Quoted in Greener (1966, 109).

45. Minutoli (1827, 5–8).

46. Lane, "Notes and views in Egypt and Nubia made during the years 1825–1828" [British Museum Add. MS 34080], quoted in Ridley (1998, 27).

47. Manley (1998, 102).

48. Paris at the time had about half a million people.

49. Quoted in ibid., 98, 100.

50. Saulnier (1822, 50).

51. Through the end of this section, the details of Lelorrain's expe-dition are taken from Saulnier's 1822 account, with exceptions as noted.

52. Ibid., 35.

53. Greener (1966, 132).

54. Saulnier (1822, 32).

55. Ibid., 39.

56. Ibid., 48, 56.

Chapter 2: Antiquity Imagined

1. The *Ezourvedam* is reprinted with an introduction in Rocher (1984a).

2. On which see Mohan (2005); Rocher (1984b); Schwab (1984, 152–55).

3. The elements of the theory are first spelled out in the *Eclair-cissemens* to Bailly (1775), without the more elaborate details that were to follow. At this point Bailly confined himself to locating Atlan-tis in the far north as the origin. On Bailly, see Sonenscher (2008, 251–64), who emphasizes his connection to Court de Gébelin (who is mentioned below); and Smith (1954).

4. See the posthumously published Bailly (1799, 2:1ff).

5. Chapin (1970).

6. Carlyle (1956 [1837], 542).

7. Bailly (1779, 1–2).

8. Ibid. A portrait of Madame du Bocage appears in Jackson (1880, 2:257–59). Although Voltaire was evidently quite taken with the widow, Jackson is less persuaded of her charms. After this Bailly wrote nothing more about antediluvian antiquity, and these letters were printed only after his death. There are several possible reasons according to Sonenscher (2008, 255), including Bailly's growing public persona in the last years of the ancien régime, but there is apparently no specific evidence to settle the question. We will see in what follows that the issue of antediluvian civilization was far from closed down at the time, despite arguments advanced by the Englishman William Jones on the comparative youth of the Indian zodiac's age. (Bailly had deployed the great antiquity of their zodiac in his own arguments, cf. Jones et al. [1792], cited as a possible reason for Bailly's subsequent silence by Sonenscher.)

9. Baudin gave a discourse on Bailly's essays to the Conseil des anciens (Baudin 1799).

10. Michaud (1843a, 18:126–27).

11. Guignes (1785 [1809]).

12. On Izouard, see Malandain (1982); Spinelli (1984).

13. Darnton (1982, 10) on Izouard. Also see Mornet (1933).

14. On this, see Manuel (1959, 25–39).

15. Cited and discussed in Grell (1993, 112).

16. Brosses (1750).

17. Manuel (1959, 193–95); cf. Brosses (1760).

18. Quoted in Grell (1993, 114); cf. Boulanger (1766, 5–6).

19. Darnton (1995, 20).

20. Boulanger (n.d.).

21. Rudwick (2005, 186).

22. Ibid., 181.

23. Ibid., 236.

24. Ibid., 237.

25. McMahon (2001).

26. McMahon (1998, 81ff).

27. McMahon (2001, 43); cf. Noé (1785).

28. On Volney, see Gillispie (2004, 514–18); Gaulmier (1951).

29. Volney (1787, 8). Volney conflated the Ottoman sultan's personal guards—the Janissaries—with the substantially independent Mamelukes, though both were in origination slave boys captured or sold, for the most part in the Caucasus and converted to Islam.

He certainly knew the difference, because the Mamelukes, he later remarked, had expanded their power in the half century or so before his visit.

30. Ibid., 12.

31. Ibid., 365.

32. The first edition is Volney (1791). Translated in Volney (1890 [1802], 110).

33. Volney (1890 [1802], 113).

34. Ibid., 223.

35. Priestley (1904 [1809], 48). His letters to Volney were printed in Priestley (1797).

36. Volney (1799, 444).

37. For Dolomieu and the translated quotation see Rudwick (2005, 320). He was a member of Napoleon's Egypt expedition, though illness forced him to leave Egypt in March 1799, after which he was captured and held prisoner under horrific conditions, eventually to be released after Napoleon's victory over the French and Italians at the Battle of Marengo.

38. On Volney's life, see Sibenaler (1992).

39. Baltrusaitis (1967, 40). Cf. Robinet (1896–98, 2:455) for the full text of the order of 2 Brumaire. The order for destruction had first been issued the previous August (Guilhermy 1855, 27).

40. "Anaxagoras" because the pre-Socratic philosopher, a materialist, had been charged with lack of respect for Athenian religion.

41. Aulard (1909, 54–55).

42. Ibid., 56–57.

43. Carlyle (1956 [1837], 554).

44. Aulard (1909, 67).

45. Guilhermy (1855, 27); Aulard (1909, 64).

46. Delaulnaye (1791); Delaulnaye and Blond (1792); cf. Rétat (1999) for details. Delaulnaye likely learned either directly from Dupuis or indirectly through Gaspard-Michel Leblond, who had been instrumental in assisting Dupuis' entry to the Académie.

47. Aulard (1909, 82).

48. Gifford (1797, vol. 2).

49. In the words of Mouna Ozouf (1998, 107), Robespierre was a man "in whom the reflections of Rousseau nourished a spontaneously religious outlook."

50. As originally argued by Aulard (1909) and further developed in Ozouf (1988, 97–102), the imagery even of the Festival of Reason

was ambiguous and elsewhere frequently involved syncretism with an underlying impulse to reproduce a deity, albeit in a new form, mostly as a female goddess.

51. Baudin (1795, 41–42).

52. Ibid., 77–80, for the text of the Year 3 decree. The Year 3 Constitution is reproduced in translation in Anderson (1904, 212–54). Article 352 specifies that "the law does not recognize religious vows nor any obligation contrary to the natural rights of man"; article 354 provides that "no one can be prevented from engaging in the worship which he has chosen, while he conforms to the laws." No mention of atheism appears.

Chapter 3: The Origin of All Religions

1. There are several short biographies, most of which copy, occasionally verbatim, from Dacier (1812): cf. Dupuis (1795a; 1822a) for others published before the entry in the 1855 Michaud *Biographie universelle*. See especially the memoir by Dupuis' wife (Dupuis 1813). On Dupuis, see Rétat (1999; 2000; 2001; 2002).

2. The Collège d'Harcourt had itself taken a stance in opposition to Jesuit influence on instruction.

3. Bailey (1979, 333–35); Palmer (1975, 25).

4. The Collège Royale was renamed the Collège Impériale under Napoleon, and in 1870 the Collège de France. However it was often referred to as the Collège de France even in the 1830s and 1840s, e.g., in Carteron (1843).

5. For an amusing account of Lalande, see Alder (2002, 76–81).

6. Cited from Dupuis (1813) in Rétat (2001, 25).

7. Dupuis (1781a).

8. Newton (1728).

9. Galasière (1782; 1784; 1785).

10. Dupuis (1822b, 6).

11. Goudemetz (1796) provides a contemporary list of the regicides.

12. Dupuis (1822a, 11).

13. Dupuis (1806, 97).

14. Cited and translated in Gillispie (2004, 293).

15. Dupuis (1806, 94).

16. Gébelin (1773, 176ff). On Gébelin, see Sonenscher (2008, 251–60), who notes that Gébelin ran a virtual research group "funded, in

part, by Physiocracy's cofounder, the marquis de Mirabeau" (260). Gébelin died while being treated for illness by mesmerism.

17. See Macrobius (1969, bk. 1, chap. 17, no. 61) for the associations with Cancer and Capricorn.

18. Dupuis (1781b, 435).

19. Ibid., 439.

20. Ibid., 378, for his first suggestions on how to avoid too great an antiquity, and p. 420, for the quotation.

21. On which see Ptolemy (1998, 404–7).

22. A separate volume of plates was printed in 1795 together with Dupuis' magnum opus, and it contained an advertisement after the title page for a precession globe built under Dupuis' direction by "Citoyen Loysel." It could be bought at Loysel's residence, Rue du Platre-Jacques, no. 9, first floor on the right (Dupuis 1795b).

23. The poem critiqued by Hipparchus was Aratus (1997). His discussion of it was first printed in Hipparchus (1567). The version used by Newton was contained in Petau (1630).

24. The first major account of Newton's chronology and its effects was Manuel (1963).

25. Dacier (1812, 10).

26. Le Coz and Roussel (1900, 1:149).

27. Ibid., 154.

28. Ibid., 209.

29. Lalande (1795) (30 Fructidor Year 3).

30. On the idéologues, see Duzer (1935); Gillispie (2004, 600–11); and especially Moravia (1974).

31. Destutt de Tracy (1799); Dupuis (1797; 1798).

32. On which see Gillispie (2004, 516–18).

33. Volney (1826 [1795]).

34. Laurens et al. (1989, 36).

35. Ibid., 42.

36. Voltaire (1753, 35). Authors' translation.

37. Laurens et al. (1989, 29).

38. Translated in Feldman and Richardson (1972, 286–87).

39. Quoted in Cole (2007, 141) from Bousquet (1968, 110n).

Chapter 4: On Napoleon's Expedition

1. The crocodile incident is recounted in Denon (1802a, 1:278–79). Details about the division with which Denon was traveling may be found in Charles-Roux (1937, 200–201). Denon's life has been amply

documented; see La Fizilère (1873); de Ris (1877); Leliévre (1942); Chatelain (1973); Ghali (1986); Leliévre (1993). In English the pickings are slimmer, but see Nowinski (1970).

2. Denon (1802a, 1:278).

3. Russell (2005, 202).

4. Denon (1802a, 1:278).

5. Darnton (1982) discusses this aspect of the era's print culture.

6. Nowinski (1970, 71); Chatelain (1973, 29).

7. Nowinski (1970, 30).

8. Ceram (1979, 88).

9. Nowinski (1970, 69). Nowinski's account is based on the story Denon told late in his life to his friend Pierre André Coupin, to whom he admitted that the memory was a "nightmarish recollection." See Coupin (1825).

10. Nowinski (1970, 81), from Savary (1828, 1:80–81).

11. Charles-Roux (1937, 220).

12. Nowinski (1970, 82), from Denon (1802a, 2:vii).

13. All three accounts are given in Nowinski (1970, 75–78), from de Ris (1877) and Leliévre (1942).

14. Denon (1802a, 1:279).

15. Ibid., 278–79.

16. Russell (2005, xviii).

17. Gillispie (1989, 448). Our discussion of the scientific aspects of Napoleon's Egyptian expedition relies heavily on Gillispie's clear and detailed account.

18. Byrd (1992, 53–55).

19. Gillispie (1989, 448).

20. Brégeon (1998, 259).

21. Ceram (1979, 88).

22. Bourrienne (1895, 142ff).

23. Ibid., 151. Jupiter was said to have been born on the mountain of Dicte in eastern Crete.

24. Bret (1998, 32–34).

25. Quoted in ibid., 34.

26. Charles-Roux (1937, 213).

27. The full list of books that Napoleon ordered for the expedition was sent on 8 Germinal VI (March 28, 1798) by Cafferelli to one J-B Say, "homme des lettres," along with permission to spend up to 10,000 francs for their procurement. The letter is reprinted in Keller (1901, 4:121–22). See also Rose (1902, 185); Byrd (1992, 45).

28. Bret (1998, 28).

29. Denon (1802a, 1:7).

30. Saint-Hilaire (1901, 34–36), letter 8; the quotation appears on p. 34.

31. Ibid., 36. Saint-Hilaire reported that, in the end, the bickering proved so intractable that Napoleon was forced to revoke his decision.

32. Laissus (1998, 63).

33. Bret (1998, 206); Byrd (1992, 78–79).

34. 6 Fructidor Year 6, according to the Revolutionary Calendar.

35. Gillispie (1989, 448–49).

36. Ibid., 449–50.

37. Laissus (1998, 110–16) reconstructs the institute's first days in detail. Sadly, virtually all of its records and minutes have been lost. A single *procés-verbaux* exists in the library of the Académie des Inscriptions et Belles-Lettres; additional facts must be gleaned from memoirs and volumes of personal correspondence like Saint-Hilaire's, as well as the institute's official publications, the *Courier d'Egypte*, the *Décade égyptienne*, and the *Mémoires sur L'Egypte*.

38. Gillispie (1989, 449).

39. Goby (1953, 95–96).

40. Goby (1987, 7).

41. Bret (1998, 208–14).

42. Al-Jabarti (1997, 109).

43. On the contemporary restoration of the Sinnari house, see Brégeon (1998, 269).

44. Saint-Hilaire (1901, 74), letter 17, 6 Fructidor Year 6 (August 3, 1798).

45. Bret (1998, 208–19).

46. Al-Jabarti (1997, 109–10); Byrd (1992, 96 n. 31). On al-Jabarti and the French, see Wielandt (1997).

47. Bret (1998, 187–89).

48. Gillispie (1989, 455); Bret (1998, 186).

49. Al-Jabarti (1997, 110–11).

50. On which see Gillispie (1989, 451–52).

51. Iversen (1993).

52. *Supper at Beaucaire*, a pamphlet written as a dialogue between a Napoleonesque soldier and two Marseilles merchants, urged an end to the civil war in the south of France.

53. The *Décade philosophique* not only recorded and publicized Italian artworks selected for transport to Paris after the Treaty of Campo Formio but also, perhaps disingenuously, devoted space to discussions about the propriety of doing so.

54. Much of what appeared in the *Décade* was also printed in the four-volume *Mémoires sur l'Égypte* (1799–1802), published in Paris by P. Didot. The *Mémoires* contained, in addition, other papers by the institute's members. See Laissus (1998, 186–89).

55. Hanley (2005, 55–71).

56. Charles-Roux (1937, 222–23).

57. Nowinski (1970, 83–85) . Denon's complaint about the lack of sturdy drawing surfaces appears in the preface to Denon (1802a, 1:xix).

58. Charles-Roux (1937, 222–23). Denon's commitment to drawing his subjects without artifice went only so far. It did not extend, for instance, to the inclusion of much detail about the lives of everyday Egyptians, which troubled otherwise sympathetic reviewers. See Spieth (2001, 480).

59. Spieth has uncovered exactly how and when Denon's *Voyage* was serialized. It appeared in six installments of the *Moniteur* in August and September 1802. See Spieth (2001, 497).

60. Ceram (1979, 90).

61. Denon (1802a, 1:180). Denon's account of his second trip to Dendera appears in ibid., 280–84.

62. Denon's reflections upon sketching the temple at Dendera, from which the quotations in this paragraph and the three subsequent paragraphs are taken, appear in ibid., 280–82 (with exceptions as noted).

63. Ibid., 96, 282.

64. Ibid., 269.

65. Ibid.

66. Al-Jabarti (1997, 38). Denon reports similar atrocities in the *Voyage*, though he is careful to leave Napoleon out of it; see Spieth (2001, 502–6).

67. Al-Jabarti (1997, 35).

68. Ibid., 67, 117–18.

69. Cole (2007, 78–79).

70. Russell (2005, 109).

71. Al-Jabarti (1997, 66, 70, 106–7).

72. Russell (2005, 98).

73. For a firsthand account of the conflict, see Jollois and Lefèvre-Pontalis (1904); Byrd (1992, 100). Denon's remark appears in Denon (1802a, 1:105–6).

74. Brégeon (1998, 256 n. 6).

75. Laurens et al. (1989, 225).

76. Spieth (2001, 511).

77. Russell (2005, 249).

78. Goby (1987, 84), meeting 58, February 8, 1801; Jollois and Lefèvre-Pontalis (1904, 21).

79. Bret (1998, 208–9).

80. Nowinski (1970, 108–9); the contents of the reliquary are listed in Richard-Desaix (1880).

81. Details of al-Jabarti's later years are given in al-Jabarti (1979, 13).

Chapter 5: One Drawing, Many Words

1. Before the revolution and after the restoration, Devilliers was known as Villiers du Terrage.

2. Devilliers had actually passed his École Polytechnique exit examinations (in Cairo) just months before.

3. Jollois and Lefèvre-Pontalis (1904, 12); Villiers du Terrage and Villiers du Terrage (1899, 213). See also Laissus (1998, 295).

4. The letter might have been carried on the *Muiron*, or it may have reached Paris later, in 1800, due to the tremendous difficulties of communication between Egypt and France at a time when the English navy, having destroyed Napoleon's fleet at Aboukir Bay barely a month after its arrival, prowled the Mediterranean in search of French frigates. We know that Louis Ripault, who had been elected by the Institut d'Égypte as its librarian, and who returned with Napoleon to France, carried letters back with him, among which were several from Devilliers to his family (Villiers du Terrage and Villiers du Terrage [1899, 111]).

5. Jollois and Devilliers were still in Upper Egypt, having reached Esneh only thirteen days before Coraboeuf sent his dispatch to Lalande.

6. The circular zodiac is plate 48 in the now rare atlas of Denon's *Voyages*; the rectangular ones at Dendera are depicted in plate 49. Fourier had certainly seen Denon's drawings; see Jollois and Devilliers (1809a).

7. The drawing of the Dendera rectangulars done a few months later by Jollois and Devilliers, which was eventually engraved for publication by the talented Charlin (who between 1802 and 1816 also engraved the magnificent floral paintings by another member of the expeditionary force, Redouté), has all the zodiacal figures facing correctly. But their depiction would not have been available to anyone

outside Egypt until the winter of 1802 at the earliest, and it was not published until 1817, though it was certainly printed as an unbound sheet before then.

8. Grobert (1800, 114–15).

9. There were in fact two rectangular zodiacs near Esneh: one at the main temple in Esneh proper (the subject of Coraboeuf's letter), and another in a smaller temple at Dair, just north of the town. At the time of Denon's visit, the smaller, northern temple was already in poor condition, while the one at Esneh lay somewhat below grade (Figure 5.5). Denon did not print the zodiac sketches from his visit to either one, though he must have either shown them to Coraboeuf or described them to him. The smaller one, in any case, was described as badly deteriorated by 1818. Even at the time of Coraboeuf's letter, it was missing about a third of the lower panel, which has today vanished altogether. We know what was there only because of Jollois and Devilliers, who sketched both ceilings during their travels in Upper Egypt. The order of zodiacal symbols in the two rectangulars near Esneh was in any case the same, and even though the one at Dair was incomplete, it seemed that it, too, began with Virgo. The depictions done by Jollois and Devilliers show the similarity; further, unlike in Denon's drawing of the Dendera rectangulars, the figures in their sketches all face in the correct direction. Of course, Coraboeuf could not have seen these drawings at the time of his letter to Lalande. See Richardson (1822, 315) for the zodiac at Dair two decades later. See Figure 8.6 for the engravings of Jollois and Devilliers' depictions of the zodiacs at Esneh.

10. Coraboeuf's letter was included in an appendix to Denon (1802a, 2:xci), but it had become available to the reading public almost at once, in Grobert (1800, 115).

11. For a discussion of these matters, see Alder (1997).

12. Langins (1987).

13. Monge (1798, 2).

14. On the art of the *Description*, see Prochaska 1994.

15. Pinpointing the chronology of these exchanges raises daunting problems, given the large but fragmentary and scattered records, but in this case there is clear evidence: see Denon (1802a, lx–lxiiff), together with Lalande (1803b, 263–65). The latter contains material that appeared originally in Lalande (1800, 36).

16. Grobert (1800, 118).

17. Ibid., 119.

18. Champollion-Figeac (1844, 50–51).

19. Garat (1800, 68). Friend of Barras and flatterer of Robespierre until Thermidor, he was a less than penetrating litterateur.

20. Champollion-Figeac (1844, 51).

21. Lalande and Maréchal (1803, 76–77).

22. Lalande (1803a, 839–40).

23. Lalande (1803b, 264–65).

24. Jollois and Lefèvre-Pontalis (1904, 26).

25. Bernard (1802, 581–82).

26. Bernard (1812a; 1812b).

27. As noted by Fourier's close friend, Champollion-Figeac (1844, 7).

28. Fourier (1800).

29. Bernard (1802, 582).

30. Fourier's letter to Berthollet was also printed at the time in Millin's comprehensive *Magasin Encyclopédique*, 7:199—see Paravey (1821, x–xii).

31. Champollion-Figeac (1844, 51–52).

32. Jollois and Lefèvre-Pontalis (1904, 180).

Chapter 6: The Dawn of the Zodiac Controversies

1. Clement XIV's religious toleration did not extend far, but it was positively enlightened compared to that of his predecessors, who had confined Jews to ghettos in the Papal States in 1555; the oppression went on from there. Clement took control of the ghetto from the Holy Office of the Inquisition and reassigned it to the cardinal vicar of Rome. He also permitted Jews to become doctors and artisans. Pius VI rescinded all of that and reinstituted as well the requirement that Jews periodically attend harangues by priests urging them to convert.

2. Kertzer (2002, chap. 1).

3. Napoleon (1846, 203, 248).

4. Desmarest (1833, 75–76), including the remark by Napoleon.

5. Cabanis (1975, 211).

6. Staël (1843, 177).

7. Staël (1818, 2:277).

8. A Dutch translation was also produced in Holland.

9. The list follows p. 283 in vol. 3 of the first French edition (Denon 1802b). It was omitted from the English editions.

10. Wilson (1803, xix–xx).

11. Visconti (1802, 567–76).

12. Chigi et al. (1775). The ostensible author was a playwright by the name of Gaetano Sertor. The play was incinerated in the Piazza Colonna (Collins 2004, 299 n. 9), though it was sufficiently popular to have generated a second Roman edition in 1775 (pub. Poggioli) and to be translated into French, German, and Dutch.

13. His major effort before the Herodotus was a nearly 400-page tome on the history of the goddess Venus (Larcher 1776).

14. Voltaire (1765).

15. Larcher (1767b).

16. Voltaire (1767).

17. Larcher (1767a).

18. Larcher (1769, 35).

19. On Larcher and Voltaire, see Moureaux (1974) and McGinnis (1997).

20. A contemporary obituary of Larcher portrays him as a sedulous defender of religious matters (Boissonade 1814). See also Moureaux (1974, 617–18).

21. See Moureaux (1974, 620–21).

22. Larcher (1829, 429), in the English translation.

23. Larcher (1802, 2:566).

24. On de Sacy and his influence, see Irwin (2006, 141–88).

25. Visconti (1802, 567–76).

26. Devilliers to Jollois in Jollois and Lefèvre-Pontalis (1904, 26).

27. This is perhaps more than conjecture because Visconti also reversed the order of the two panels as printed in Denon's *Voyage*, calling the one printed as upper the lower and vice versa, suggesting that he had seen a drawing that differed from the printed version. Still, the engraver would have to have reversed all the figures, or as many as Denon had correctly depicted.

28. Visconti (1802, 569). This is not strictly correct. For example, in 2000 BCE, the summer solstice occurred in the asterism Leo, while the autumnal equinox occurred toward the part of Libra that overlaps the claws of Scorpio.

29. Ibid.

30. Which of course presumes that the Egyptians could not have transmitted the zodiac to the Greeks.

31. We now know that, as a result, the Egyptians maintained a separate, lunar calendar that was kept in sync with the civil one by adding an extra month whenever the beginning of the lunar year would occur before the start of the civil year. In fact the Egyptians

had not two but three calendars, since they also kept the very first one that had been devised, which was also lunar.

32. Visconti (1802, 570). All dates that follow are Julian.

33. As corrected by Scaliger, Censorinus's date for the first day of the Egyptian calendar year (1 Thoth) occurred on July 20 in 139 CE— not 132 CE, though the difference is nugatory for present purposes. The dating of the civil calendar's origin has since early modern times relied on Censorinus's claim, which continues to raise complex problems. O'Mara (2003) continues the now centuries-long attempt to extract absolute dating information from Censorinus's remark.

34. Visconti did not say how he came to this conclusion. However, on Denon's drawing (Figure 4.3), the distance from the farthest part of Virgo to the circumference of the circle is greater than the distance from the farthest part of any of the other zodiacal symbols to it, with the exception of Cancer, which is in any event markedly out of place. So if the sequence begins with Virgo, it would thereafter almost (though not quite) seem to spiral outward.

35. So called because Julius Caesar learned of it while in Alexandria, though hardly from Egyptian priests, despite legends to the contrary. Roman priests used it without proper understanding until Augustus (Larcher apparently disliked this alternative). Visconti essayed a related argument, according to which the equinoxes and solstices correspond to the points of the "circle that most closely approach the four sides of the square." He did not specify which figures these were, though he did assert that it seems the "beginning of the zodiac is at Libra; which is to say at the one of these four points which is closest to the beginning of the year." Running counterclockwise from the top of Denon's sketch, it would seem that Sagittarius, Virgo, Gemini, and Pisces are the closest to the sides. Libra is physically closest to Virgo, and Virgo begins the year on Visconti's reasoning. Circa 150 CE the autumnal equinox had entered Virgo from Libra, while the summer solstice was just entering Gemini from Cancer.

36. See the excellent account of the period in McMahon (2001, chap. 4). Few if any among the returnees and the revivified apostles of religion accepted Napoleon's growing hold on power. Indeed, many hoped that he would be merely a stop-gap on the way to reintroducing the monarchy—a French General Monk, as it were.

37. Cited and translated in ibid., 127.

38. The characterization of the right made here derives especially from ibid., chap. 4. The quotation is translated on p. 139.

39. Voltaire had, we have seen, attacked Larcher, but Larcher's choice of words nevertheless seemed designed to stimulate the aged but still aggressive philosophe's pen.

40. Anonymous (1803, 3).

41. The dissertation appeared in French translation in 1807. Its later publishing history is a testament to the persistence of interest in the zodiacs at the highest levels of the Catholic hierarchy, for it was republished in 1822, again in French translation, at the height of the controversy provoked by the Dendera circular's arrival in Paris. This later, Genoese edition was dedicated to Luigi Lambruschini, the archbishop of Genoa from 1819. Lambruschini was also the domestic prelate for Pius VII (Barnaba Niccolò Maria Luigi Chiaramonti), who had been elected on March 21, 1800, at the conclave that began on November 30, 1799, following Pius VI's death in Valence the previous August.

42. Piolanti (1977, 24, 142).

43. Montani (1844).

44. Denon was himself intrigued by the Monte Bolca fossils; unlike Testa, he had visited, and drawn, Bozza's collection: see Panta (1998, 47–50). We thank Darius Spieth for the reference.

45. Gaudant (1999).

46. Ibid., 199, for Testa's remark, and 197 for Brochhi's.

47. Biot (1858, 453). On the virulently reactionary policies of Leo XII, see Kertzer (2002, chap. 3).

48. Testa (1802, 5).

49. Lalande (1800, 36), reprinted three years later in Lalande (1803b, 264–75).

50. Macrobius's "fact" about the ram can be found in the Roman rhetorician Claudius Aelianius's *De natura animalium* (written originally in Greek).

51. Macrobius (1969, 144), which is book 1, chap. 21. Testa cites chap. 18 instead, which does deal with the sun but not with Aries.

52. Testa (1807, 13–14).

53. Ibid., 12–16. It's unclear where Testa obtained his values for the difference between the sidereal and tropical years. In any case his calculation is at best correct only for the return of coincidence between the position of the sun among the stars and the beginning of a 365-day calendar—not for the return to coincidence of Sirius's heliacal rise.

54. Ibid., 16.

55. Ibid., 16–23.

56. Ibid., 29–34. Testa cited various authorities for the importance of Sirius in ancient Egypt, including Pliny, Porphyry, Plutarch, Theon of Alexandria, and Solon. None of these, however, explicitly links the heliacal rise of Sirius to the flood. Rather, they insist only on the importance of Sirius, which is, after all, the brightest star in the sky.

57. Testa's 1322 BCE is a Julian date that was based on Censorinus's *De die natali*, to which we will return in discussing Visconti below.

58. Testa (1807, 35–36).

59. To the effect that the Egyptians would add five days to the year "between August and September," as indeed did take place regularly after the reform of the calendar under Augustus.

60. Memoirs that did not do so, including ones on the antiquity of Egypt, passed muster nonetheless. For example, a French translation of Samuel Henley's brief for zodiacal youth, directed specifically against Denon and Fourier, was printed in April 1803 in Millin's *Magasin Encyclopédique* (Henley 1803 [Floréal Year 11]). Henley was a member of the London Society of Antiquaries who, born in England, became professor of moral and intellectual philosophy at William and Mary College in Virginia. He became embroiled in religious and political controversy and returned to England in 1775. Thomas Jefferson bought several of his books. In England itself, Testa's counterassault on zodiacal antiquity was publicized in the *Monthly Magazine* (Anonymous 1802b); the same volume contains Henley's remarks.

Chapter 7: Ancient Skies, Censored

1. There are many accounts of Fouché's career, including a novel by Stefan Zweig. See, among others, Arnold (1979); Cole (1971); Forssell (1925); Madelin (1960).

2. Cole (1971, 41).

3. On Fouché and the police under Napoleon, as well as Fouché's opportunism and early depredations, see especially Arnold (1979).

4. Welschinger (1882, 10ff). On censorship under the empire, also see Coffin (1917).

5. Castres (1810, 107).

6. Cabanis (1975, 197).

7. Laissus (1998, 319).

8. Ripauld (1800, 10).

9. Ibid., 33.

10. According to the editor of the *Mémoire*, Diesbach (1987), Louis enjoyed retrieving strawberries from Talon's bosom.

11. The fabrication was demonstrated by Norman Cohn and Richard Kieckhefer: see Cohn (2000, 187–93). On Lamothe-Langon, see Switzer (1955). Lamothe-Langon also fabricated spurious memoirs by Napoleon and even by Louis XVIII. In the latter he insinuated that the Dauphin had not been executed and was still alive, possibly in the person of a Prussian army deserter by the name of Naundorff (Cohn 2000, 190).

12. Diesbach (1987, 223–24).

13. Carnot (1837, 10).

14. A contemporary obituary of David Hume quoted the then forty-five-year-old Denina to the effect that, had Hume "not shown so much eagerness to insinuate his pernicious opinions, he would have escaped the just censures of the religious, added greater weight to his history, and rendered it at once more interesting and spirited." The obituary, dated August 25, 1776, is pasted into a copy of Hume's autobiography (1777) in the possession of the McGill University David Hume Collection (B1497 A3 1777, copy 3).

15. Madelin (n.d., 117–18).

16. Arnold (1979, 37).

17. Fouché (1903, 185–86).

18. Meurthe (1882, 136).

19. Desmarest (1833, 77).

20. See Byrnes (1991, 318–19).

21. Chateaubriand (1823, 302–3).

22. Chateaubriand (1877, 202).

23. Chateaubriand (1823, 135).

24. Ibid.

25. Enghien was incorrectly thought to be involved in a conspiracy to assassinate Napoleon. On Napoleon's orders, he was taken, tried, and executed, though Napoleon had meanwhile found that Enghien was not involved.

26. D'Hauterive (1908, 1:125, entry 391).

27. On Napoleon's complicated relations with authors, see Polowetzky (1993).

28. Staël (1818, 575–76).

29. Lalande and Maréchal (1803, iv); Diesbach (1987).

30. Welschinger (1882, 268–69).

31. D'Hauterive (1908).

32. Lalande (1803a; 1804 [Nivose Year 12]). The 1803 is much shorter than the 1804.

33. Lalande (1804 [Nivose Year 12], 365).

34. Champollion-Figeac (1844, 52–53).

35. McMahon (2001, 150–1).

36. On the history and fate of the Second Class, see Staum (1980; 1996).

37. These quotations are from the sympathetic biography by Pierre-René Auguisin in Dupuis (1822c).

38. Dupuis (1806, 56).

39. Dupuis (1822 [1806]).

40. Lalande (1803a, 878).

41. Dupuis (1822 [1806], 563ff).

42. Plutarch, *De Iside*, chap. 1, sec. 11.

43. *Stromata*, bk. 5, and *Georgics*, bk. 1, verse 217, which reads: "With the spring comes bean-sowing; thee, too, Lucerne, the crumbling furrows then receive, and millet's annual care returns, what time the white bull with his gilded horns, opens the year, before whose threatening front, routed the dog-star sinks."

44. Dupuis (1822 [1806], 545).

45. Ibid., 549.

46. Horapollo (1993, 73, bk. 2, chap. 3).

47. The colures are the two great circles drawn perpendicular to the equator that include either the solstices or the equinoxes. To assume that the ancient Egyptians deployed the colures requires assuming that they had, in effect, something very like the Greek geometrical model of the celestial sphere. Of course, Dupuis and many contemporary savants were convinced that the Greeks had learned astronomy from the Egyptians.

48. Dupuis (1806, 59).

49. Dupuis (1822 [1806], 566).

Chapter 8: Egypt Captured in Ink and Porcelain

1. The account of Fourier's early career that follows is drawn from three major sources, two modern and one written shortly after his death: Cousin (1838, 360ff); Dhombres and Dhombres (1989, chaps. 3–4); and Herivel (1975, part 1).

2. Adolphus (1799, 2:1–11) for an interesting contemporary biography by an Englishman.

3. On Ichon, see Bénétrix (1894, 47–78), which, however, makes no mention of the Fourier affair.

4. Maure reputedly told the Jacobins that lard sent to provision the army would serve nicely to "grease the guillotine," according to Fréron. Cf. Michaud (1843b, 334).

5. Cited and translated in Herivel (1975, 54–55).

6. Ibid., 57–58.

7. Jollois and Lefèvre-Pontalis (1904, 25–26).

8. Ibid.

9. Ibid., 26–27.

10. Ibid., 27–28.

11. The history of the publication plans that follows derives for the most part from Gillispie and Dewachter (1987, 23–29).

12. On Jomard's almost numerological attachment to cartography and measurement, see Godlewska (1999, 132–47).

13. Humbert, Pantazzi, and Ziegler (1994, 211).

14. See ibid., 220–25, for the material in this paragraph.

15. On which see Herivel (1975, 76–82).

16. Dhombres and Robert (1998, 288). The account that follows of Fourier's interactions with the brothers Champollion and of his life in Grenoble is drawn from Dhombres, from the three major biographies of Champollion (Champollion and Hartleben 1909; Faure 2004; Lacouture 1988), and from Champollion-Figeac's reminiscences (1844). See also the account in Adkins and Adkins (2000, 42ff).

17. Dhombres and Robert (1998, 295).

18. See Crosland (1967) for an extensive discussion of Arcueil.

19. A considerable literature exists on Fourier's work, and significant points of controversy concerning it remain. See especially Darrigol (2007, 401ff); Grattan-Guinness (1990, 2:587–632); Grattan-Guinness and Ravetz (1972); Herivel (1975, 151–91); Truesdell (1980, 46–78). Darrigol remarks the presence in older work by Lagrange of results that, given a particular condition, are entirely similar to Fourier's series.

20. Champollion-Figeac (1844, 76–88).

21. Ibid., 83.

22. Ibid., 92–93.

23. Ibid., 119.

24. Ibid., 55.

25. Ibid., 166, for the material in this paragraph.

26. Ibid., 54.

27. Ibid.

28. Champollion-Figeac (1806, 2).

29. Years later a critic of just about everyone who had ever thought the zodiacs to be representations of the sky also demurred concerning Champollion-Figeac's (and others') translation: the historical geographer Letronne dated the inscription to the thirtieth year of Augustus's reign, and not to his age (Letronne 1823, 155ff).

30. Champollion-Figeac (1806, 17).

31. Champollion-Figeac (1844, 85).

32. Ibid., 81.

33. A comparative list of the contents of both editions of the *Description* is provided in the indispensable Munier (1943).

34. Jollois and Devilliers (1809b); Raige (1809).

35. Fourier (1809, 803–24). We will refer to the more easily found, unaltered second edition, Fourier (1829 [1809], 1–42), printed by the publisher Panckoucke, which for this article of Fourier's (though not for his *préface*) is identical to the 1809 version.

36. The young Champollion, who had often discussed things with Fourier, remarked in 1814 in his first published work that insight into the zodiacs as then known would have to await Fourier's forthcoming work, which could not be the memoir we discuss here since that had already appeared (Champollion 1814, 231).

37. As indeed they do, given appropriate interpretations of the constellations. Fourier wrote nothing at all about that, but it is not hard to see what he had in mind since Dupuis had long before charted the path, provided that the associations are made between a rising constellation and the setting sun. See below for Jollois and Devilliers' associations, which would have been the same as Fourier's, and indeed they had been suggested as an alternative possibility by Dupuis himself, who had not favored the idea in his *Origine*.

38. Fourier (1829 [1809], 26–27, 35).

39. Ibid., 26.

40. See Young (2006) for the complexities involved.

41. For Biot's remarks, see Biot (1823, xxxi).

42. See Biot (1811, 2:326) for his discussion of the heliacal rise and the *arcus visionis*.

43. The Egyptians divided the seasonal year into three parts: *Akhet*, when the flood occurred, *Peret*, after its subsidence, allowing the land to be sown, and *Shemu*, when the crops could be harvested. On the Egyptian calendar, see Parker (1981); Clagett (1995).

44. There was another problem with Denon's depiction, though Fourier chose simply to leave it unremarked: namely, that in Denon's drawing Cancer appears in both panels. In fact the figure near the panel that ends with Leo is a scarab, whereas the figure in the panel that contains the solar rays is indubitably a crab. This might well have affected Dupuis' interpretation of the zodiac, but, unlike Fourier, he had only Denon's drawing to rely upon.

45. Fourier (1829 [1809], 18).

46. Ibid., 24.

47. Widely used in antiquity, the ancient cubit spans the length of a man's forearm.

48. Girard (1824 [1809]).

49. Fourier (1829 [1809], 30). The calculation is not trivial, which again proves Fourier's knowledge of computational astronomy.

50. Reybaud (1832, 159). However, in 12,000 BCE at Dendera, Sirius's heliacal rise took place about two months after the summer solstice, and so it would have been useless for the flood.

51. Jollois and Devilliers (1822 [1809], 476).

52. See the account in the introduction to Villiers du Terrage and Villiers du Terrage (1899).

53. Ibid., xxi–xxii.

54. Jollois and Devilliers (1822 [1809], 371).

55. Ibid., 366.

56. On which see Halma (1821, 34 n. 1).

57. Jollois and Devilliers (1822 [1809], 473–74).

58. Fourier also had it that the solstice corresponding to the Esneh zodiac took place near the center of the constellation Leo, albeit in his case because the inundation occurred in Virgo (the first constellation fully passed through after Sirius's heliacal rise in Leo). Jollois and Devilliers learned about this from Fourier after they had submitted their memoir, but in time to add a note to the effect that Fourier had used different reasoning but had come to the same conclusion about dates (ibid., 476 n.).

59. Ibid., 476–78.

60. Jollois and Devilliers (1821 [1809], 512–13).

61. *Courier de l'Égypte*, 26 Messidor, Year 7, 3–4.

62. Raige (1822 [1809], 392).

63. The Egyptian months were Thoth, Phaophi, Athyr, Choiak, Tybi, Mechir, Phamenoth, Pharmuthi, Pachons, Payni, Epiphi, and Mesore.

64. Outram (1984, 78).

65. Rudwick (2005, 353–56).

66. Outram (1984, 53).

67. Rudwick (2005, 356–63).

68. Quoted in ibid., 333; see his chapter 6 for a full account of De Luc.

69. Ibid., 367

70. Anonymous (1802a).

71. This and subsequent remarks by Cuvier are from the translations in Rudwick (1997, 174 and elsewhere), as noted.

72. Ibid., 260.

73. Translated in ibid., 239–40.

74. Ibid., 246, for the translation, here altered a bit to capture Cuvier's meaning that the last clause is the reasonable alternative to the first one. The relevant part of the original reads "précession des equinoxes, et ne réponde pas simplement au . . . " (Cuvier 1812, 106).

75. See, e.g., Cuvier (1825, 242ff). For Visconti, see 264–65. By 1825 Cuvier was well versed in the zodiac controversies that erupted following the circular's arrival in Paris.

76. Rudwick (1997, 247).

77. Dupuis, we have seen, had also noted the possibility of reducing the time period for Egypt drastically by invoking constellation risings as the sun set. Cuvier did not mention the latter point, which is not surprising since it was buried in Dupuis' monumentally boring account.

Chapter 9: Egyptian Stars under Paris Skies

1. Cited in Mansel (2003, 58), on whose excellent account of restoration culture and politics we have relied.

2. Fellowes (1815, 81–82, 92).

3. Scott (1816, 198).

4. Cited in Mansel (2003, 211).

5. Hobhouse (1816, 2:231).

6. Méjan (1820, 1:2). Méjan was a lawyer at the Royal Court, so his account is infected with the advocate's usual rhetorical pleading. Years later the son of Louvel's lawyer gave his own brief history. It includes a transcription of Louvel's statement, which was not communicated at the time to his lawyers (Bonnet 1864, 68–71).

7. McMahon (1998, 172).

8. Discussed and translated in Kroen (1998, 31–33ff).

9. Ibid., 32, 49–56.

10. Furet (2000, 296).

11. Mansel (2003, 211–12). Rauzan was first superior of the missions: see the adulatory Delaporte (1892).

12. Kroen (1998, 27).

13. Bonald (1819, 2:115–23). The italics are Bonald's.

14. Bonald (1802, 31).

15. Locré (1819, 65–67).

16. Ibid., 102–4.

17. Francoeur (1838 [1812], 298). The Paris editions were printed by Courcier (1812, 1818, and 1821) and by Béchet (1812), Bachelier (1837), and Mallet-Bachelier (1853).

18. Picot (1815, 373).

19. Picot (1816, viii).

20. Duclot (1824, xii).

21. He even adduced a passage, allegedly from Dupuis, that seems to make no astronomical sense whatsoever and that we have been unable to find: ibid., 151.

22. Ibid., 169, 198.

23. Ibid., 161.

24. Anonymous (1816, 257).

25. Ibid., 263.

26. For the events of the period, see Mansel (2003, chap. 6).

27. Anonymous (1822a).

28. Anonymous (1850, 143).

29. The trail to prove this is a bit complex but runs as follows. There were three editions in 1822, including a third and a fourth. The fourth edition notes that the zodiac had been bought for 150,000 francs. This means that the fourth appeared after April 4, since the letter that Saulnier wrote to the interior minister trying to push a sale to the government bore that date (Saulnier 1822, 67). The police had seized the *Abrégé* at Chasseriau's before March 16 (Anonymous 1822a, 155), which must therefore have been the third edition, and Chasseriau nearly immediately produced a fourth, evidently not fearing the consequences—which he suffered the following year.

30. Referring of course to communion (Dupuis 1822a, 293).

31. Ibid., 364.

32. The original pamphlet is now almost impossible to find, but it was soon reprinted and has been many times since. The original title was *Comme quoi Napoleon n'a jamais existé, ou Grand*

Erratum, Source d'un Nombre Infini d'Errata a Noter dans l'Histoire du 19e Siècle; the first reprint in French seems to be Pérès (1836). It appeared that same year in German translation, the next year in Polish and again in German. Although Pérès likely did not know it, Bishop Richard Whately in England had already written a satire doubting Napoleon's existence on evidentiary grounds, though his target had not been Dupuis in particular but contemporary religious skepticism (Whately 1819). Whately was nevertheless himself highly skeptical of mystical belief, in particular as represented by high-church Tractarians and his former friend, the Catholic convert Cardinal Henry Newman.

33. Pérès (1836, 36–37).

34. Saulnier (1822, 57).

35. Ibid., 60–61.

36. Only six years later, the same journey would be made by another celebrated Egyptian curiosity, a giraffe who walked the nearly 900 kilometers to the capital, through throngs of ardent fans, wearing specially made shoes and draped in a yellow blanket emblazoned with royal fleurs-de-lis and accompanied by none other than Saint-Hilaire, the famous zoologist and supervisor of the *ménagerie royale* (Dardaud 1951; Allin 1998, 137–69). The exotic charm of these objects and their itineraries should not obscure the fact that they were made part of a discourse about the East that, among other things, made the French invasion, in 1830, and subsequent occupation of Algeria seem necessary and inevitable (Said 1978, 122–23).

37. Lagier (1923, 57).

38. The complaints about the "ugly black stone" and the stormy chamber session are recounted in Champollion-Figeac (1844, 58) and reappear in Lagier.

39. Prendergast (1996, 75).

40. Mansel (2003, 282, 86).

41. Ibid., 295.

42. Prendergast (1996, 78).

43. Ibid., 93.

44. Bryan (2003, 183–85).

45. Ibid., 186.

46. Delvau (1860), quoted in Prendergast (1996, 3). Burton (1988) elaborates on the idea of Paris as a sphinx.

47. Mansel (2003, 291); Prendergast (1996, 86).

48. Jones (2004, 66); see also Chevalier (1973).

49. The narrative's success was ensured by the destruction of official documents concerning the campaigns in Egypt and Syria. Spieth (2007, 17) discusses Napoleon's destruction of these records.

50. Kroen (1998, 40–41).

51. Lindsay (2000, 476–78).

52. The letter was published in the *Revue des Deux Mondes* under the title "Du Vandalisme en France." See Rudolph (2006, 22).

53. Jones (2004, 267).

54. Ibid.

55. Kroen (1998, 50).

56. Jones (2004, 267).

57. Constant's quip is reproduced in Mansel (2003, 173). On the religious revival, see Jones (2004, 266–70).

58. Spieth (2007, 20–21).

59. Ibid., 109.

60. Ibid., 23–26.

61. Lenoir (1822). Lenoir dated the Dendera circular to 770 BCE (Baltrusaitis 1967, 30–43). On Lenoir, see Greene (1981).

62. Spieth (2007, 46, 55–56, and 162 n. 1, for the quotation).

63. According to Spieth, Denon only "frequented Sophisian circles" (2007, 161 n.7), while Sollers (1995, 34–35) claims Denon was a member of the order, based on the list given in Ligou (1987). The anecdote by Lady Morgan is related in Spieth (2007, 44–45).

64. Spieth (2007, 113) notes that the Compagnie Franche turns up repeatedly in memoirs by Sophisians.

65. Ibid., 212.

66. Ibid., 69, 174 n. 31.

67. Ibid., 116.

68. Ibid., 112.

69. McCormick (1993, 15).

70. Pao (1997, 29).

71. McCormick (1993, 144, 53–54).

72. Pao (1997).

73. Albert (1902, 280).

74. Kroen (2000, 173).

75. McCormick (1993, 113–15).

76. Albert (1902, 296).

77. Spieth (2007, 111).

78. Mitchell (1991, 4).

79. Spieth (2007, 115).

80. Ibid., 21–22.

81. Ibid., 66, 178 n. 3.

82. Ibid., 68.

83. Ibid., 122–23.

84. Letter from Emile Deschamps to Guilbert de Pixerécourt, May 4, 1841. Quoted in Pao (1997, 19).

85. Albert (1902, 303).

86. Quoted in Pao (1997, 130).

87. Spieth (2007, 126, 33–34).

88. Quoted in Pao (1997, 130–31).

89. Brisset, Ferdinand, and Théaulon (1822).

90. Ayers (1994, 48).

91. Galignani (1827, 653).

92. Wicks (1950).

93. Curmer (1839–1842).

94. On de Trie's *Othello*, see Spieth (2007, 125).

95. For a general study of the reception of Shakespeare in Paris in the nineteenth century, see Pemble (2005); on the fortunes of *Othello* in Paris, 93–117.

96. Haig (1964, 53); Jusserand (1899, 451).

97. While the printed version of the play published in 1822 contained no directions for the staging of *Othello*, a contemporary review of the play described this twist on the original (D'Argé 1824, 434–36: "Le Zodiaque de Paris").

98. Comment (1999, 57–58).

99. On the connection between the use of a chorus in Parisian theaters in this time and in Sophisian initiation rituals, see Spieth (2007, 125); on de Trie's adaptations of Shakespeare's plays, see McCormick (1993, 140–41). De Trie enjoyed mounting loose adaptations of Shakespeare, such as his *Macbeth* of 1817, a four-act pantomime with actors on horseback and a quick dispatch of Lady Macbeth.

100. Krakovitch (1985, 32); Pao (1997, 63).

101. Pao (1997, 72–73).

102. Brisset, Ferdinand, and Théaulon (1822).

103. Lamothe-Langon (1830, 1:47–48). Chazet was also the author of "Le fils par hazard, ou ruse et folie" (1809), which inspired Rossini's comic opera *Il Signor Bruschino*, and of a long, treacly poem celebrating the marriage of Napoleon to Marie-Louise, printed (despite Chazet's position in the restoration government) in 1821.

104. Pao (1997, 72–73).

105. "Censeur" appears (under a heavy black line) in the manuscript original on p. 14; "Frondeur" appears in the printed version on p. 12.

106. Although Saint-Martin gave a talk at Inscriptions et Belles-Lettres on February 22 which, in its printed version, quoted extensively from Lelorrain's account as provided by Saulnier (Saint-Martin 1822, 14–16). However, there is no doubt that Saulnier's account appeared after mid-April, in which case either Saint-Martin's passages from it were added after that date, which means the Inscriptions talk was printed even later, or Saint-Martin obtained a copy of the manuscript directly from Saulnier before it reached print. We will return to Saint-Martin below.

107. Chabert and Ferlus (1822).

108. Ferlus was also, however, the author five years later of a brief pamphlet on the giraffe *Zarafa* that was brought to Paris (Ferlus 1827); on *Zarafa*, see Allin (1998).

109. Saulnier (1822, 88).

110. Such as that the figures had a relief of about "5 or 6 millimeters."

111. Saulnier (1822, 11).

112. Dupuis (1822b).

113. Cuvier (1821). See also the differently titled Cuvier (1825).

114. D'Argé (1825).

115. D'Argé (1824, preface).

116. Ibid., 436.

Chapter 10: The Zodiac Debates

1. Saint-Martin (1822, 10). Castex had accompanied the expedition to Upper Egypt under the leadership of Costaz in the summer of 1799.

2. Laurens and Laurens (1995, 301): Kléber to General Lanusse, 19 Frimaire Year 7 (December 10, 1799): "I pray you therefore to send them [the marble] to me as soon as possible." Either Lanusse never sent the marble or it did not exist in the Alexandrian storehouses.

3. Anonymous (1825, 7–8).

4. Paravey's career is discussed in Drouin (1970).

5. Spitzer (1987, 41).

6. The professor in question was Lefébure de Fourcy, so-called— originally Louis Lefebvre, he had added the (false) claim of nobility following the restoration. He was apparently abrupt and domineering with students. On the contretemps, see Pickering (1993, 29–30).

7. Among other effects of these changes, Louis Poinsot, a liberal but frequently absent professor of mechanics, was replaced by the young, highly religious, and conservative mathematician Augustin Louis Cauchy, following the joining together of the classes in analysis and mechanics at the Polytechnique.

8. See the brief account in Belhoste (1991, 47–49).

9. LaMennais (1820, 2:iv–v). The Right was incensed by one professor in particular, François Andrieux. A lawyer, poet, and classical playwright, he was an apostle of Enlightenment views and strongly opposed to Romanticism. An anonymous author asserted of him in 1816 that his "hatred of religion exceeded all bounds. He cites the *Vicaire Savoyard*, the *Lettres persanes*, speaks of Voltaire and Volney, treats the early years of the Middle Ages as a period of barbarism, and praises the horrible correspondance of Voltaire. Frenetically impious, he indecently calumniates religion" (quoted in Pinet [1887, 96]).

10. He later claimed to have "upheld the rights of this useful establishment from [the otherwise oppressive actions of] a poorly enlightened minister," i.e., Viénot.

11. Paravey's increasingly bitter recollections are scattered, frequently reprinted and updated, through works printed from 1821 through 1835. Delambre's report was quickly printed by Paravey together with important additional remarks by Delambre that were not part of the official original in Paravey (1821). The report was preceded there by an "extract" of various earlier remarks about the zodiacs, and by an "aperçu" by Paravey himself of his unpublished memoirs. The next year he printed an "explication" that also considered Biot's interventions (Paravey 1822). All of this was republished years later, together with other material, by an angry and bitter Paravey, who took the opportunity somewhat to alter his original words without always signaling the changes (Paravey 1835b). That year, he also reprinted the same collection but included as well a refutation of Biot (Paravey 1835c). And six years after his presentations before the Académie des Sciences, Paravey published a work specifically on the origins of writing, which included a "rapid glance at the history of the world between the time of Creation and that of Nabonassar" (Paravey 1826).

12. Godlewska (1999, 220).

13. Herivel (1975, 118–24); and Dhombres and Robert (1998, 363–71).

14. Paravey (1835b), the *avertissement* following a reprint of the Delambre report.

15. Ibid., the introductory *avant-propos*.

16. Ibid., 3ff, the *avertissement*.

17. Ibid., 8, note by Paravey following the reprint of the official report.

18. Ibid., 7, note by Delambre to the report.

19. Ibid., 9.

20. See, e.g., the first note to the aperçu at ibid., 25, and (especially) Paravey (1826, v&ff), where he states unequivocally that all peoples had "their civilization from the same origin, and in the same country where Moses places the family of Noah after the Deluge."

21. Paravey (1826, 5).

22. Fourier (1818, 65–66).

23. Figure 10.5 was printed for the first time in this form in Biot (1823).

24. Paravey (1821).

25. Paravey (1835b, 22), of the Delambre report.

26. Paravey (1835d, 9), bound in at least one instance with Paravey (1835b).

27. Anonymous (1821, 70).

28. Reprinted in Paravey (1835a, 12–13).

29. Ibid., 19.

30. On Walckenaer's geographic interests in particular see Godlewska (1999, 273–75).

31. Saulnier (1822, 63–65).

32. Biot (1823, v).

33. Ibid., vi.

34. On Arago, see Daumas (1987). There is no book-length biography of Biot, but see Frankel (1972; 1978); Lapparent (1894); Lefort (1867); Martius (1862); Picard (1927); Sainte-Beuve (1864).

35. Arago and Barral (1854–59, 2:702–3). See also Arago and Biot (1806a; 1806b).

36. On which see Alder (2002).

37. For these events see Daumas (1987) and Arago's own account in Arago and Barral (1854–59, 1:1–102).

38. Arago and Barral (1854–59, vol. 1).

39. For details of these events, and the content of Fresnel's work, alone and with Arago, see Buchwald (1989).

40. Fresnel, Senarmont, and Verdet (1866–70, 2:853–54); Buchwald (1989, 237–38).

41. Crosland (1967, 333).

42. Paravey (1835 [1821], 21, *Rapports*, n. 1).

43. The king paid half the price, which Biot could report because he was "charged with signing the act of acquisition in the name of the government" (Biot 1823, vi).

44. Biot (1844, 4).

45. Herivel (1975, 122–23); Dhombres and Robert (1998, 369–72).

46. Cited and translated in Herivel (1975, 125).

47. Henry (1854, 167).

48. Biot (1823, 282). Biot later argued otherwise and claimed, not implausibly, that he had shown the dying Delambre his work to the latter's great interest and encouragement.

49. For the method's elements see ibid., 2&ff.

50. Ibid., 20.

51. Ibid., 25&ff.

52. Ibid., 41.

53. Ibid., 43&ff.

54. Ibid., 48. Biot in fact asserted that the probability is only half as much as this for reasons having to do with distance measurements.

55. Though it is admittedly hard to square their rather vague statement with the specificity of Biot's method.

56. Biot (1823, 252–53).

57. A facsimile of the letter follows page 49 in Jollois and Devilliers (1834).

58. Paravey (1822, 6).

59. Jomard (1822, 5).

60. Ibid., 12.

61. Biot (1823, xii).

62. Ibid., xiii–xxii.

63. Halma (1822a).

64. Halma (1822b, 12). There was no evidence whatsoever in the images Halma referred to for human sacrifice. This was a common, and antique, accusation thrown by monotheists at pagans in their time. Human sacrifice certainly had been practiced, though it had likely disappeared by the eighth century BCE or thereabouts in the Near East, whereas monotheists ever prefer to exterminate stubborn heretics.

65. Leprince (1822). Leprince was particularly adamant that Dupuis' *Origine* had failed utterly, prompting an acerbic reply by Dupuis' defender, the astronomer Francoeur, who remarked that Leprince's critique on this point "is completely unintelligible, and that system [Dupuis'] remains the victor of this attack" (Francoeur 1822, 576).

66. Leprince (1819).

67. Chabrol and Jomard (1809, 7); Chabrol (1822, 524). Chabrol's two other memoirs for the *Description* were Chabrol and Lancret (1813a; 1813b).

68. The five pamphlets included in Chabrol's volume were Dalmas (1823); Jomard (1822); Leprince (1822); Letronne (1824); and Saint-Martin (1822). Chabrol annotated Saint-Martin, Leprince, and Dalmas, the latter of which was presented to him by the author.

69. Saint-Martin (1822, 9, 37, 43, 51). This article was translated into English and printed as Saint-Martin (1823).

70. In vol. 2, plate 7, no. 2.

71. The cartouche is no. 14 in vol. 4, plate 28.

72. Contemporary remarks on the zodiacs based primarily on stylistic and historical considerations include—and the list is by no means exhaustive—Dalmas (1823); Lenoir (1822); Letronne (1824); and Saint-Martin (1822). Dumersan (1824) is particularly interesting as he canvasses everyone's opinion.

Chapter 11: Champollion's Cartouche

1. Hartleben (1906, 1:172).

2. Adkins and Adkins (2000, 103–4).

3. Ibid., 105, on Coptic place-names, and 94, on *Scholasticomanie*. See Hartleben (1906, 1:chap. 4) for a full account of the period.

4. Hartleben (1906, 1:chap. 4).

5. Lacouture (1988, 271).

6. Hartleben (1906, 1:206).

7. Ibid., 238, provides the original quotation, which we have translated. Also see Adkins and Adkins (2000, 128–29) on Champollion's anger with De Sacy, who by this time was jealous as well of another former protégé, Étienne Quatremère, whom he had earlier come to prefer over Champollion.

8. Hartleben (1906, 1:309).

9. Ibid., 267–90.

10. Adkins and Adkins (2000, 144–6); Hartleben (1906, 1:286–87, 297).

11. Hartleben (1906, 1:282–84).

12. Quoted in Lacouture (1988, 430).

13. Adkins and Adkins (2000, 157).

14. Noted in Faure (2004, 426–27). The remarks follow the printing of Saulnier's letter, dated October 4, announcing the zodiac's

arrival at Marseilles, and immediately after Dacier's congratulatory letter to Saulnier dated October 5. Although the remarks in question are signed "note of the editors," they do certainly sound like Champollion, who is one of the very few likely to have protested the zodiac's removal (Champollion 1821, 470).

15. Anonymous (1823).

16. Hartleben (1906, 1:49).

17. Champollion (1822c, 232–33).

18. Ibid., 256&ff.

19. Biot (1844, 68).

20. These events and Champollion's preceding work are engagingly recounted most recently in Adkins and Adkins (2000, 155–210). See also Faure (2004, 413–67); Lacouture (1988, 280–307); and especially Hartleben (1906, 1:345–500).

21. Champollion (1822a, 620–28).

22. Italics as in the original: Champollion (1822b, 3).

23. Of these, the first (Champollion's no. 48) does not appear, strictly speaking, on plate 28 in volume 4 of the *Description* (note also that the Dacier letter misprinted 27 for 28). It is, however, a variant of the others.

24. Champollion (1822b, 25).

25. Hartleben (1906, 1:417). There seems to be no source beyond Hartleben for Jomard's immediate reactions, and she provided no references.

26. Saint-Martin (1822, 47).

27. On Letronne, see Godlewska (1999, 283–303). The *éloge* of Letronne by Walckenaer is particularly illuminating about him since Walckenaer did not share Letronne's broad approach to historical geography (Letronne 1850, i–xxv).

28. Translated in Godlewska (1995, 298) from Letronne (1842, 1:xliii).

29. Letronne (1823, xxxvii).

30. Ibid., xv.

31. Ibid., xxxv.

32. Ibid., xxxviii.

33. Letronne (1824, 9).

34. Letronne (1823, 108).

35. Letronne (1837 [1824]). The memoir had been read in 1824 at Inscriptions et Belles-Lettres but reached print only in 1837 when Letronne had the time to perfect his arguments.

36. Carteron (1843).

37. Letronne (1846, 118–19).

38. Ibid., 122. Letronne's attack on the presumption of mathematical system was quite extensive, yet he succumbed to system of a different kind. For he was convinced that the zodiac had been invented by the Greeks and not by the Egyptians or the denizens of ancient Mesopotamia. He was wrong about that, since the constellations originated in Mesopotamia toward the beginning of the first millennium BCE (on which see Brack-Bernsen and Hunger 1999), though at the time the evidence did seem to him to indicate otherwise.

39. Anonymous (1822b, 445).

40. Champollion (1823).

41. Champollion (1824).

42. Biot (1858, 2:456).

43. Hartleben (1909, 1:226–29).

44. Hartleben (1906, 1:574–75).

45. Hartleben (1909, 1:230).

Chapter 12: Epilogue

1. Champollion (1833, 89).

2. The phrase is from Horace's *Ars Poetica*, 5.

3. Hartleben (1909, 153–54). The letter in question is dated November 24.

4. Champollion (1833, 91–92).

5. Chéronnet-Champollion (1868, 75); Champollion-Figeac (1887, 174).

6. Champollion and Champollion-Figeac (1835–1845). The volumes are now rare but were reprinted in toto in Champollion and Champollion-Figeac (1970).

7. Anonymous (1825, 7–8).

8. Letronne (1837 [1824], 11).

9. On which see Aubourg (1995); Cauville (1997). For a full translation of the inscriptions at Dendera, with commentary, see Cauville (2009), as well as her fictional diaries of the priest and architect of the temple (Cauville 1999).

10. We thank Sally-Ann Ashton of the Fitzwilliam Museum for noting that, though the depiction is probably of a goddess (since its forehead bears the vulture rather than the royal cobra), it was common to represent deities in the image of a living ruler. The temple has

many images of Cleopatra. This one has not to our knowledge been previously identified and published.

Acknowledgments

1. In the initial drafting, Buchwald wrote chapters 2 and 3, while Greco Josefowicz wrote chapters 1 and 4. He continued with 5 through 8, while she wrote the last four sections of 9 and the first one of 11. Buchwald then drafted the first two sections of 9, followed by 10, the last two sections of 11, the introduction, and the epilogue.

BIBLIOGRAPHY

Primary Sources

Adolphus, John (1799). *Biographical Memoirs of the French Revolution*. London, T. Cadell, Jun. and W. Davies.

Al-Jabarti, Abd-al-Rahman (1979). *Journal d'un Notable du Caire durant l'Expédition Française, 1798–1801*, trans. Joseph Cuoq, Paris, Albin Michel.

—— (1997). *Napoleon in Egypt. Al-Jabarti's Chronicle of the French Occupation, 1798*, trans. Shmuel Moreh. Princeton, Markus Wiener.

Anderson, Frank Maloy (1904). *The Constitutions and Other Select Documents Illustrative of the History of France 1789–1901*. Minneapolis, H. W. Wilson.

Anonymous (1802a). "[De Luc on the zodiacs]," *The Edinburgh Magazine or Literary Miscellany* 20: 370–71.

—— (1802b). "[Testa on the zodiacs]," *The Monthly Mirror* 14: 139.

—— (1803). "Variétés [review of Larcher's 1802 *Hérodote*]," *Journal des Débats*: 2–3.

—— (1816). "Review of the Abbé Duclot's 'La sainte Bible vengée . . . ,'" *L'Ami de la Réligion et du Roi; Journal Ecclésiastique, Politique et Littéraire* 9: 257–63, 337–43.

—— (1821). "Sur les zodiaques d'Egypte," *L'Ami de la Réligion et du Roi; Journal Ecclésiastique, Politique et Littéraire* 29: 65–74.

—— (1822a). "Nouvelles politiques," *L'Ami de la Réligion et du Roi; Journal Ecclésiastique, Politique et Littéraire* 31: 155 (Dupuis), 310–13 (Frayssinous).

—— (1822b). "The Zodiac of Dendera," *European Magazine and London Review* 82: 441–46.

—— (1823). "[Review of] Notice sur le Zodiaque de Denderah, par M. J. Martin," *North American Review* 41: 233–42.

———— (1825). *Exhibition, 47, Leicester Square. Zodiac of Dendera. Epitome of the Celebrated Sculptured Zodiac of Dendera, so
Famous in Egyptian Antiquity, On Which It Is Conjectured the Present System of Astronomy Was Founded*. London, J. Haddon.

———— (1850). *Catalogue des Écrits, Gravures et Dessins condamnés
depuis 1814 jusqu'au 1er janvier 1850 suivi de la liste des indivi-
dus condamnés pour délits de presse*. Paris, Adolphe Delahays.

Arago, François, and J. A. Barral (1854–59). *Œuvres complètes de
François Arago*. Paris, Gide et J. Baudry.

Arago, François, and Jean-Baptiste Biot (1806a). "Mémoire sur les
affinités des corps pour la lumière et particulièrement sur les forces
réfringents des différens gaz," *Mémoires de l'Institut* 7: 301–85.

———— (1806b). *Mémoire sur les affinités des corps pour la lumière:
et particulièrement sur les forces réfringentes des différens gaz*.
Paris, Bachelier.

Aratus (1997). *Phaenomena*, trans. Douglas Kidd. Cambridge, Cambridge University Press.

Aulard, A. (1909). *Le Culte de la Raison et le Culte de l'Être Supreme
(1793–1794)*. Paris, Félix Alcan et Guillaumin Réunies.

Bailly, Jean Sylvain (1775). *Histoire de l'astronomie ancienne depuis
son origine jusqu'à l'établissement de l'école d'Alexandrie*. Paris,
Frères Debure.

———— (1779). *Lettres sur l'Atlantide de Platon et sur l'ancienne
histoire de l'Asie: pour servir de suite aux Lettres sur l'origine des
sciences, adressées à M. de Voltaire*. Paris, Frères Debure.

———— (1799). *Essai sur les fables, et sur leur histoire, adressé a la
citne Du Bocage*. Paris, G. de Bure l'aîné an VII.

Baudin, P. C. L. (1795). *Du fanatisme et des cultes*. Paris, Leclere.

———— (1799). *Discours prononcé par P.C.L. Baudin (des Ardennes),
en présentant au Conseil l'Essai sur les fables et sur leur histoire, ouvrage posthume de Jean-Sylvain Bailly*. Paris, Imprimerie
nationale.

Bénétrix, Paul (1894). *Les Conventionnels du Gers*. Auch, J. Capin.

Bernard, Samuel (1802). "Copie d'une lettre du citoyen S.B., membre
de la commission des sciences et arts d'Égypte, au citoyen Morand,
membre du corps-legislatif. Marseille, le . . . nivoise an 10," *Gazette
Nationale ou le Moniteur Universel*: 581–82.

———— (1812a). "Mémoire sur les monnoies d'Égypte," in *Description de l'Égypte*, edited by E. Jomard, vol. 2: État Moderne. Paris,
Imprimerie Impériale, 321–24.

———— (1812b). "Notice sur les poids arabes anciens et modernes," in *Description de l'Égypte*, edited by E. Jomard, vol. 2: État Moderne. Paris, Imprimerie Impériale, 229–48.

Biot, Jean-Baptiste (1811). *Traité élémentaire d'astronomie physique.* Paris, J. Klostermann.

———— (1823). *Recherches sur plusieurs points de l'astronomie Égyptienne appliquées aux monumens astronomiques trouvés en Égypte.* Paris, F. Didot.

———— (1844). *Mémoire sur le zodiaque circulaire de Denderah.* Paris, Imprimerie Royale.

———— (1858). *Mélanges scientifiques et littéraires.* Paris, M. Lévy.

Boissonade, J. F. (1814). "Notice sur la vie et les ecrits de Larcher," *Classical Journal* 10: 130–44.

Bonald, Louis-Gabriel-Ambroise de (1802). *Législation primitive: considérée dans les derniers temps par les seules lumières de la raison suivie de plusieurs traités et et discours politiques.* Paris, Chez Le Clere.

———— (1819). *Mélanges Littéraires, Politiques et Philosophiques,* Paris. Chez Le Clere.

Bonnet, Jules (1864). *Mes Souvenirs du Barreau Depuis 1804.* Paris, Auguste, Durand.

Boulanger, Nicolas Antoine (1766). *L'antiquité dévoilée par ses usages, ou examen critique des principales, opinions, cérémonies & institutions religieuses & politiques des différens peuples de la terre.* Amsterdam, Chez M. M. Rey.

———— (n.d.). *Anecdotes de la nature.* Paris, Muséum de l'histoire naturelle.

Bourrienne, Louis Antoine Fauvelet de (1895). *Memoirs of Napoleon Bonaparte,* edited by R. W. Phipps. New York, Charles Scribner's Sons.

Brisset, Ferdinand, and Théaulon (1822). *Le Zodiaque de Paris, A propos du Zodiaque de Denderah, Vaudeville-Épisodique en un Acte.* Paris, Duvernois.

———— (1822). *Gymnase dramatique. Le Zodiaque de Paris. Tableau vaudeville.* August 2. Box 642, #626, Archives Nationales, Paris.

Brosses, Charles de (1750). *Lettres sur l'État Actuel de la Ville Souterraine d'Herculée.* Dijon, F. Desventes.

———— (1760). *Du Culte des Dieux Fétiches, ou Parallèle de l'Ancienne Réligion de l'Égypte avec la Réligion Actuelle de Nigritie.* Paris, n.p.

Carlyle, Thomas (1956 [1837]). *The French Revolution: A History.* New York, Heritage Press.

Carnot, Hippolyte (1837). *Notice Historique sur Henri Grégoire.* Paris, P. Baudouin.

Carteron, Èdouard (1843). *Analyse des Recherches de M. Letronne sur les Représentations Zodiacales.* Paris, Bureau des Annales de Philosophie Chrétienne.

Castres, Sabatier de (1810). *Apologie de Spinosa et du Spinosisme, contre les Athées, les incrédules et contre les Théologiens scolastiques-platoniciens.* Paris, Fournier Frères.

Chabert, J., and L. D. Ferlus (1822). *Explication du zodiaque de Denderah (Tentyris). Observations curieuses sur ce monument précieux et sur sa haute antiquité.* Paris, Guidaudet.

Chabrol, Volvic de (1822). "Essai sur les moeurs des habitans modernes d'Égypte," in *Description de l'Égypte*, edited by E. Jomard, vol. 7: État Moderne. Paris, Imprimerie Royale, 361–524.

Chabrol, and Edmé Jomard (1809). "Description d'Ombos et des Environs," in *Description de l'Égypte*, edited by E. Jomard, vol. 1: Antiquités, Description. Paris, Imprimerie Impériale, 2–26.

Chabrol and Lancret (1813a). "Mémoire sur le canal d'Alexandrie," in *Description de l'Égypte*, edited by E. Jomard, vol. 6: État Moderne. Paris, Imprimerie Impériale, 185–94.

——— (1813b). "Notice topographique sur la partie de l'Égypte comprise entre Rahmanyeh et Alexandrie et sur les environs du Lac Mareotis," in *Description de l'Égypte*, edited by E. Jomard, vol. 6: État Moderne. Paris, Imprimerie Impériale, 482–90.

Champollion, Jean-François (1814). *L'Égypte sous les Pharaons, ou Recherches sur la géographie, la réligion, la langue, les écritures et l'histoire de l'Égypte avant l'invasion de Cambyse.* Paris, Chez de Bure Frères.

——— (1821). "Note de rédacteurs [on Saulnier's announcement of the zodiac's arrival at Marseilles]," *Revue encyclopédique* 12: 232–39.

——— (1822a). "Extrait d'un Mémoire relatif à l'Alphabet des Hiéroglyphes phonétiques égyptiens," *Journal des Savans*: 620–28.

——— (1822b). *Lettre à M. Dacier, secrétaire perpetuel de l'Académie Royale des Inscriptions et Belles-Lettres, relative à l'alphabet des hiérroglyphes phonétiques employés par les Égyptiens pour inscrire sur leurs monuments les titres, les noms et les surnoms des souverains Grecs et Romains.* Paris, Firmin Didot Père et Fils.

———— (1822c). "Lettre à M. le Rédacteur de la Revue encyclopédique, relative au Zodiaque de Denderah," *Revue encyclopédique*: 232–39.

———— (1823). *Panthéon Égyptien, Collection des Personnages
Mythologiques de l'Ancienne Égypte, d'Après les Monuments; avec
un Texte Explicatif*. Paris, Firmin Didot.

———— (1824). *Précis du système hiéroglyphique des anciens Égyptiens: ou, Recherches sur les élémens premiers de cette écriture
sacrée*. Paris, Treuttel et Würtz.

———— (1833). *Lettres écrites d'Égypte et de Nubie en 1828 et 1829*.
Paris, Firmin Didot Frères.

Champollion, Jean-François, and Jacques-Joseph Champollion-Figeac
(1835–45). *Monuments de l'Égypte et de la Nubie, d'après les
dessins exécutés sur les lieux sous la direction de Champollion-
le-jeune, et les descriptions autographes qu'il en a rédigés*. Paris,
Firmin Didot.

———— (1970). *Monuments de l'Égypte et de la Nubie*. Genève, Éditions de Belles-Lettres.

Champollion, Jean-François, and Hermine Hartleben (1909). *Lettres
de Champollion le jeune*. Paris, E. Leroux.

Champollion-Figeac, Jacques-Joseph (1806). *Lettre sur l'inscription
Grecque de temple de Dendera: addressée a Monsieur Fourier*.
Grenoble, J. H. Peyronard.

———— (1844). *Fourier et Napoléon. L'Égypte et les Cent Jours.
Mémoires et Documents Inédits*. Paris, Firmin Didot Frères.

Chateaubriand, F. R. de (1823). *Génie du christianisme ou beautés
de la réligion chrétienne*. Paris, LeNormont.

———— (1877). *Génie du christianisme ou beautés de la réligion
chrétienne*. Tours, Alfred Mame et Fils.

Chéronnet-Champollion, Zoë (1868). *Lettres écrites d'Égypte et de
Nubie en 1828 et 1829*. Paris, Didier.

Chigi, Sigismondo, Pietro Metastasio, Niccolò Piccinni, and [Gaetano]
Sertor (1775). *Il concalve dell'anno MDCCLXXIV; dramma per
musica da recitarsi nel Teatro delle Dame nel carnevale de
MDCCLXXV*. Roma, Cracas.

Coupin, Pierre André (1825). "Notice nécrologique sur M. le Baron
Denon," *Revue encyclopédique* 28: 30–41.

Cousin, Victor (1838). *Fragments Philosophiques*. Paris, Ladrange.

Curmer, Henri Léon (1840–42). *Les Français peints par eux-mêmes:
encyclopédie morale du dix-neuvième siècle*. Paris, L. Curmer.

Cuvier, Georges (1812). *Recherches sur les ossemens fossiles de quadrupèdes ou, l'on rétablit les caractères de plusieurs espèces d'animaux que les révolutions du globe paroissent avoir détruites.* Paris, Deterville.

——— (1821). *Discours sur la théorie de la terre.* Paris, G. Dufour and E. D'Ocagne.

——— (1825). *Discours sur les Révolutions de la Surface du Globe.* Paris, G. Dufour and E. D'Ocagne.

D'Argé, A. P. Chaalons (1824). *Histoire Critique et Littéraire des Théatres de Paris, Année 1822.* Paris, Pollet.

——— (1825). *Voyage du Capitaine Hiram Cox dans l'Empire des Birmans, avec des Notes et un Essai Historique sur cet Empire.* Paris, Bertrand.

D'Hauterive, Ernest (1908). *La Police Secrète du Premier Empire, Bulletins Quotidiens Addressés par Fouché à l'Empereur.* Paris, Perrin.

Dacier, Bon-Joseph (1812). "Notice historique sur la vie et les ouvrages de M. Dupuis," *Le Moniteur* nos. 216/17: 20.

Dalmas, V. de (1823). *Mémoire sur Le Zodiaque, en faveur de la Réligion Chrétienne.* Paris, Adrien Leclerc.

Delaulnaye, François (1791). *Histoire générale et particulière des réligions et du culte de tous les peuples du monde, tant anciens que modernes, par M. Delaulnaye: ouvrage proposé par souscription libre, et orné de plus de 300 figures gravées sur les dessins de M. Moreau le jeune, et sous sa direction, par les meilleurs artistes de Paris.* Paris, Fournier le Jeune.

Delaulnaye, François, and Gaspard Le Blond (1792). *Histoire générale et particulière des religions et du culte de tous les peuples du monde, tant anciens que modernes.* Paris, Fournier le Jeune.

Delvau, Alfred (1860). *Les dessous de Paris.* Paris, Poulet-Malassis et de Broise.

Denon, Vivant (1802a). *Voyage dans la Basse et la Haute Égypte: pendant les campagnes du général Bonaparte.* Londres, Cox, Fils, et Baylis.

——— (1802b). *Voyage dans la Basse et la Haute Égypte, pendant les campagnes du général Bonaparte.* Paris, P. Didot l'Aîné.

Desmarest, Pierre (1833). *Témoignages Historiques, ou Quinze Ans de Haute Police sous Napoléon.* Paris, Alphonse Levasseur.

Destutt de Tracy, Antoine Louis Claude (1799). *Analyse de l'origine de tous les cultes, par le citoyen Dupuis, et de l'abrégé qu'il a donné de cet ouvrage.* Paris, H. Agasse.

Diesbach, Ghislain de, ed. (1987), *Mémoires d'une femme de qualité sur le Consulat et l'Empire*. Paris, Mercure de France.

Drovetti, Bernardino (1985). *Epistolario*. Milan, Istituto Editoriale Cisalpino-La Goliardica.

Duclot, Abbé Joseph-François (1824). *La Sainte Bible vengée des attaques de l'incrédulité: et justifiée de tout reproche de contradiction avec la raison, avec les monumens de l'histoire, des sciences et des arts: avec la physique, la géologie, la chronologie, la géographie, l'astronomie*. Lyon, Rusand.

Dumersan, Théophile Marion (1824). *Notice sur le zodiaque de Dendera et sur son transport en France: avec un résumé des principales opinions et des systêmes les plus remarquables des antiquaires, des géomêtres et des astronomes, sur ce monument*. Paris, Chez M. Journé.

Dupuis, Charles (1781a). *Mémoire sur l'origine des constellations et sur l'explication de la fable par le moyen astronomique*. Paris, Veuve Desaint.

―――― (1781b). "Mémoire sur l'Origine des Constellations, et sur l'Explication de la Fable, par le Moyen de l'Astronomie," in *Astronomie*, edited by Jérome Lalande, vol. 4, Paris, La Veuve Desaint, 350–576.

―――― (1795a). *Origine de tous les cultes, ou, Réligion universelle*. Paris, H. Agasse.

―――― (1795b). *Planches de l'Origine de Tous les Cultes, du Citoyen Dupuis, avec leur Explication*. Paris, H. Agasse.

―――― (1797 and 1798). *Abrégé de l'origine de tous les cultes*. Paris, H. Agasse.

―――― (1798). *Abrégé de l'origine de tous les cultes*. Paris, André, Latour, Delaunay.

―――― (1806). *Mémoire explicatif du zodiaque chronologique et mythologique. Ouvrage contenant le tableau comparatif des maisons de la lune chez les diff'erens peuples de l'Orient, et celui des plus anciennes observations qui s'y lient, d'après les 'Égyptiens, les Chinois, les Perses, les Arabes, les Chald'eens et les calendriers grecs*. Paris, Courcier.

―――― (1822a). *Abrégé de l'origine de tous les cultes*. Paris, Chasseriau.

―――― (1822b). *L'origine de tous les cultes. Édition populaire complète*. Paris, Librairie Anti-Cléricale.

―――― (1822c). *Origine de tous les cultes ou réligion universelle*. Paris, Émile Babeuf.

———— (1822 [1806]). *Dissertation sur le zodiaque de Dendera* [Originally *Astronomie. Observations . . .*]. Paris, Chasseriau.

Dupuis, Mme (1813). *Notice historique sur la vie littéraire et politique de M. Dupuis, par Madame sa Veuve.* Paris, Brasseur Ainé.

Fellowes, W. D. (1815). *Paris; During the Interesting Month of July, 1815. A Series of Letters, Addressed to a Friend in London.* London, Gale and Fenner.

Ferlus, L. D. (1827). *Nouvelle notice sur la Girafe envoyée au Roi de France par le Pacha d'Égypte et arrivée à Paris le 30 Juin 1827. Observations curieuses sur le caractère, les habitudes et l'instinct de ce quadrupède.* Paris, Moreau.

Fouché, Joseph (1903). *Memoirs of Joseph Fouché.* New York, Merrill and Baker.

Fourier, Joseph (1800). "Le 10 Frimaire, VIIIe Année de la République (no. 47)," *Courier de l'Égypte.*

———— (1809). "Recherches sur les sciences et le gouvernment de l'Égypte," in *Description de l'Égypte*, edited by E. Jomard, vol. 1: Antiquités, Mémoires. Paris, Imprimerie Impériale, 803–24.

———— (1818). "Premier Mémoire sur les Monumens Astronomiques de l'Égypte," in *Description de l'Égypte*, edited by E. Jomard, vol. 2: Antiquités, Mémoires. Paris, Imprimerie Royale, 71–86.

———— (1829 [1809]). "Recherches sur les sciences et le gouvernment de l'Égypte," in *Description de l'Égypte*, edited by E. Jomard, vol. 9: Antiquités, Mémoires. Paris, Panckoucke, 1–42.

Francoeur, Louis Benjamin (1822). "Review of Leprince, *Essai d'interpretation du Zodiaque circulaire de Denderah*," *Revue Encyclopédique* 16: 575–77.

———— (1838 [1812]). *Uranographie ou Traité Élémentaire d'Astronomie.* Brussels, Meline, Cans.

Fresnel, Léonor, Henri de Senarmont, and Émile Verdet, eds. (1866–70). *Oeuvres Complètes d'Augustin Fresnel.* Paris, Imprimerie Impériale.

Galasière, Guillaume-Joseph Le Gentil de la (1782). "Dissertation sur l'origine du zodiaque, et sur l'explication des douze signes," *Histoire de l'Académie Royale des Sciences*: 368–456.

———— (1784). "Remarques et observations sur l'astronomie des Indiens, & sur l'ancienneté de cette Astronomie," *Histoire de l'Académie Royale des Sciences*: 482–91.

———— (1785). "Mémoire sur l'origine du zodiaque, l'explication de ses douze signes, et sur le Système chronologique de Newton," *Histoire de l'Académie Royale des Sciences*: 9–16.

Galignani (1827). *New Paris Guide, or Stranger's Companion through the French Metropolis.* Paris, A. and W. Galignani.

Garat, Dominique-Joseph (1800). *Éloge funèbre des généraux Kléber et Desaix: prononcé le 1er vendémaire an 9, à la Place des Victoires.* Paris, L'Imprimerie de la République.

Gébelin, Court de (1773). *Allégories orientales ou le fragment de Sanchoniaton, quit contient l'histoire de Saturne, suivie de celles de Mercure et d'Hercule, et de sez douze travaux, avec leur explication, pour servire à l'intelligence du Génie symbolique de l'Antiquité.* Paris, Boudet; Valleyre l'ainé.

Gifford, John, ed. (1797). *A Residence in France, during the Years 1792, 1793, 1794, and 1795; Described in a Series of Letters from an English Lady: with General and Incidental Remarks on the French Character and Manners,* 2 vols. London, T. N. Longman.

Girard, Pierre Simon (1824 [1809]). "Observations su la Vallée d'Égypte et sur l'Exhaussement Séculaire du Sol qui la Recouvre," in *Description de l'Égypte,* edited by E. Jomard, vol. 20. Paris, Pancoucke, 130ff. (sec. 4); 1809, vol. 1, 343ff.

Goudemetz, H. (1796). *Historical Epochs of the French Revolution. Translated from the French of H. Goudemetz, a French Clergyman in England.* Bath, R. Cruttwell.

Grobert, J. F. L. (1800). *Description des pyramides de Ghizé, de la ville du Kaire, et ses environs.* Paris (Rémont, libraire, quai des Augustins, no. 41), Chez Logerot-Petiet imprimeur rue et maison des Capucines vis-à-vis la place Vendôme.

Guignes, Joseph de (1785 [1809]). "Mémoire concernant l'origine du zodiaque et du calendrier des orientaux, et celles de différentes constellations de leur ciel astronomique," *Histoire de l'Académie des Inscriptions et des Belles-Lettres, Mémoires de littérature* 47: 378–434.

Guilhermy, M. F. de (1855). *Itinéraire Archéologique de Paris.* Paris, A. Morel.

Halma, Abbé Nicolas-B. (1821). *Les Phénomènes, d'Aratus de Soles, et de Germanicus César.* Paris, Merlin.

——— (1822a). *Commentaire de Théon d'Alexandrie sur le livre III de l'Almageste de Ptolemée; Tables manuelles des mouvemens des astres.* Paris, A. Bobée.

——— (1822b). *Examen et explication du zodiaque de Denderah Comparé au globe Farnese.* Paris, Merlin.

Hartleben, Hermine (1909). *Lettres de Champollion le Jeune.* Paris, Ernest Leroux.

Henley, Samuel (1803 [Floréal an 11]). "Remarques sue le Zodiaque de Dendera, autrefois Tentyra," *Magasin Encyclopédique, ou Journal des Sciences, des Lettres et des Arts* 6: 433–51.

Henniker, Frederick (1823). *Notes during a Visit to Egypt, Nubia, the Oasis, Mount Sinai and Jerusalem.* London, John Murray.

Henry, William Charles, ed. (1854). *Memoirs of the Life and Scientific Researches of John Dalton.* London, Cavendish Society.

Hipparchus (1567). *Hipparchi Bithyni In Arati et Evdoxi Phænomena libri III: Eiusdem Liber asterismorum. Achillis Statii In Arati Phaenomena. Arati vita, & fragmenta aliorum veterum in eius poema.*Florentiae, In officina Ivntarvm, Bernardi filiorum.

Hobhouse, John Carn (1816). *The Substance of Some Letters, Written by an Englishman Resident at Paris during the Last Reign of the Emperor Napoleon,* London.

Horapollo (1993). *The Hieroglyphics of Horapollo,* trans. George Boas. Princeton, Princeton University Press.

Jollois, Jean Baptiste Prosper, and René Edouard Devilliers (1809a). "Description des Monumens Astronomiques Découverts en Égypte," in *Description de l'Égypte,* edited by E. Jomard, vol. 1: Antiquités, Descriptions. Paris, Imprimerie Impériale, 1–16 (Appendice no. II).

——— (1809b). "Recherches sur les bas-reliefs astronomiques des Égyptiens, et parallèle de ces bas-reliefs avec les différens monumens astronomiques de l'antiquité, d'où résulte la connoissance de la majeure partie des constellations égyptiennes," in *Description de l'Égypte,* edited by E. Jomard, vol. 1: Antiquités, Mémoires. Paris, Imprimerie Impériale, 427–94.

——— (1821 [1809]). "Description générale de Thebes," in *Description de l'Égypte,* edited by E. Jomard, vol. 2: Antiquités, Mémoires. Paris, Imprimerie Royale.

——— (1822 [1809]). "Recherches sur les bas-reliefs astronomiques . . . ," in *Description de l'Égypte,* edited by E. Jomard, vol. 8: Antiquités, Mémoires. Paris, Imprimerie Royale, 357–489.

——— (1834). *Appendice aux Recherches sur les Bas-Reliefs Astronomiques des Égyptiens.* Paris, Carilian-Goeury.

Jollois, Jean Baptiste Prosper, and Pierre Lefèvre-Pontalis (1904). *Journal d'un ingénieur attaché à l'expédition d'Égypte, 1798–1802.* Paris, Ernest Leroux.

Jomard, E. (1822). "Examen d'une opinion nouvelle sur le Zodiaque circulaire de Dendéra," *Revue encylopédique ou analyses et annonces raisonnées* 15: 433–51.

Jones, William, et al. (1792). *Dissertations and Miscellaneous Pieces Relating to the History and Antiquities, the Arts, Sciences, and Literature of Asia.* London, G. Nicol, J. Walter, and J. Sewell.

Keller, Alexander, ed. (1901). *Correspondence, Bulletins, et Orders de Napoléon Bonaparte.* Paris, Albert.

Lalande, Joseph Jérome Le Français de (1795 [30 Fructidor Year 3]). "Review of Dupuis, Origine de tous les cultes," in *Supplément à la Gazette Nationale,* i–iii.

—— (1800). *Connaissance des Tems, ou des Mouvemens Célestes à l'Usage des Astronomes et des Navigateurs pour l'An XIII de l'Ère de la République Française.* Paris, Imprimerie de la République.

—— (1803a). *Bibliographie astronomique avec l'histoire de l'astronomie depuis 1781 jusqu'à 1802.* Paris, Imprimerie de la République.

—— (1803b). "Histoire de l'astronomie pour l'année VIII [1800]," in *Connaissance des Tems, ou des Mouvemens Célestes à l'Usage des Astronomes et des Navigateurs pour l'An XIII de l'Ère de la République Française.* Paris, Bureau des Longitudes, 235ff.

—— (1804 [Nivose an 12]). "Histoire de l'astronomie pour l'année X [1802]," in *Connaissance des Tems, ou des Mouvemens Célestes à l'Usage des Astronomes et des Navigateurs pour l'An XIV de l'Ère de la République Française.* Paris, Bureau des Longitudes, 342ff.

Lalande and Sylvain Maréchal (1803). *Dictionnaire des athées anciens et modernes.* Paris, Chez Grabit.

LaMennais, H. F. de (1820). *Essai sur l'Indifférence en Matière de Réligion.* Paris, Tournachon-Molin et H. Séguin.

Lamothe-Langon, E. L. (1830). *Private Memoirs of the Court of Louis XVIII, By A Lady.* London, Henry Colburn and Richard Bentley.

Larcher, Pierre-Henri (1767a). *Réponse à la Défense de Mon Oncle.* Amsterdam, Changuion.

—— (1767b). *Supplément à la Philosophie de l'Histoire de Seu M. l'Abbé Bazin.* Amsterdam, Changuion.

—— (1769). *Supplément à la Philosophie de l'Histoire de Seu M. l'Abbé Bazin. Nouvelle Édition, considérablement augmentée.* Amsterdam, Changuion.

—— (1776). *Mémoire sur la Déesse Vénus, auquel l'Académie Royale des Inscriptions et Belles-Lettres a adjugé le Prix de la Saint Martin 1775.* Paris, Valade.

—— (1802). *Histoire d'Hérodote: traduite du grec, avec des remarques historiques et critiques, un essai sur la chronologie d'Hérodote, et une table géographique.* Paris, G. Debure l'aîné (C. Crapelet).

———— (1829). *Larcher's Notes on Herodotus: Historical and Critical Remarks on the Nine Books of the History of Herodotus, with a Chronological Table.* London, John R. Priestley.

Laurens, Jacques, and Henry Laurens, eds. (1995). *Kléber en Égypte 1798–1800. Kléber, commandant en chef 1799–1800*, vol. 3. Paris, Institut Français d'Archéologie Orientale.

Le Coz, Claude, and Alfred Roussel (1900). *Correspondance de Le Coz, évêque constitutionnel d'Ille-et-Vilaine.* Paris, A. Picard et Fils.

Lenoir, Alexandre (1822). *Essai sur le zodiaque circulaire de Denderah, maintenant au Musée du roi.* Paris, Librairie des annales françaises.

Leprince, H. S. (1819). *Nouvelle Croagénésie ou Réfutation du Traité d'Optique de Newton. Première Partie.* Paris, Leblanc.

———— (1822). *Essai d'Interprétaton du Zodiaque Circulaire de Denderah.* Paris, Panthieu et Delaunai.

Letronne, Antoine Jean (1823). *Recherches pour servir à l'histoire de l'Egypte pendant la domination des Grecs et des Romains, tirés des inscriptions grecques et latines relatives à la chronologie, à l'état des arts, aux usages civils et réligieux de ce pays.* Paris, Boulland-Tardieu.

———— (1824). *Observations Critiques et Achéologiques sur l'Objet des Représentatons Zodiacales qui Nous restent de l'Antiquité; a l'Occasion d'un Zodiaque Égyptien Peint Dans Une Caisse de Momie quie porte une Inscription Grecque du temps de Trajan.* Paris, Auguste Bolland.

———— (1837 [1824]). "Sur l'origine Grecque des zodiaques prétendus Égyptiens," *Revue des Deux Mondes (Extrait).*

———— (1842). *Receuil des inscriptions grecques et latines d'Egypte.* Paris, Imprimerie Royale.

———— (1846). "Analyse critique des représentations zodiacales de Dendéra et d'Esné, où l'on établit, 1. que ces représentations ne sont point astronomiques; 2. que les figures, autres que celles des signes du zodiaque, ne sont pas des constellations; 3. que le zodiaque circulaire de Dendéra n'est point un planisphère soumis à une projection quelconque," *Mémoires de l'Institut Royal de France. Académie des inscriptions et belles-lettres* 16: 102–210.

———— (1850). *Mélanges d'erudition et de critique.* Paris, Ducrocq.

Locré, Baron (1819). *Discussions sur la Liberté de la Presse, la Censure, la Propriété Littéraire, L'Imprimerie, et la Librairie, qui ont lieu dans le Conseil de l'État, pendant les Années 1808, 1809, 1810 et 1811.* Paris, Garnery & H. Nicolle.

Macrobius (1969). *The Saturnalia,* trans. Percival Vaughan Davies. New York, Columbia University Press.

Méjan, Maurice (1820). *Histoire du Procès de Louvel.* Paris, J. G. Dentu.

Minutoli, Baroness W. M. von (1827). *Recollections of Egypt 1820–1821.* London, Treuttel & Wurtz.

Monge, Gaspard (1798). *Géométrie descriptive: leçons données aux Écoles normales, l'an 3 de la République.* Paris, Baudouin.

Montani, Francesco Fabi (1844). "Elogio storico di Monsignor Gian Domenico Testa," in *Continuazione delle Memorie di Religione di Morale di Letteratura,* vol. 18. Modena, Eredi Soliani.

Montulé, Édouard de (1821). *Voyage en Amérique, en Italie, en Sicile et en Égypte, pendant les années 1816, 1817, 1818, et 1819.* Paris, Delaunay.

Mountnorris, George Annesley, Earl of (1809). *Voyages and Travels in India, Ceylon, the Red Sea, Abyssinia, and Egypt, in 1802, 1803, 1804, 1805, and 1806.* London, W. Miller.

Napoleon (1846). *The Bonaparte Letters and Despatches, Secret, Confidential, and Official; from the Originals in His Private Cabinet.* London, Saunders and Otley.

Newton, Isaac (1728). *The Chronology of Ancient Kingdoms Amended. To Which Is Prefix'd, a Short Chronicle from the First Memory of Things in Europe, to the Conquest of Persia by Alexander the Great.* London, Printed for J. Tonson [etc.].

Noé, Marc-Antoine de (1788). *Discours de Monseigneur l'éveque de Lescar sur l'état futur de l'Église.* France, n.p.

Panta, Elena Del, ed. (1998). *Pages d'un Journal de Voyage en Italie (1788).* Paris, Gallimard.

Paravey, Charles Hippolyte chevalier de (1821). *Rapport de M. le Chevalier Delambre sur les mémoires relatifs a l'origine commune des sphères de tous les anciens peuples . . . Mémoires Lus et Présentés à l'Académie par M. de Paravey.* Paris, Belin.

——— (1822). *Explication du Zodiaque de Denderah (Tentyris). Observations Curieuses sur ce Monument Précieux et sur sa Haute Antiquité.* Paris, Guiraudet.

——— (1826). *Essai sur L'Origine Unique et Hiéroglyphique des Chiffres et de Lettres de Tous les Peuple . . . Précédé D'un Coup*

*d'Oeil Rapide sur l'Histoire du Monde, entre l'Époque de la Créa-
tion et l'Époque de Nabonassar.* Paris, Treuttel et Wurtz.

——— (1835 [1821]). "Aperçu des Mémoires encore Manuscrits," in
*Illustrations de l'Astronomie Hiéroglyphique et des Planisphères
et Zodiaques Retrouvés en Égypte, en Chaldée, dans l'Inde et au
Japon.* Paris, Adolphe Delahays.

——— (1835a). "Connaissances Astronomiques des Anciens Peuples
de l'Égypte et de l'Asie, sur les Satellites de Jupiter et l'Anneau de
Saturne ou Lettres Adressées a l'Académie des Sciences en 1835,"
in *Illustrations de l'Astronomie Hiéroglyphique et des Planisphères
et Zodiaques Retrouvés en Égypte, en Chaldée, dans l'Inde et au
Japon.* Paris, Adolphe Delahays.

——— (1835b). *Illustrations de l'Astronomie Hiéroglyphique et des
Planisphères et Zodiaques Retrouvés en Égypte, en Chaldée, dans
l'Inde et au Japon.* Paris, Adolphe Delahays.

——— (1835c). *Illustrations de l'Astronomie Hiéroglyphique et des
Planisphères Zodiaques Retrouvés, en Égypte, en Chaldée dans
l'Inde, et au Japon; et Réfutation des Mémoires Astronomiques de
Dupuis, de Volney, de Fourier, et de M. Biot.* Paris, Treuttel & Wurtz
and Bachelier.

——— (1835d). *Réfutation des Anciens et des Nouveaux Mémoires
de M. Biot, sur les Zodiaques Égyptiens, et sur l'Astronomie
Comparée de l'Égypte, de la Chaldée et de l'Asie Orientale.* Paris,
Bachelier and Treuttel & Wurtz.

Pérès, Jean-Baptiste (1836). *Grand Erratum, Source d'Un Nombre
Infini d'Errata A Noter Dans l'Histoire du 19e Siècle.* Paris, J. J. Risler.

Petau, Denis (1630). *Uranologion, sive, Systema variorum authorum
qui de sphaera ac sideribus eorumque motibus graecè commen-
tati sunt. Sunt autem horum libri: Gemini, Achillis Tatii Isagoge
ad Arati Phaenomena; Hipparchi Libri tres ad Aratum; Ptolemaei
De apparentiis; Theodori Gazae De mensibus; Maximi, Isaaci
Argyri . . . S. Andreae Cretensis Computi.* Lutetiae Parisiorum,
sumptibus Sebastiani Cramoisy.

Picot, Michel Pierre Joseph (1815). *Mémoires pour servir À l'Histoire
Ecclésiastique, pendant le Dix-Huitième Siècle.* Paris, Adrien le
Clere.

——— (1816). *L'Ami de la Réligion et du Roi, ou L'Order Rétabli,
Suivi de quelques considérations sur les avantages de la Religion,
et terminé par les deux Tetsamens de LL. MM. Louis XCI, et Marie-
Antoinette, de glorieuse mémoire.* Lyon, Guyot Frères.

Priestley, Joseph (1797). *Letters to Mr. Volney: Occasioned by a Work of His Entitled Ruins, and by His Letter to the Author.* Philadelphia, Thomas Dobson.

———— (1904 [1809]). *Memoirs of Dr. Joseph Priestley, Written by Himself (To the Year 1795) with a Continuation to the Time of His Decease by His Son, Joseph Priestley.* London, H. R. Allenson.

Ptolemy (1998). *Ptolemy's Almagest,* trans. G. J. Toomer. Princeton, Princeton University Press.

Raige, Remi (1809). "Mémoire sur le zodiaque nominal et primitif des anciens Égyptiens," in *Description de l'Égypte,* edited by E. Jomard, vol. 1: Antiquités Mémoires. Paris, Imprimerie Impériale, 169–80.

———— (1822 [1809]). "Mémoire sur le zodiaque . . . ," in *Description de l'Égypte,* edited by E. Jomard, vol. 6: Antiquités Mémoires. Paris, Imprimerie Royale, pp. 391–412.

Reybaud, Louis (1832). *Histoire Scientifique et Militaire de l'Expédition Française en Égypte.* Paris, A.-J. Dénain.

Rhind, A. Henry (1862). *Thebes, Its Tombs and Their Tenants.* London, Longman.

Richardson, Robert (1822). *Travels along the Mediterranean, and Parts Adjacent; in Company with the Earl of Belmore, during the Years 1816–17–18: Extending as far as the Second Cataract of the Nile, Jerusalem, Damascus, Balbec, &c. &c.* London, T. Cadell.

Rifaud, J.-J. (1830). *Tableau de L'Égypte, de la Nubie et des Lieux Circonvoisins.* Paris, Truettel & Wurtz.

Ripauld, Louis (1800). *(Ripault) Report of the Commission of Arts to the First Consul Bonaparte, on the Antiquities of Upper Egypt.* London, J. Debret.

Rocher, Ludo, ed. (1984). *Ezourvedam, a French Veda of the Eighteenth Century.* Amsterdam/Philadelphia: John Benjamins.

Roy, J. J. E. (1855). *Les Français en Égypte ou Souvenirs des Campagnes d'Égypte et de Syrie par un Officier de l'Expédition, receuillis et mis en order par J. J. E. Roy.* Tours, Mame.

Saint-Hilaire, Etienne-Geoffroy (1901). *Lettres Ecrits d'Egypte.* Paris, Librairie Hachette.

Saint-Martin, Antoine Jean (1822). *Notice sur le Zodiaque de Denderah.* Paris, Delaunay.

———— (1823). "Notice sur le Zodiaque de Denderah," *North American Review* 41: 233–42.

Saulnier, Sebastien (1822). *Notice sur le voyage de M. Lelorrain en Égypte et observations sur le zodiaque circulaire de Denderah.* Paris, n.p.

Savary, M. (1828). *Memoirs of the Duke of Rovigo, (M. Savary,) Written by Himself: Illustrative of the History of the Emperor Napoleon.* London, Henry Colburn.

Scott, John (1816). *Paris Revisited in 1815, By Way of Brussels.* Boston, Wells and Lilly.

Sonnini, C. S. (1799). *Voyage dans la Haute et Basse Égypte, fait par ordre de l'Ancien Gouvernement.* Paris, F. Buisson.

Staël, Madame de (1818). *Considérations sur les Principaux Événements de la Révolution Française.* Paris, Delaunay.

——— (1843). *Dix Années d'Exil.* Paris, Charpentier.

Testa, Abbé Domenico (1802). *Dissertazione sopra due Zodiaci novellamente scoperti nell'Egitto letta in una adunanza straodinaria dell'Accademia di Religione Catolica*, trans. Accademia di Religione Catolica. Rome.

——— (1807). *Dissertation sur les deux zodiaques nouvellement découverts en Égypte*, trans. Gauthier de Chaubry. Paris, Le Clere.

Villiers du Terrage, René-Edouard, and Marc Villiers du Terrage (1899). *Journal et souvenirs sur l'expédition d'Égypte (1798–1801).* Paris, Plon Nourrit.

Visconti, Ennio Quirino, 1802, "Notice Sommaire des Deux Zodiaques de Tentyra; Supplément a la Notice Précédente," in Larcher, *Histoire d'Hérodote: traduite du grec, avec des remarques historiques et critiques, un essai sur la chronologie d'Hérodote, et une table géographique.* Paris, G. Debure l'aîné (C. Crapelet), 567–76.

Volney, C. F. (1787). *Voyage en Syrie et en Égypte, pendant les années 1783, 1784 et 1785.* Paris, Desenne.

——— (1791). *Les ruines, ou, Méditation sur les révolutions des empires.* Paris, Desenne.

——— (1799). "Miscellanies. The Wrangling Philosophers. Volney's Answer to Dr. Priestley," *Anti-Jacobin Review and Magazine; or, Monthly Political and Literary Censor* 2: 331–34 and 443–46.

——— (1826 [1795]). *Leçons d'Histoire prononcés à l'École Normale.* Paris, Baudoin Frères.

——— (1890 [1802]). *The Ruins; or, Meditation on the Revolutions of Empires: and The Law of Nature.* New York, Peter Eckler.

Voltaire (1753). *Le Fanatisme, ou Mahomet le Prophète, Tragédie.* Amsterdam, Estienne Ledet.

——— (1765). *La Philosophie de l'Histoire par seu l'Abbé Bazin.* Amsterdam, Changuion.

——— (1768). *La Défense de Mon Oncle.* Geneva, n.p.

Whately, Richard (1819). *Historic Doubts Relative to Napoleon Buonaparte.* London, Longmans, Green.

Wilson, Erasmus (1877). *Cleopatra's Needle: With Brief Notes on Egypt and Egyptian Obelisks.* London, Brain.

Wilson, Robert Thomas (1803). *History of the British Expedition to Egypt.* London, C. Roworth.

Secondary Sources

Adkins, Lesley, and Roy Adkins (2000). *The Keys of Egypt. The Obsession to Decipher Egyptian Hieroglyphics.* New York, HarperCollins.

Albert, Maurive (1902). *Les Théatres des Boulevards (1789–1848).* Paris, Société Française d'Imprimerie et de Librairie.

Alder, Ken (1997). *Engineering the Revolution. Arms and Enlightenment in France, 1763–1815.* Princeton, Princeton University Press.

——— (2002). *The Measure of All Things. The Seven-Year Odyssey and Hidden Error That Transformed the World.* New York, Free Press.

Allin, Michael (1998). *Zarafa: A Giraffe's True Story, from Deep in Africa to the Heart of Paris.* New York, Walker.

Arnold, Eric A. (1979). *Fouché, Napoleon, and the General Police.* Washington, DC, University Press of America.

Aubourg, E. (1995). "La date de conception du zodiaque du temple d'Hator à Dendera," *Bulletin de l'Institut Français d'Archéologie Orientale:* 647–58.

Ayers, Andrew (1994). *The Architecture of Paris.* Paris, Édition Axel Menges.

Bailey, Charles R. (1979). "Educational Administration and Politics: The Collège of Louis-le-Grand, 1763–1790," *History of Education Quarterly* 19: 333–50.

Baltrusaitis, Jurgis (1967). *La Quête d'Isis. Essai sur la Légende d'un Mythe. Introduction à l'Égyptomanie.* Paris, Olivier Perrin.

Behdad, Ali (1994). *Belated Travelers: Orientalism in the Age of Colonial Dissolution.* Durham, Duke University Press.

Belhoste, Bruno (1991). *Augustin-Louis Cauchy: A Biography.* New York, Springer-Verlag.

Bhabha, Homi (1986). "Signs Taken for Wonders: Questions of Ambivalence and Authority under a Tree Outside Delhi, May 1817," in *"Race," Writing and Difference*, edited by Henry Louis Gates Jr. Chicago, University of Chicago Press, 163–84.

Bousquet, G.-H. (1968). "Voltaire et l'Islam," *Studia Islamica* 28: 109–26.

Brack-Bernsen, Lis, and Hermann Hunger (1999). "The Babylonian Zodiac: Speculations on Its Invention and Significance," *Centaurus* 41: 280–92.

Brégeon, Jean-Joël (1998). *L'Egypte de Bonaparte*. Paris, Perrin.

Bret, Patrice (1998). *L'Egypte au temps de l'expédition de Bonaparte 1798–1801*. Paris, Hachette.

Bryan, Cathie (2003). "Napoleon's Savants: The Return to France," in *Ancient Egypt: the History, People & Culture of the Nile Valley* 3: 14–21.

Buchwald, Jed Z. (1989). *The Rise of the Wave Theory of Light*. Chicago, University of Chicago Press.

Burleigh, Nina (2007). *Mirage. Napoleon's Scientists and the Unveiling of Egypt*. New York, HarperCollins.

Burton, Richard D. E. (1988). *Baudelaire in 1859: A Study in the Sources of Poetic Creativity*. Cambridge, Cambridge University Press.

Byrd, Melanie (1992). "The Napoleonic Institute of Egypt," Ph.D dissertation, Florida State University.

Byrnes, Joseph F. (1991). "Chateaubriand and Destutt de Tracy: Defining Religious and Secular Polarities in France at the Beginning of the Nineteenth Century," *Church History* 60: 316–30.

Cabanis, André (1975). *La Presse sous le Consulat et l'Empire*. Paris, Société des Études Robespierristes.

Cauville, S. (1997). *Le Zodiaque d'Osiris*. Louvain, Peters.

———— (1999). *L'Oeil de Re. Histoire de la Construction du Temple d'Hathor à Dendara (du 16 Juillet av. J.-C. au Printemps 64 ap. J.-C.)*. Paris, Pygmalion.

———— (2009). *Dendara. Le Temple d'Isis. Traduction* (vol. 1) and *Analyse à la lumière du temple d'Hathor* (vol. 2). Louvain, Peeters.

Ceram, C. W. (1979). *Gods, Graves & Scholars*. New York, Vintage Books, Random House.

Champollion-Figeac, Aimé Louis (1887). *Les deux Champollion; leur vie et leurs œuvres, leur correspondance archéologique relative au Dauphiné et à l'Égypt. Étude complète de biographie et de bibliographie, 1778–1867, d'après des documents inédits*. Grenoble, X. Drevet.

Chapin, Seymour L. (1970). "Bailly, Jean-Sylvain," in *Dictionary of Scientific Biography*, edited by Charles C. Gillispie, vol. 1. New York, Charles Scribner's Sons, 400–2.

Charles-Roux, François (1910). *Les Origines de l'Expédition d'Égypte*. Paris, Plon-Nourrit.

—— (1925). *L'Angleterre et l'expédition française en Égypte*. Cairo, Société Royale de Géographie d'Égypte.

—— (1934). "Emotions et travaux de l'Institut d'Egypte," *La Revue de Paris*: 829–52.

—— (1935). "Une Académie coloniale au Caire," *La Revue de Paris*: 49–78.

—— (1937). *Bonaparte: Governor of Egypt*, trans. E. W. Dickes. London, Methuen.

Chatelain, Jean (1973). *Vivant Denon et le Louvre de Napoléon*. Paris, Librairie Académique Perrin.

Chevalier, Louis (1973). *Laboring Classes and Dangerous Classes*. New York, Howard Fertig.

Clagett, Marshall (1995). *Ancient Egyptian Science: A Source Book*, vol. 2: *Calendars, Clocks, and Astronomy*. Philadelphia, American Philosophical Society.

Clifford, James (1988). *The Predicament of Culture*. Cambridge, Harvard University Press.

Coffin, Victor (1917). "Censorship and Literature under Napoleon I," *American Historical Review* 22: 288–308.

Cohn, Norman (2000). *Europe's Inner Demons*. Chicago, University of Chicago Press.

Cole, Hubert (1971). *Fouché. The Unprincipled Patriot*. New York, McCall.

Cole, Juan (2007). *Napoleon's Egypt. Invading the Middle East*. New York, Palgrave Macmillan.

Collins, Jeffrey (2004). *Papacy and Politics in Eighteenth-Century Rome: Pius VI and the Arts*. Cambridge, Cambridge University Press.

Comment, Bernard (1999). *The Painted Panorama*. New York, Harry N. Abrams.

Crosland, Maurice (1967). *The Society of Arceuil*. London, Heinemann Educational Books.

Dardaud, Gabriel (1951). "L'extraordinaire aventure de la giraffe du pacha d'Égypte," *Revue des conferences françaises en Orient* 14: 1–72.

Darnton, Robert (1982). *The Literary Underground of the Old Regime.* Cambridge, Harvard University Press.

——— (1995). *The Corpus of Clandestine Literature in France 1769–1789.* New York, W. W. Norton.

Darrigol, Olivier (2007). "The acoustic origins of harmonic analysis," *Archive for History of Exact Sciences* 61: 343–424.

Daumas, Maurice (1987). *Arago. La jenuesse de la science.* Paris, Belin.

De Ris, Clement (1877). *Les Amateurs des Autrefois.* Paris, E. Plon.

Delaporte, P. A. (1892). *Vie du Très Révérend Père Jean-Baptiste Rauzan Fondateur et Premier Supérieure Général de la Société des Missions de France.* Paris, Maison de la Bonne Presse.

Dhombres, Jean, and Nicole Dhombres (1989). *Naissance d'un pouvoir: sciences et savants en France, 1793–1824.* Paris, Payot.

Dhombres, Jean, and Jean-Bernard Robert (1998). *Fourier. Créateur de la Physique-Mathématique.* Paris, Belin.

Drouin, Jean-Claude (1970). "Un esprit original du XIXe siècle: le chevalier de Paravey (1787–1871)," *Revue d'histoire de Bordeaux et du département de la Gironde*: 65–78.

Duzer, Charles van (1935). *Contributions of the Ideologues to French Revolutionary Thought.* Baltimore, Johns Hopkins University Press.

Fagan, Brian M. (1975). *The Rape of the Nile: Tomb Robbers, Tourists, and Archaeologists in Egypt.* New York, Charles Scribner's Sons.

Fahmy, Khaled (2009). *Mehmed Ali: From Ottoman Governor to Ruler of Egypt.* Oxford, Oneworld Publications.

Faure, Alain (2004). *Champollion. Le Savant Déchiffré.* Paris, Fayard.

Feldman, Barton, and Robert D. Richardson, eds. (1972). *The Rise of Modern Mythology.* Bloomington, Indiana University Press.

Forssell, Nils (1925). *Fouché. The Man Napoleon Feared*, trans. Anna Barwell. London, George Allen and Unwin.

Forster, E. M. (1961 [1922]). *Alexandria: A History and a Guide.* Garden City, NY, Anchor Books.

Frankel, Eugene (1972). "Jean-Baptiste Biot: The Career of a Physicist in Nineteenth Century France," Ph.D dissertation, Princeton University.

——— (1978). "Career-Making in Post-Revolutionary France: The Case of Jean-Baptiste Biot," *British Journal for the History of Science* 37: 36–48.

Furet, François (2000). *Revolutionary France, 1770–1880*, trans. Antonia Nevill. Oxford, Blackwell.

Gaudant, Jean (1999). "La querelle des trois abbés (1793–1795): le débat entre Domenico Testa, Alberto Fortis, et Giovanni Serafino Volta sur la Signification des Poissons Pétrifiées du Monte Bolca (Italie)," *Miscellanea Paleontologica (Verona)* 8: 159–206.

Gaulmier, Jean (1951). *L'idéologue Volney: Contribution à l'histoire de l'orientalisme en France.* Beirut, Imprimerie Catholique le Deux.

Ghali, Ibrahim Amin (1986). *Vivant Denon ou La Conquête du Bonheur.* Cairo, Institut Français d'Archaeologie Orientale.

Gillispie, Charles C. (1989). "Scientific Aspects of the French Egyptian Expedition 1798–1801," *Proceedings of the American Philosophical Society* 133: 447–74.

——— (2004). *Science and Polity in France at the End of the Old Regime. The Revolutionary and Napoleonic Years.* Princeton, Princeton University Press.

Gillispie, Charles C., and Michel Dewachter (1987). *Monuments of Egypt: The Napoleonic Edition: The Complete Archaeological Plates from la Description de l'Egypte.* Princeton, Princeton Architectural Press in association with Architectural League of New York and J. Paul Getty Trust.

Goby, Jean-Édouard (1953). "Où vécurent les savants de Bonaparte en Egypte?," *Cahiers d'histoire égyptienne* 5: 290–301.

——— (1987). *Premier Institut d'Égypte, Restitution des Comptes Rendus des Séances.* Paris, Institut de France.

Godlewska, Anne (1995). "Map, Text and Image: The Mentality of Enlightened Conquerors: A New Look at the Description de l'Égypte," *Transactions of the Institute of British Geographers* 20: 5–28.

——— (1999). *Geography Unbound: French Geographic Science from Cassini to Humboldt.* Chicago, University of Chicago Press.

Grattan-Guinness, Ivor (1990). *Convolutions in French Mathematics, 1800–1840.* Basel, Birkhäuser Verlag.

Grattan-Guinness, Ivor, and J. R. Ravetz (1972). *Joseph Fourier, 1768–1830.* Cambridge, MIT Press.

Greene, Christopher M. (1981). "Alexandre Lenoir and the Musée des monuments français during the French Revolution," *French Historical Studies* 12: 200–22.

Greener, Leslie (1966). *The Discovery of Egypt.* London, Cassell.

Grell, Chantal (1993). *L'histoire entre érudition et philosophie; étude sur la connaissance historique à l'age des lumières.* Paris, Presse Universitaires de France.

Haig, Stirling (1964). "Vigny and Othello," *Yale French Studies* 33: 53–64.

Hanley, Wayne (2005). *The Genesis of Napoleonic Propaganda, 1796–1799.* New York, Columbia University Press.

Hartleben, Hermine (1906). *Champollion; sein Leben und sein Werk.* Berlin, Weidmann.

Hatin, Eugéne (1866). *Bibliographie historique et critique de la presse périodique française, ou catalogue systématique et raisonné de tous les écrits périodiques de quelque valeur publiés ou ayant circulé en France depuis l'origine du journal jusqu'à nos jours, avec extraits, notes historiques, critiques, et morales, indication des prix que les principaux journaux ont atteints dans les ventes publiques, etc.; précédé d'un essai historique et statistique sur la naissance et les progrès de la presse périodique dans les deux mondes.* Paris, Firmin-Didot Fréres, Fils.

Herivel, John (1975). *Joseph Fourier, the Man and the Physicist.* Oxford, Oxford University Press.

Humbert, Jean-Marcel, Michael Pantazzi, and Christiane Ziegler (1994). *Egyptomania. l'Égypte dans l'art occidental. 1730–1930.* Paris, Spadem.

Irwin, Robert (2006). *Dangerous Knowledge. Orientalism and Its Discontents.* Woodstock, Overlook Press.

Iversen, Erik (1993). *The Myth of Egypt and Its Hieroglyphs.* Princeton, Princeton University Press.

Jackson, Catherine Charlotte (1880). *The Old Régime: Courts, Salons and Theatres.* London, R. Bentley and Son.

Jeffreys, D. G. (2003). *Views of ancient Egypt since Napoleon Bonaparte: Imperialism, Colonialism and Modern Appropriations.* London, UCL.

Jones, C. (2004). *Paris: The Biography of a City.* New York, Vintage Books.

Jonquière, Clément de la (1899–1907). *L'Expédition d'Égypte.* Paris, H. Charles-Lavauzelle.

Jusserand, J. J. (1899). *Shakespeare in France under the Ancien Régime.* London, T. Fisher Unwin.

Kertzer, David I. (2002). *The Popes Against the Jews. The Vatican's Role in the Rise of Modern Anti-Semitism.* New York, Vintage Books.

Krakovitch, Odile (1985). *Hugo Censuré: La Liberté du Théatre au XIXe Siècle.* Paris, Calmann-Lévy.

Kroen, Sheryl T. (1998). "Revolutionizing Religious Politics during the Restoration," *French Historical Studies* 21: 27–53.

————— (2000). *Politics and Theater: The Crisis of Legitimacy in Restoration France.* Berkeley, University of California Press.

Lacouture, Jean (1988). *Champollion. Une vie de lumières.* Paris, Bernard Grasset.

La Fizilère, A. (1873). *L'oeuvre originale de Vivant Denon.* Paris, A. Barraud.

Lagier, Camille (1923). "La querelle des zodiaques," *Revue des Questions Scientifiques* 4: 48–71.

Laissus, Yves (1998). *L'Égypte, une aventure savante 1798–1801.* Paris, Fayard.

Langins, Janis (1987). *La Républlique avait besoin des savants; les débuts de l'École Polytechnique: L'École Centrale des Travaux Publics et les Cours Révolutionnaires de l'an III.* Paris, Belin.

Lapparent, A. de (1894). "Biot (1774–1862)," in *Livre du Centenaire.* Paris, École Polytechnique.

Laurens, Henry (1997). *L'expédition d'Égypte.* Paris, Éditions de Seuil.

Laurens, Henry, Charles Coulston Gillispie, Jean-Claude Golvin, and Claude Traunecker (1989). *L'expédition d'Égypte, 1798–1801.* Paris, Armand Colin.

Lefort, F. (1867). "Un Savant Chrétien. J.-B. Biot," *Le Correspondant*: 955–95.

Lelièvre, Pierre (1942). *Vivant Denon, directeur des beaux-arts de Napoléon.* Paris, Librairie Floury.

————— (1993). *Vivant Denon.* Paris, Picard.

Ligou, Daniel (1987). *Dictionnaire de la Franc-Maçonnerie.* Paris, Presses Universitaires de France.

Lindsay, S. G. (2000). "Mummies and Tombs: Napoleon and the Death Ritual," *Art Bulletin* 82: 476–502.

Lowe, Lisa (1991). *Critical Terrains: French and British Orientalisms.* Ithaca, Cornell University Press.

McCormick, John (1993). *Popular Theatres of Nineteenth-Century France.* London, Routledge.

McGinnis, Reginald (1997). "L'histoire Prostituée: Voltaire Contre Larcher, et Contre Lui-Même," *The Romaic Review* 88.

————— (2001). *Enemies of the Enlightenment: The French Counter-Enlightenment and the Making of Modernity.* Oxford, Oxford University Press.

MacKenzie, John M. (1995). *Orientalism: History, Theory, and the Arts.* New York, Manchester University Press.

McMahon, Darrin (1998). "The Counter-Enlightenment and the Low-Life of Literature in Pre-Revolutionary France," *Past and Present* 159: 78–112.

Madden, R. R. (1840). *Egypt and Mohammed Ali.* London, H. Adams.

Madelin, Louis (1960). *Fouché. 1759–1820.* Paris, Club des Éditeurs.

Malandain, Pierre (1982). *Delisle de Sales, philosophe de la nature (1741–1826).* N.p., Voltaire Foundation.

Manley, Deborah (1998). "Two Brides: Baroness von Minutoli and Mrs. Colonel Elmwood," in *Travellers in London*, edited by Paul Starkey and Janet Starkey. London, I. B. Tauris.

Manley, Deborah, and Peta Rée (2001). *Henry Salt: Artist, Traveller, Diplomat, Egyptologist.* London, Libri.

Mansel, Philip (2003). *Paris Between Empires.* New York, St. Martin's Press.

Manuel, Frank Edward (1959). *The Eighteenth Century Confronts the Gods.* Cambridge, Harvard University Press.

——— (1963). *Isaac Newton, Historian.* Cambridge, Belknap Press of Harvard University Press.

Marsot, Afaf Lutfi al-Sayyid (1984). *Egypt in the Reign of Muhammad Ali.* Cambridge, Cambridge University Press.

Martius, C. F. P. von (1862). *Zum Gedächtniss an Jean Baptiste Biot.* Munich, Verlag der k. Akademie.

Meurthe, Boulay de la (1882). *La Négociation du Concordat.* Paris, Jules Gervais.

Meyerson, Daniel (2004). *The Linguist and the Emperor: Napoleon and Champollion's Quest to Decipher the Rosetta Stone.* New York, Random House.

Michaud, ed. (1843a). *Biographie universelle ancienne et moderne: histoire par ordre alphabétique de la vie publique et privée de tous les hommes*, vol. 2. Paris, Desplaces.

——— (1843b). *Biographie universelle ancienne et moderne: histoire par ordre alphabétique de la vie publique et privée de tous les hommes*, vol. 27. Paris, Desplaces.

Mitchell, Timothy (1991). *Colonising Egypt.* Berkeley, University of California Press.

Mohan, Jyoti (2005). "La civilization la plus antique: Voltaire's images of India," *Journal of World History* 16: 173–85.

Moravia, S. (1974). *Il pensiero degli idéologues: scienza et filosofia in Francia, 1780–1812.* Firenze, La Nuova Italia.

Mornet, Daniel (1933). *Les Origines Intellectuelles de la Révolution Française (1715–1787).* Paris, Armand Collin.

Moureaux, J. M. (1974). "Voltaire et Larcher, ou le faux 'mazarinier'," *Revue d'Histoire Littéraire de la France* 74: 600–26.

Munier, Henri (1943). *Tables de la Description de l'Égypte, suivies d'une bibliographie sur l'expédition française de Bonaparte.* Cairo, Institut Français d'Archéologie Orientale.

Nowinski, Judith (1970). *Baron Dominique Vivant Denon (1747–1825): Hedonist and Scholar in a Period of Transition.* Rutherford, NJ, Fairleigh Dickinson University Press.

O'Mara, Patrick F. (2003). "Censorinus, the Sothic Cycle, and Calendar Year One in Ancient Egypt: The Epistemological Problem," *Journal of Near Eastern Studies* 62: 17–26.

Outram, Dorinda (1984). *Georges Cuvier: Vocation, Science, and Authority in Post-Revolutionary France.* Manchester, Manchester University Press.

Ozouf, Mona (1988). *Festivals and the French Revolution*, trans. Alan Sheridan. Cambridge, Harvard University Press.

Palmer, R. R., ed. (1975). *The School of the French Revolution: A Documentary History of the College of Louis-le-Grand and Its Director, Jean-François Champagne 1762–1814.* Princeton, Princeton University Press.

Pao, Angela Chia-Yi (1997). *The Orient of the Boulevards: Exoticism, Empire, and Nineteenth-Century French Theater.* Philadelphia, University of Pennsylvania Press.

Parker, Richard A. (1981). "Egyptian Astronomy, Astrology, and Calendrical Reckoning," in *Dictionary of Scientific Biography*, edited by Charles Coulston Gillispie, vol. 15. Supplement 1: Topical Essays. New York, Charles Scribner's Sons, 706–27.

Pemble, Jonathan (2005). *Shakespeare Goes to Paris.* London, Hambledon.

Picard, Émile (1927). *La Vie et l'Oeuvre de Jean-Baptiste Biot.* Paris, Académie des Sciences.

Pickering, Mary (1993). *Auguste Comte. An Intellectual Biography.* Cambridge, Cambridge University Press.

Pinet, G. (1887). *Histoire de l'École Polytechnique.* Paris, Librairie Polytechnique, Baudry.

Piolanti, Mons. Antonio (1977). *L'Accademia di Religione Cattolica. Profilo della sua Storia e del Suo Tommismo.* Vatican City, Libreria Editrica Vaticana.

Polowetzky, Michael (1993). *A Bond Never Broken: The Relations Between Napoleon and the Authors of France.* London, Associated University Presses.

Prendergast, Christopher (1996). *Paris and the Nineteenth Century.* Oxford, Blackwell.

Prochaska, David (1994). "Art of Colonialism, Colonialism of Art: The Description de l'Égypte (1809–1828)," *L'Esprit Créateur* 34: 69–91.

Rétat, Claude (1999). "Lumières et ténèbres du citoyen Dupuis," *Chroniques d'histoire maçonnique* 50: 5–68.

——— (2000). "La teinte de la Nature. L'Isis du citoyen Dupuis," in *Isis, Narcisse, Psyché entre lumières et romantisme*, edited by P. Jonchière. Clermont-Ferrand, Presses Blaise-Pascal, 71–79.

——— (2001). "La machine a poème. Charles-Francois Dupuis, Alexandre Lenoir," *Politica hermetica* 15: 4–37.

——— (2002). "Traduction/ réfection du poème antique à la fin du XVIIIe siècle: Charles-François Dupuis," *Cahiers Roucher-André Chénier* 21: 133–39.

Richard-Desaix, Ulric (1880). *La Relique de Moliére du Cabinet du Baron Vivant Denon.* Paris, Vignères.

Ridley, Ronald (1998). *Napoleon's Proconsul in Egypt: The Life and Times of Bernardino Drovetti.* London, Rubicon.

Robinet (1896–98). *Le Mouvement Réligieux à Paris Pendant la Révolution.* Paris, Leopold Cerf, Charles Noblet, Maison Quantin.

Rocher, Ludo (1984). *Ezourvedam: A French Veda of the Eighteenth Century.* Philadelphia, John Benjamins.

Rose, John Holland (1902). *The Life of Napoleon I Including New Materials from the British Official Records.* London, George Bell and Sons.

Rudolph, Conrad, ed. (2006). *A Companion to Medieval Art: Romanesque and Gothic in Northern Europe.* Cornwall: TJ International.

Rudwick, M. (1997). *Georges Cuvier, Fossil Bones, and Geological Catastrophes: New Translations and Interpretations of the Primary Texts.* Chicago, University of Chicago Press.

——— (2005). *Bursting the Limits of Time: The Reconstruction of Geohistory in the Age of Revolution.* Chicago, University of Chicago Press.

Russell, Terence M. (2005). *The Discovery of Egypt. Vivant Denon's Travels with Napoleon's Army.* Stroud, Sutton.

Said, Edward W. (1978). *Orientalism.* New York, Vintage Books.

Sainte-Beuve, C. A. (1864). "M. Biot," *Nouveaux lundis* 2: 71–110.

Schwab, Raymond (1984). *The Oriental Renaissance: Europe's Rediscovery of India and the East, 1680–1880.* New York, Columbia University Press.

Sibenaler, Jean (1992). *Il se faisait appeler Volney: approche biographique de Constantin-François Chassebeuf, 1757–1820.* Paris, Maulévier Hérault.

Smith, Edwin Burrow (1954). "Jean-Sylvain Bailly—Astronomer, Mystic, Revolutionary—1736–1793," *Transactions of the American Philosophical Society* 44: 427–538.

Sollers, Philippe (1995). *Le Cavalier du Louvre: Vivant Denon.* Paris, Pion.

Sonenscher, Michael (2008). *Sans-Culottes: An Eighteenth-Century Emblem in the French Revolution.* Princeton, Princeton University Press.

Spieth, Darius A. (2001). "The Printed Work of Vivant Denon (1747–1825)," Ph.D. dissertation, University of Illinois at Urbana-Champaign.

——— (2007). *Napoleon's Sorcerers: The Sophisians.* Newark, University of Delaware Press.

Spinelli, Donald C. (1984). "Review of Pierre Malandrain, Delisle de Sales, philosophe de la nature," *French Review* 58: 291–92.

Spitzer, Alan B. (1987). *The French Generation of 1820.* Princeton, Princeton University Press.

Staum, Martin S. (1980). "The Class of Moral and Political Sciences, 1795–1803," *French Historical Studies* 11: 371–97.

——— (1996). *Minerva's Message: Stabilizing the French Revolution.* Montreal, McGill-Queen's University Press.

Strathern, Paul (2008). *Napoleon in Egypt.* New York, Bantam.

Switzer, Richard (1955). *Etienne-Léon de Lamothe-Langon and the French Popular Novel, 1800–1830.* Berkeley, University of California.

Truesdell, Clifford (1980). *The Tragicomical History of Thermodynamics, 1822–1854.* New York, Springer-Verlag.

Varisco, Daniel Martin (2007). *Reading Orientalism. Said and the Unsaid.* Seattle, University of Washington Press.

Welschinger, Henri (1882). *La Censure sous le Premier Empire. Avec Documents Inédits.* Paris, Charavay Frères.

Wicks, Charles Beaumont (1950). *The Parisian Stage: Alphabetical Indexes of Plays and Authors.* Birmingham, University of Alabama Press.

Wielandt, Rotraud (1997). *Das Bild der Europäer in der Modernen Arabischen Erzähl- und Theaterliteratur.* N.p., Franz Steiner Verlag.

Young, Andrew (2006). "Understanding Astronomical Refraction," *Observatory* 126: 82–115.

FIGURE SOURCES

Title Page. Entrance to the chapel containing the Dendera circular zodiac. Photo by JZB. Introduction. Cover of the *Zodiaque* pamphlet collection. Photo by JZB.

Figure 1.1. The Dendera temple. Top photo by JZB; bottom by Francis Frith, courtesy of George Eastman House, International Museum of Photography.

Figure 1.2. The Dendera circular zodiac, plaster cast. Photo by JZB.

Figure 1.3. The Dendera circular zodiac in the Louvre. Réunion des Musés Nationaux / Art Resource, NY.

Figure 2.1. Notre Dame interior. Photo by JZB.

Figure 2.2. Zodiacal and agricultural carvings on the façade of Notre Dame. Photos by JZB.

Figure 3.1. Frontispiece to C. Dupuis (1822a).

Figure 3.2. Dupuis' depiction of carvings on the façade of Notre Dame. C. Dupuis (1795a, 3:48, plate 18).

Figure 3.3. Precession of the summer solstice. Graphic by JZB using *TheSky* astronomy software by SofwareBisque.

Figure 3.4. Positions of the summer solstice among the constellations. Graphic by JZB.

Figure 3.5. Flemish precession globe. Photos by JZB from the Museum of the History of Science, Oxford.

Figure 3.6. Frontispiece to Dupuis' *Origine*, based on a similar design printed originally in Delaulnaye's 1791 *Histoire générale*. C. Dupuis (1795b).

Figure 4.1. Monge and Berthollet by Dutertre. R.-E. Villiers du Terrage and M. Villiers du Terrage (1899, 81, 7).

Figure 4.2. The Sinnari house in Cairo. Photos by JZB.

Figure 4.3. Denon's sketch of the circular Dendera zodiac. V. Denon (1802b, 3:242–43 [description of graphic], plate 130, no. 2). Courtesy of the John Hay Library, Brown University.

Figure 4.4. Frontispiece to the *Description de l'Égypte*, vol. 1: drawn by Cécile, engraved by Girardet and Sellier. Courtesy of the John Hay Library, Brown University.

Figure 5.1. Map of Upper Egypt. Map by Bonne (July 1781), JZB print collection.

Figure 5.2. Jollois and Devilliers by Dutertre, in R.-E. Villiers du Terrage and M. Villiers du Terrage (1899, frontispiece and xix).

Figure 5.3. The temple of Isis at Philae. Photos by JZB.

Figure 5.4. The great rectangular zodiac at Dendera today, and as drawn by Denon. Photo by JZB (*top*); V. Denon (1802b, 3:246 [description of graphic], plate 132).

Figure 5.5. Denon's drawings of the temples at Esneh. V. Denon (1802b, 3:87 [description of graphic], plate 53, no. 1 [*top*] and no. 2). Courtesy of the John Hay Library, Brown University.

Figure 5.6. Jollois and Devilliers' drawings of the temples at Esneh. Jollois and Devilliers, *Description de l'Égypte* (1809, vol. 1, plate 73 [*top*, engraved by Louvet] and plate 88 [engraved by Lorieux]).

Figure 6.1. The rectangular zodiac at Dendera, sections from Figure 5.4.

Figure 6.2. The sun entering and exiting Leo. Graphic by JZB using *StarryNight* astronomy software by Imaginova Canada.

Figure 6.3. The sun in Virgo. Graphic by JZB using *StarryNight* astronomy software by Imaginova Canada.

Figure 6.4. The solstice and the heliacal rise of Sirius. Graphic by JZB using *StarryNight* astronomy software by Imaginova Canada.

Figure 7.1. Ripault and Napoleon by Dutertre. R.-E. Villiers du Terrage and M. Villiers du Terrage (1899, 113, 18).

Figure 7.2. Fontanes and Chateaubriand, lithographs by Delpeche. JZB print collection.

Figure 7.3. The zodiacal signs on Denon's sketch of the Dendera rectangular (see Figure 5.4).

Figure 7.4. A representation of what Dupuis envisioned. Graphic by JZB using *StarryNight* astronomy software by Imaginova Canada.

Figure 7.5. The sun next to Aries on the Dendera rectangular: section from Figure 8.7.

Figure 7.6. The single-legged figure on the Dendera rectangular: sections from Denon's (Figure 5.4) and Jollois and Devilliers' (Figure 8.7) depictions. The Denon by courtesy of the John Hay Library, Brown University.

Figure 8.1. Jomard and Fourier by Dutertre. R.-E. Villiers du Terrage and M. Villiers du Terrage (1899, 224, 74).

Figure 8.2. Dendera in Sèvres porcelain (*top*), by permission of the Victoria and Albert Museum, London. Jollois and Devilliers' drawing of the Dendera temple (*bottom*). Jollois and Devilliers, *Description de l'Égypte* (1817, vol. 4, plate 29) (engraved by Leisnier). By permission of the Department of Special Collections, Charles E. Young Research Library, UCLA. Jollois and Devilliers' plate 7 in vol. 4 shows the temple half-buried in sand.

Figure 8.3. Sonnini's depiction of the sky goddess at Dendera. C. S. Sonnini (1799, vol. 4, plate 36).

Figure 8.4. Fourier's friend, Jacques-Joseph Champollion-Figeac. A. L. Champollion-Figeac (1887, frontispiece).

Figure 8.5. Locations of the sun at Sirius's heliacal rise. Graphic by JZB.

Figure 8.6. Jollois and Devilliers' depictions of the zodiacs at Esneh. Jollois and Devilliers, *Description de l'Égypte* (1809, vol. 1, plate 79) (engraved by Pomel); top two by permission of the Department of Special Collections, Charles E. Young Research Library, UCLA; and (1809, vol. 1, plate 87) (engraved by Allais).

Figure 8.7. Jollois and Devilliers' drawing of the Dendera rectangular. *Description de l'Égypte* (1817, vol. 4, plate 20) (engraved by Charlin). By permission of the Department of Special Collections, Charles E. Young Research Library, UCLA.

Figure 8.8. The visibility of Sirius over the millennia. Graphic based on map by Bonne (Figure 5.1).

Figure 8.9. The constellation Corvus on a sky chart and on Jollois and Devilliers' drawing of the Dendera circular (Figure 10.1). Graphic by JZB using *StarryNight* astronomy software by Imaginova Canada.

Figure 9.1. A page with censor's marks from the manuscript of the *Zodiaque de Paris*. Photo by DGJ from F18, Box 642, #626, at Archives Nationales, Paris.

Figure 10.1. The shaded engraving of the circular zodiac and surrounds. Jollois and Devilliers, *Description de l'Égypte* (1817, vol. 4, plate 21) (engraved by Allais). By permission of the Department of Special Collections, Charles E. Young Research Library, UCLA.

Figure 10.2. The marble sculpture of the Dendera circular, carved by Castex. By permission of the Fitzwilliam Museum, University of Cambridge.

Figure 10.3. The "hook" in Chinese and on the Dendera circular. C. H. c. d. Paravey (1835 [1821], 29) (*left*); section of Figure 10.1.

Figure 10.4. Paravey's suggested projection. Graphic by JZB.

Figure 10.5. Plan of the Dendera temple as it would be seen from above if the ceiling were transparent. J.-B. Biot (1844, plate 2).

Figure 10.6. Paravey's axis, drawn across the engraving by Jollois and Devilliers, Figure 10.1.

Figure 10.7. Delambre's bird-man highlighted at center, section from the engraving by Jollois and Devilliers, Figure 10.1.

Figure 10.8. Laplace, lithograph by Delpeche. JZB print collection.

Figure 10.9. Biot and Arago. Biot lithograph by Delpeche. JZB print collection.

Figure 10.10. The "projection by development." Graphic by JZB.

Figure 10.11. (Left) Biot's location for Antares on the actual Dendera circular, section from the Dendera circular in the Louvre. Photo by JZB. (Right) Biot's cow and hawk for Sirius, section from the engraving by Jollois and Devilliers, Figure 10.1.

Figure 10.12. Biot's table of probabilities. J.-B. Biot (1823, 45).

Figure 10.13. How Biot found the zodiac's pole. Graphic by JZB.

Figure 10.14. Chabrol's cartouches, drawn in the left-hand margin of his copy of A. J. Saint-Martin (1822, 47); section from the engraving by Jollois and Devilliers, Figure 10.1.

Figure 11.1. The young Jean-François Champollion. A. L. Champollion-Figeac (1887, 52).

Figure 11.2. Biot's modification of Gau's engraving. J.-B. Biot (1823), reprinted in J.-B. Biot (1844).

Figure 11.3. Champollion's variant hieroglyphs for *autocrator*. J.-F. Champollion (1822b, plate 2).

Figure 11.4. Denon's drawing of the Dendera cartouches. V. Denon (1802b, 243 [*description*], plate 130, no. 2).

Figure 11.5. Champollion's renditions of the sky goddess. J.-F. Champollion (1823, 357, plate for "Uranie, la Désse Ciel").

Figure 12.1. "Champoléon" inscribed on a column at Karnak. Photo by JZB.

Figure 12.2. The empty cartouches of the Dendera zodiac and the drawing of them done during Champollion's expedition. Photo by JZB; J.-F. Champollion and J. J. Champollion-Figeac (1835–1845, vol. 4, plate 349).

Figure I.1. The festival of the Supreme Being by Naudet. Musée de la Ville de Paris, Musée Carnavalet, Paris. Erich Lessing / Art Resource, NY.

Figure I.2. The Coronation of Napoleon at Notre Dame by David, section of original in the Louvre, Paris. Erich Lessing / Art Resource, NY.

Figure I.3. The great rectangular zodiac at Dendera today. Photo by JZB.

Figure I.4. Section of the Dendera rectangular. Photo by JZB.

Figure I.5. *Théatre du Gymnase* by Adolph von Menzel. The National-galerie, Staatliche Museen zu Berlin. Bildarchiv Preussischer Kulturbesitz / Art Resource, NY.

Figure I.6. The Franco-Tuscan Expedition in Egypt by Giuseppe Angelelli. Museo Archeologico, Florence. Scala / Art Resource, NY.

Figure I.7. One of the many empty cartouches at Dendera. Photo by JZB.

Figure I.8. Cleopatra VII in the guise of an Egyptian goddess (Mut) at Dendera. Photo by JZB.

SUBJECT INDEX

NAME INDEX